Electrical Engineering

Problems and Solutions

Electrical Engineering

Problems and Solutions

Eighth Edition

Lincoln D. Jones, P.E.
Internet:73125.1013@CompuServe.Com
San Jose State University

Engineering Press Austin, Texas 78720-0129

Printed in the United States of America.

ISBN 1-57645-033-3

Engineering Press **P.O. Box 200129 Austin, Texas 78720-0129**

Contents

Electrical Engineering

Problems and Solutions

Exam Files

Professors around the country have opened their exam files and revealed their examination problems and solutions. These are actual exam problems with the complete solutions prepared by the same professors who wrote the problems. Exam Files are currently available for these topics:

Calculus I
Calculus II
Calculus III
Circuit Analysis
College Algebra
Differential Equations
Dynamics
Engineering Economic Analysis
Fluid Mechanics
Linear Algebra
Materials Science
Mechanics of Materials
Organic Chemistry
Physics I Mechanics
Physics III Electricity and Magnetism
Probability and Statistics
Statics
Thermodynamics

Becoming A Professional Engineer

To achieve registration as a Professional Engineer there are four distinct steps: education, fundamentals of engineering (engineer-in-training) exam, professional experience, and finally, the professional engineer exam. These steps are described in the following sections.

Education

The obvious appropriate education is a B.S. degree in electrical engineering from an accredited college or university. This is not an absolute requirement. Alternative, but less acceptable, education is a B.S. degree in something other than electrical engineering, or from a non-accredited institution, or four years of education but no degree.

Fundamentals of Engineering (FE/EIT) Exam

Most people are required to take and pass this eight-hour multiple-choice examination. Different states call it by different names (Fundamentals of Engineering, E.I.T., or Intern Engineer) but the exam is the same in all states. It is prepared and graded by the National Council of Examiners for Engineering and Surveying (NCEES). Review materials for this exam are found in other books like Newnan: Engineer-In-Training License Review and Newnan and Larock: Engineer-In-Training Examination Review.

Experience

Typically one must have four years of acceptable experience before being permitted to take the Professional Engineer exam, but this requirement may vary from state to state. Both the length and character of the experience will be examined. It may, of course, take more than four years to acquire four years of acceptable experience.

Professional Engineer Exam

The second national exam is called Principles and Practice of Engineering by NCEES, but probably everyone else calls it the Professional Engineer or P.E. exam. All states, plus Guam, the District of Columbia, and Puerto Rico use the same NCEES exam.

Electrical Engineering Professional Engineer Exam Background

The reason for passing laws regulating the practice of electrical engineering is to protect the public from incompetent practioners. Beginning about 1907 the individual states began passing title acts regulating who could call themselves an electrical engineer. As the laws were strengthened, the practice of certain aspects of electrical engineering was limited to those who were registered electrical engineers, or working under the supervision of a registered electrical engineer. There is no national registration law; registration is based on individual state laws and is administered by boards of registration in each of the states. A listing of the State Boards is in Table 1.1.

Examination Development

Initially the states wrote their own examinations, but beginning in 1966 the NCEES took over the task for some of the states. Now the NCEES exams are used by all states. This greatly eases the ability of an electrical engineer to move from one state to another and achieve registration in the new state. About 2500 electrical engineers take the exam each year. As a result about 23% of all electrical engineers are registered professional engineers.

The development of the electrical engineering exam is the responsibility of the NCEES Committee on Examinations for Professional Engineers. The committee is composed of people from industry, consulting, and education, plus consultants and subject matter experts. The starting point for the exam is an electrical engineering task analysis survey that NCEES does at roughly five to ten year intervals. People in industry, consulting and education are surveyed to determine what electrical engineers do and what knowledge is needed. From this NCEES develops what they call a "matrix of knowledges" that form the basis for the electrical engineering exam structure described in the next section.

The actual exam questions are prepared by the NCEES committee members, subject matter experts, and other volunteers. All people participating must hold professional registration. Using workshop meetings and correspondence by mail, the questions are written and circulated for review. The problems relate to current professional situations. They are structured to quickly orient one to the requirements, so the examinee can judge whether he or she can successfully solve it. While based on an understanding of engineering fundamentals, the problems require the application of practical professional judgement and insight. While four hours are allowed for four problems, probably any problem can be solved in 20 minutes by a specialist in the field. A professionally competent applicant can solve the problem in no more than 45 minutes. Multi-part questions are arranged so the solution of each succeeding part does not depend on the correct solution of a prior part. Each part will have a single answer that is reasonable.

Examination Structure

The twelve problems in the morning four-hour session are regular computation "essay" problems. In the afternoon four-hour session all twelve problems are multiple choice. In each category (Generation Systems, Transmission Systems, Rotating Machines, and so on) about half of the problems will be in the morning session and half in the afternoon. Engineering economics may appear as a component within one or two of the problems.

Generation Systems
A. Fundamental Design of Large Scale Power Plants - One Problem

B. Final Design and Applications of Large Scale Power Plants - One Problem

Transmission and Distribution Systems
C. Fundamental Design of Transmission and Distribution Systems - One Problem
Transformers, protection, safety, overhead and underground lines, metering, batteries, relays, substations, circuit protectors, and intrinsic safety devices.

D. Final Design and Applications of Transmission and Distribution Systems - Two Problems
The description is the same as C.

Rotating Machines
E. Final Design and Applications of Motors and Generators - One Problem

Lightning Protection and Grounding
F. Final Design and Applications of Lightning Protection and Physical Grounding of Equipment and Structures - One Problem

Control Systems
G. Design of Industrial Process and Operations Controls, Feedback Controls and Transient Theory - Two Problems

Electronic Devices
H. Design of Electronic Devices - Two Problems
Digital storage devices, integrated circuit components, and operational amplifiers.

I. Applications of Electronic Devices - One Problem
The description is the same as H.

Instrumentation
J. Design of Instrumentation - One Problem
Instrument transformers, metering systems, measurement systems, test procedures, and transducers.

K. Applications of Instrumentation - One Problem
The description is the same as J.

L. Final Design and Applications of Instrumentation - One Problem
The description is the same as J.

Digital Systems
M. Design of Digital Systems,including Interfaces, protocols, and Standards - Two Problems

N. Design of Digital and Analog Computer Systems - One Problem

O. Applications of Digital and Analog Computer Systems - Two Problems

Communication Systems
P. Design of Communications Systems - Two Problems
Broadcast, voice and data communication; fiber optics, antennas, microwave systems and HF transmission lines.

Q. Applications of Communication Systems - One Problem
The description is the same as P.

Biomedical Systems
R. Design of Biomedical Systems
Electrical applications in living systems.

Note: The examination is developed with problems that will require a variety of approaches and methodologies including design, analysis, application, economic aspects, and operations.

The New Breadth/Depth Exam

At this writing, NCEES has not announced when the new exam format will be implemented. The examination will probably consist of 80 items in the morning and another 80 items in the afternoon. All items to be objectively scored with a,b,c,d multiple-choice answers.

The morning exam will probably consist of 80 problems, distributed about the knowledge clusters as shown in the attached chart. All problems will be scored by computer. The afternoon exam will probably consist of 80 problems. Five different afternoon exams will be prepared with the problems distributed as shown in the table. The new exam will not have any essay type questions.

Knowledge Clusters	AM Test	PM Test				
		OPTION NUMBER				
		1	2	3	4	5
MATHEMATICS	15%			5%	5%	
MEASUREMENT	5%	5%	5%	5%	5%	10%
CODES & STANDARDS		10%	10%	5%	10%	10%
CIRCUIT THEORY	30%	10%	5%	10%	10%	15%
FIELDS	5%	15%			5%	10%
ELECTRONICS	20%	10%	20%	5%	30%	5%
COMPUTERS	5%	10%	45%	10%	15%	
COMMUNICATIONS	5%	35%	5%	10%	10%	
CONTROL SYSTEMS	5%	5%	10%	50%	10%	
POWER	10%					50%
TOTAL PERCENTAGE	100%	100%	100%	100%	100%	100%

Taking The Exam

Exam Dates
The National Council of Examiners for Engineering and Surveying (NCEES) prepares Electrical Engineering Professional Engineer exams for use on a Friday in April and October each year. Some state boards administer the exam twice a year in their state, while others offer the exam once a year. The scheduled exam dates are:

	April	October
1999	23	29
2000	14	27
2001	20	26
2002	19	25

People seeking to take a particular exam must apply to their state board several months in advance.

Exam Procedure
Before the morning four-hour session begins, the proctors will pass out an exam booklet and solutions pamphlet to each examinee. There are likely to be civil, chemical, and mechanical engineers taking their own exams at the same time. You must solve four of the twelve electrical engineering problems.

The solution pamphlet contains grid sheets on right-hand pages. Only work on these grid sheets will be graded. The left-hand pages are blank and are for scratch paper. The scratch work will not be considered in the scoring.

If you finish more than 30 minutes early, you may turn in the booklets and leave. In the last 30 minutes, however, you must remain to the end to insure a quiet environment for all those still working, and to insure an orderly collection of materials.

The afternoon session will begin following a one-hour lunch break. The afternoon exam booklet will be distributed along with an answer sheet. The booklet will have twelve 10-part multiple choice questions. You must select and solve four of them. An HB or #2 pencil is to be used to record your answers on the scoring sheet.

Exam-Taking Suggestions
People familiar with the psychology of exam-taking have several suggestions for people as they prepare to take an exam.

1. Exam taking is really two skills. One is the skill of illustrating knowledge that you know. The other is the skill of exam-taking. The first may be enhanced by a systematic review of the technical material. Exam-taking skills, on the other hand, may be improved by practice with similar problems presented in the exam format.

2. Since there is no deduction for guessing on the multiple choice problems, an answer should be given for all ten parts of the four selected problems. Even when one is going to guess, a logical approach is to attempt to first eliminate one or two of the five alternatives. If this can be done, the chance of selecting a correct answer

obviously improves from 1 in 5 to, say, 1 in 3.

3. Plan ahead with a strategy. Which is your strongest area? Can you expect to see one or two problems in this area? What about your second strongest area? What will you do if you still must find problems in other areas?

4. Have a time plan. How much time are you going to allow yourself to initially go through the entire twelve problems and grade them in difficulty for you to solve them? Consider assigning a letter, like A, B, C and D, to each problem. If you allow 15 minutes for grading the problems, you might divide the remaining time into five parts of 45 minutes each. Thus 45 minutes would be scheduled for the first - and easiest - problem to be solved. Three additional 45 minute periods could be planned for the remaining three problems. Finally, the last 45 minutes would be in reserve. It could be used to switch to a substitute problem in case one of the selected problems proves too difficult. If that is unnecessary, the time can be used to check over the solutions of the four selected problems. A time plan is very important. It gives you the confidence of being in control, and at the same time keeps you from making the serious mistake of misallocation of time in the exam.

5. Read all five multiple choice answers before making a selection. The first answer in a multiple choice question is sometimes a plausible decoy - not the best answer.

6. Do not change an answer unless you are absolutely certain you have made a mistake. Your first reaction is likely to be correct.

7. Do not sit next to a friend, a window, or other potential distractions.

Exam Day Preparations

There is no doubt that the exam will be a stressful and tiring day. This will be no day to have unpleasant surprises. For this reason we suggest that an advance visit be made to the examination site. Try to determine such items as:

1. How much time should I allow for travel to the exam on that day? Plan to arrive about 15 minutes early. That way you will have ample time, but not too much time. Arriving too early, and mingling with others who also are anxious, will increase your anxiety and nervousness.

2. Where will I park?

3. How does the exam site look? Will I have ample work space? Where will I stack my reference materials? Will it be overly bright (sunglasses) or cold (sweater), or noisy (earplugs)? Would a cushion make the chair more comfortable?

4. Where is the drinking fountain, lavoratory facilities, pay phone?

5. What about food? Should I take something along for energy in the exam? A bag lunch during the break probably makes sense.

What To Take To The Exam

The NCEES guidelines say you may bring the following reference materials and aids into the examination room for your personal use only:

1. Handbooks and textbooks
2. Bound reference materials, provided the materials are and remain bound during the entire examination. The NCEES defines "bound" as books or materials fastened securely in its cover by fasteners which penetrate all papers. Examples are ring binders, spiral binders and notebooks, plastic snap binders, brads, screw posts, and so on.
3. Battery operated, silent non-printing calculators.

At one time NCEES had a rule that did not permit "review publications directed principally toward sample questions and their solutions" in the exam room. This set the stage for restricting some kinds of publications from the exam. <u>State boards may adopt the NCEES guidelines, or adopt either more or less restrictive rules</u>. Thus an important step in preparing for the exam is to know what will - and will not - be permitted. We suggest that if possible you obtain a written copy of your state's policy for the specific exam you will be taking. Recently there has been considerable confusion at individual examination sites, so a copy of the exact applicable policy will not only allow you to carefully and correctly prepare your materials, but also will insure that the exam proctors will allow all proper materials that you bring to the exam.

As a general rule we recommend that you plan well in advance what books and materials you want to take to the exam. Then they should be obtained promptly so you use the same materials in your review that you will have in the exam.

License Review Books

There are two rules that we suggest you follow in selecting license review books to insure that you obtain up-to-date materials:

1. Consider the purchase only of materials that have a 1993 or more recent copyright. The exam used to be 25 essay questions, including a full scale engineering economics problem. The engineering economics problem is gone (at least as a separate question) and half the remaining 24 problems are now multiple choice. This 1993 date (or later) is especially important for the National Electric Code handbook.
2. Even if a license review book has a recent copyright date, is the content up to date? Books with older content probably will also lack the afternoon multiple choice problems.

Textbooks

If you still have your university textbooks, we think they are the ones you should use in the exam, unless they are too out of date. To a great extent the books will be like old friends with familiar notation.

Bound Reference Materials

The NCEES guidelines suggest that you can take any reference materials you wish, so long as you prepare them properly. You could, for example, prepare several volumes of bound reference materials with each volume intended to cover a particular category of problem. Maybe the most efficient way to use this book would be to cut it up and insert portions of it in your individually prepared bound materials. Use tabs so specific material can be located quickly. If you do a careful and systematic review of electrical engineering, and prepare a lot of well organized materials, you just may find that you are so well prepared that you will not have left anything of value at home.

Other Items

Calculator - NCEES says you may bring a battery operated, silent, non-printing calculator. You need to determine whether or not your state permits pre-programmed calculators. Extra batteries for your calculator are essential, and many people feel that a second calculator is also a very good idea.

Clock - You must have a time plan and a clock or wristwatch.

Pencils - You should consider mechanical pencils that you twist to advance the lead. This is no place to go running around to sharpen a pencil, and you surely do not want to drag along a pencil sharpener.

Eraser - Try a couple to decide what to bring along. You must be able to change answers on the multiple choice answer sheet, and that means a good eraser. Similarly you will want to make corrections in the essay problem calculations.

Exam Assignment Paperwork - Take along the letter assigning you to the exam at the specified location. To prove you are the correct person, also bring something with your name and picture.

Items Suggested By Advance Visit - If you visit the exam site you probably will discover an item or two that you need to add to your list.

Clothes - Plan to wear comfortable clothes. You probably will do better if you are slightly cool.
Box For Everything - You need to be able to carry all your materials to the exam and have them conveniently organized at your side. Probably a cardboard box is the answer.

Exam Scoring

Essay Questions

The exam booklets are returned to Clemson, SC. There the four essay question solutions are removed from the morning workbook. Each problem is sent to one of many scorers throughout the country.

For each question an item specific scoring plan is created with six possible scores: 0, 2, 4, 6, 8, and 10 points. For each score the scoring plan defines the level of knowledge exhibited by the applicant. An applicant who is minimally qualified in the topic is assigned a score of 6 points. The scoring plan shows exactly what is required to achieve the 6 point score. Similar detailed scoring criteria are developed for the two levels of superior performance (8 and 10 points) and the three levels of inferior performance (0, 2, and 4 points). Every essay problem submitted for grading receives one of these six scores. The scoring criteria may be based on positive factors, like identifying the correct computation approach, or negative factors, like improper assumptions or calculation errors, or a mixture of both positive and negative factors. After scoring, the graded materials are returned to NCEES, which reassembles the applicants work and tabulates the scores.

Multiple Choice Questions
Each of the four multiple choice problems is 10 points, with each of the ten questions of the problem worth one point. The questions are machine scored by scanning. The input data are evaluated by computer programs to do error checking. Marking two answers to a question, for example, would be detected and no credit given. In addition, the programs identify those questions with statistically unlikely results. There is, of course, a possibility that one or more of the questions is in some way faulty. In that case a decision will be made by subject matter experts on how the situation should be handled.

Passing The Exam

In the exam you must answer eight problems, each worth 10 points, for a total raw score of 80 points. Since the minimally qualified applicant is assumed to average six points per problem, a raw score of 48 points is set equal to a converted passing score of 70. Stated bluntly, you must get 48 of the 80 possible points to pass. The converted scores are reported to the individual state boards in about two months, along with the recommended pass or fail status of each applicant. The state board is the final authority of whether an applicant has passed or failed the exam.

Although there is some variation from exam to exam, the following gives the approximate passing rates:

Applicant's Degree	Percent Passing Exam
Engineering from accredited school	62%
Engineering from non-accredited school	50
Engineering Technology from accredited school	42
Engineering Technology from non-accredited school	33
Non-Graduates	36
All Applicants	56

Although you want to pass the exam on your first attempt, you should recognize that if necessary you can always apply and take it again.

These Two Volumes

The two books are organized to cover the electrical engineering professional engineer (principles and practice) exam. The books contain:
* A review of each topic on the exam.

* Example problems to illustrate the topic discussion along with solutions.
* Essay problems for the morning session, with detailed solutions.
* Multiple choice problems for the afternoon session, also with detailed solutions.
* A complete eight-hour sample exam, together with a complete solution for each problem.

Book One - Review
Each chapter begins with a review of the particular topic including example problems where appropriate. Following this there are essay and multiple choice problems. NCEES does not allow their problems to be reproduced, so none of the problems in this book came from them. Each one is structured to approximate the scope and difficulty of the actual exam problems you will encounter. Book One provides the content that might appear in a textbook or reference book. But just as in textbooks, the solutions to the end-of-chapter problems have been omitted. The Book One end-of-chapter problems and the sample exam are reprinted in Book Two, along with their detailed step-by-step solutions.

Book Two - Problems and Solutions
The National Council of Examiners for Engineering and Surveying (NCEES), which prepares the electrical engineering examination, calls it an open book examination. Most states accept this and allow applicants to bring textbooks, handbooks and any bound reference materials to the exam. A few states, however, do not permit review publications directed principally toward sample problems and their solutions. To insure that Book One of Electrical Engineering License Review will not be banned in any state, the solutions to the end-of-chapter problems and the sample exam have been omitted. In Book Two all the Book One problems are reprinted, but now with detailed step-by-step solutions. Thus in some states both books may be taken into the exam; in others, only Book One. This division of the material also has a positive aspect. Some people will want a lot of problems and solutions without a lot of text discussion. These people will find Book Two is an ideal independent problem and solution book.

Table 1.1. State Boards of Registration for Engineers

State	Mail Address	Phone
AL	301 Interstate Park Drive, Montgomery 36109	205-242-5568
AK	P.O. Box 110806, Juneau 99811	907-465-2540
AZ	1951 W. Camelback Rd, Suite 250, Phoenix 85015	602-255-4053
AR	P.O. Box 2541, Little Rock 72203	501-324-9085
CA	2535 Capitol Oaks Dr, #300, Sacramento 95833-2926	916-920-7466
CO	1560 Broadway, Ste. 1370, Denver 80202	303-894-7788
CT	165 Capitol Ave., Rm G-3A, Hartford 06106	203-566-3386
DE	2005 Concord Pike, Wilmington 19803	302-577-6500
DC	614 H Street NW, Rm 923, Washington 20001	202-727-7454
FL	1940 N. Monroe St., Tallahassee 32399	904-488-9912
GA	166 Pryor Street SW,Rm 504, Atlanta 30303	404-656-3926
GU	P.O. Box 2950, Agana, Guam 96910	671-646-1079
HI	P.O. Box 3469, Honolulu 96801	808-586-2702
ID	600 S. Orchard, Ste. A, Boise 83705	208-334-3860

IL	320 W. Washington St, 3/FL,Springfield 62786	217-785-0820
IN	100 N. Senate Ave, Rm 1021, Indianapolis 46204	317-232-2980
IA	1918 S.E. Hulsizer, Ankeny 50021	515-281-5602
KS	900 Jackson, Ste 507, Topeka 66612-1214	913-296-3053
KY	160 Democrat Drive, Frankfort 40601	502-564-2680
LA	1055 St. Charles Ave, Ste 415, New Orleans 70130	504-568-8450
ME	State House, Sta. 92, Augusta 04333	207-287-3236
MD	501 St. Paul Pl, Rm 902, Baltimore 21202	410-333-6322
MA	100 Cambridge St, Rm 1512, Boston 02202	617-727-9956
MI	P.O. Box 30018, Lansing 48909	517-335-1669
MN	133 E. Seventh St, 3/Fl, St. Paul 55101	612-296-2388
MS	P.O. Box 3, Jackson 39205	601-359-6160
MO	P.O. Box 184, Jefferson City 65102	314-751-0047
MT	111 N. Jackson Arcade Bldg, Helena 59620-0407	406-444-4285
NE	P.O. Box 94751, Lincoln 68509	402-471-2021
NV	1755 E. Plumb Lane, Ste 135, Reno 89502	702-688-1231
NH	57 Regional Dr., Concord 03301	603-271-2219
NJ	P.O. Box 45015, Newark 07101	201-504-6460
NM	1010 Marquez Pl, Santa Fe 87501	505-827-7561
NY	Madison Ave, Cult Educ Ctr., Albany 12230	518-474-3846
NC	3620 Six Forks Rd., Raleigh 27609	919-781-9499
ND	P.O. Box 1357, Bismarck 58502	701-258-0786
MP	P.O. Box 2078, Siapan, No Mariana Is. 96950	670-234-5897
OH	77 S. High St. 16/Fl, Columbus 43266-0314	614-466-3650
OK	201 NE 27th St, Rm 120, Oklahoma City 73105	405-521-2874
OR	750 Front St, NE, Ste 240, Salem 97310	503-378-4180
PA	P.O. Box 2649, Harrisburg 17105-2649	717-783-7049
PR	P.O. Box 3271, San Juan 00904	809-722-2122
RI	10 Orms St, Ste 324, Providence 02904	401-277-2565
SC	P.O. Drawer 50408, Columbia 29250	803-734-9166
SD	2040 W. Main St, Ste 304, Rapid City 57702	605-394-2510
TN	Volunteer Plaza, 3/Fl, Nashville 37243	615-741-3221
TX	P.O. Drawer 18329, Austin 78760	512-440-7723
UT	P.O. Box 45805, Salt Lake City 84145	801-530-6628
VT	109 State St., Montpelier 05609-1106	802-828-2875
VI	No. 1 Sub Base, Rm 205, St. Thomas 00802	809-774-3130
VA	3600 W. Broad St., Richmond 23230-4917	804-367-8514
WA	P.O. Box 9649, Olympia 98507	206-753-2548
WV	608 Union Bldg., Charleston 25301	304-348-3554
WI	P.O. Box 8935, Madison 53708-8935	608-266-1397
WY	Herschler Bldg., Rm 4135E, Cheyenne 82002	307-777-6155

CIRCUITS PROB. 2.1

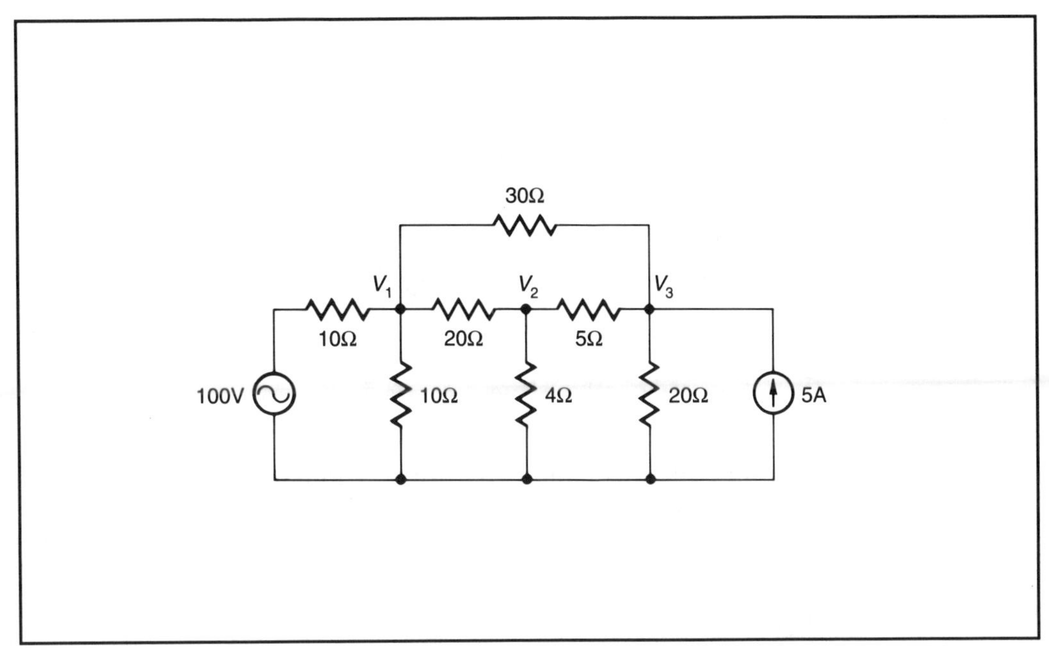

FigureP2.1

Determine the node voltages, V1, V2, and V3 by setting up the proper nodal equations in matrix form, manipulating the matrices to form the matrix solution for the voltages.

SOLUTION

To simplify the equations we may first replace the voltage source with its series resistance with a Norton equivalent circuit.

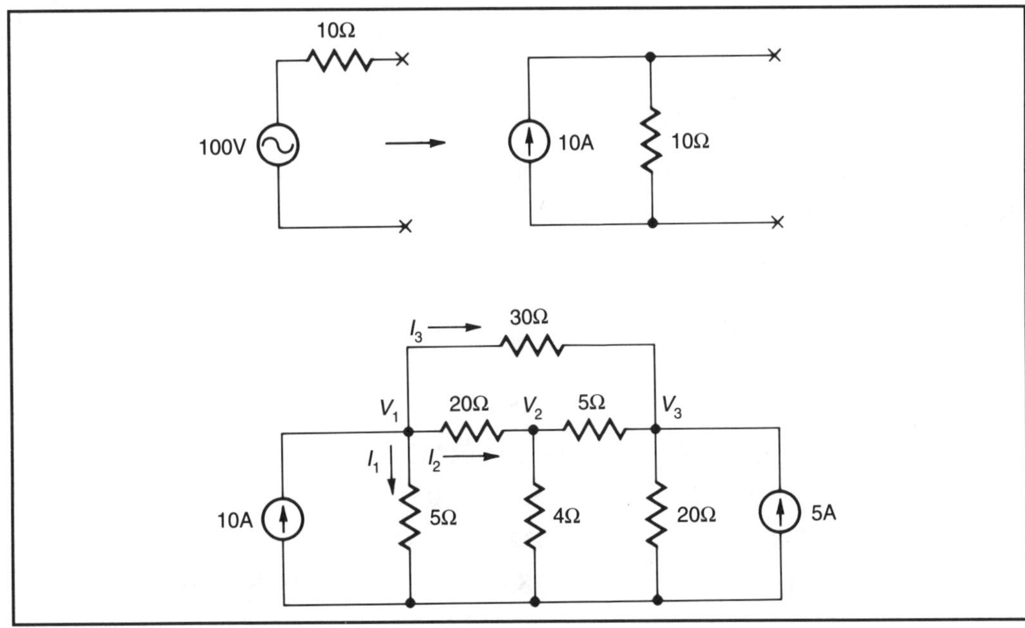

Solution 2.1

Note that the 10Ω resistor from the source is now in parallel with the 10Ω resistor of the network, making it effectively a 5Ω parallel branch.

The nodal equations are:

$$10 = V_1\left(\frac{1}{5}+\frac{1}{20}+\frac{1}{30}\right) - V_2\left(\frac{1}{20}\right) - V_3\left(\frac{1}{30}\right)$$

$$0 = -V_1\left(\frac{1}{20}\right) + V_2\left(\frac{1}{20}+\frac{1}{4}+\frac{1}{5}\right) - V_3\left(\frac{1}{5}\right)$$

$$5 = -V_1\left(\frac{1}{30}\right) - V_2\left(\frac{1}{5}\right) + V_3\left(\frac{1}{20}+\frac{1}{5}+\frac{1}{30}\right)$$

One would normally solve these equations by use of Cramer's rule; however the problem specifies the use of matrices.
This is in matrix form:

$$[I] = [Y][V]$$

Where:

$$\begin{bmatrix} 10 \\ 0 \\ 5 \end{bmatrix} = \begin{bmatrix} 0.2833 & -0.05 & -0.0333 \\ -0.05 & +0.5 & -0.2 \\ -0.0333 & -0.2 & 0.2833 \end{bmatrix} \begin{bmatrix} V_1 \\ V_2 \\ V_3 \end{bmatrix}$$

The matrix solution is:

$$[V] = [Y^{-1}][I]$$

$$\text{where: } [Y^{-1}] = \frac{\begin{bmatrix} A_{11} & A_{21} & A_{31} \\ A_{12} & A_{22} & A_{32} \\ A_{13} & A_{23} & A_{33} \end{bmatrix}}{|Y|}$$

Where A_{ij} = Signed minor of Y_{ij} (cofactor)

(Reference: Lipshutz: Theory and Problems of Linear Algerbra. Schaum's Outline Series, McGraw - Hill, Chapter 8.)

and $|Y|$ is the determinant of the Y matrix.

$$|Y| = 0.0269$$

So

$$Y^{-1} = \frac{\begin{bmatrix} +\begin{vmatrix} .5 & -.2 \\ -.2 & .2833 \end{vmatrix} & -\begin{vmatrix} -.05 & -.0333 \\ -.2 & .2833 \end{vmatrix} & +\begin{vmatrix} -.05 & -.0333 \\ .5 & -.2 \end{vmatrix} \\ -\begin{vmatrix} -.05 & -.2 \\ -.0333 & .2833 \end{vmatrix} & +\begin{vmatrix} .2833 & -.0333 \\ -.0333 & .2833 \end{vmatrix} & -\begin{vmatrix} .2833 & -.0333 \\ -.05 & -.2 \end{vmatrix} \\ +\begin{vmatrix} -.05 & .5 \\ -.0333 & -.2 \end{vmatrix} & -\begin{vmatrix} .2833 & -.05 \\ -.0333 & -.2 \end{vmatrix} & +\begin{vmatrix} .2833 & -.05 \\ -.05 & .5 \end{vmatrix} \end{bmatrix}}{0.0269}$$

$$Y^{-1} = \begin{bmatrix} 3.7807 & .7732 & .9926 \\ .7732 & 2.9442 & 2.1673 \\ .9926 & 2.1673 & 5.1747 \end{bmatrix}$$

$$\begin{bmatrix} V_1 \\ V_2 \\ V_3 \end{bmatrix} = \left[Y^{-1}\right] \begin{bmatrix} 10 \\ 0 \\ 5 \end{bmatrix}$$

Multiplying row by column:

$V_1 = 10(3.7807) + 0(0.7732) + 5(0.9926) = 42.77 volts$

$V_2 = 10(0.7732) + 0(2.9442) + 5(2.1673) = 18.57 volts$

$V_3 = 10(0.9926) + 0(2.1673) + 5(5.1747) = 35.80 volts$

Check node no. 1

$I_1 = \frac{42.8}{5} = 8.56 Amps \quad I_2 = \frac{42.8 - 18.57}{20} = 1.21 Amps$

$I_3 = \frac{42.8 - 35.8}{30} = 0.23 Amps$

CIRCUITS PROB 2.2

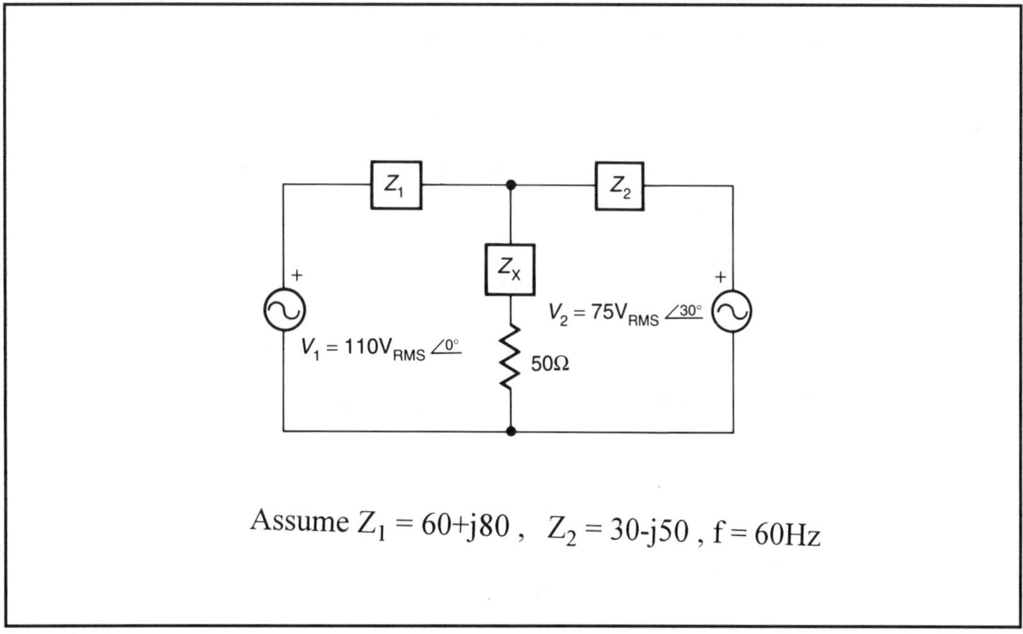

Figure P2.2

For 50 Ω load is to receive maximum, power from this system. A small series reactance, Z_x is to be placed in series with the load to accomplish this. Find the proper element for Z_x and calculate the power into the 50Ω load when this element is placed in the circuit.

SOLUTION

This problem may be simplified by first finding a Thevenin equivalent circuit for the two sources.

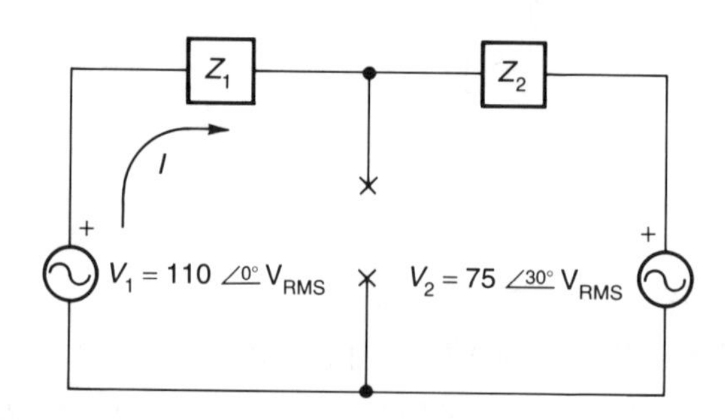

$$V_{xx} = V_1 - IZ_1 = V_{OC} \text{ and } I = \frac{V_1 - V_2}{Z_1 + Z_2}$$

$$I = \frac{110\angle 0° - 75\angle 30°}{60 + j80 + 30 - j50} = \frac{110 + j0 - (64.95 + j37.5)}{90 + j30}$$

$$I = \frac{58.61\angle -39.78°}{94.87\angle 18.43°} = 0.62\angle -58.2° \; AmpsRms$$

$$\text{Then } V_{OC} = 110\angle 0° - 0.62\angle -58.2°(60 + j80)$$

$$= 110\angle 0° - (0.62\angle -58.2°)(100\angle 53.1°)$$

$$= 110\angle 0° - 62\angle -5.1°$$

$$= 48.8\angle 6.4° V_{RMS} = V_{Thev} = V_{OC}$$

To find the Thevenin impedance, we look into terminals xx with all sources disabled.

$$Z_{eq} = Z_{Thev} = \frac{Z_1 Z_2}{Z_1 + Z_2} = \frac{(60 + j80)(30 - j50)}{90 + j30}$$

$$= \frac{(100\angle 53.1°)(58.3\angle -59°)}{94.9\angle 18.4°}$$

$$= 61.46\angle -24.3° = 56 - j25.3$$

The circuit now looks like this,

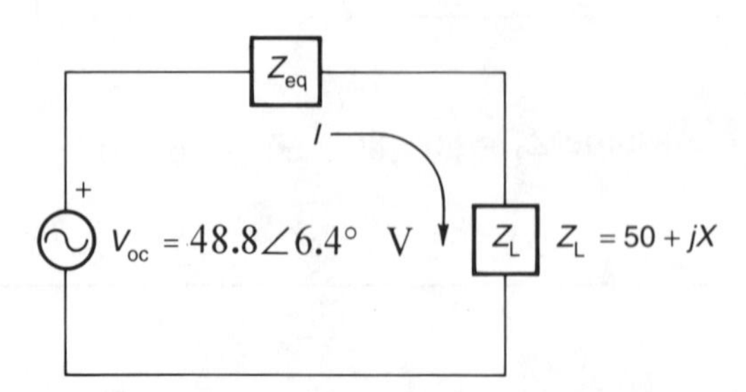

The phase angle of the source is only a reference to the phase of the original source and may be ignored while finding jX. To maximize power, $Z_L = Z^*_{Thev}$, that is, the load impedance should be the complex conjugate of the source. Since we cannot make the real parts equal, $56 \neq 50$, we must recognize that the maximum current through R_L will deliver maximum power.

$$I = \frac{V_{OC}}{Z_{eq} + Z_L}$$

$$Z_{eq} + Z_L = 56 - j25.3 + 50 + jX$$

The denominator will be minimum when $|-j25.3| = |+jX|$

$$\text{Then} \quad X = \omega L = 25.3$$

$$L = \frac{25.3}{\omega} \quad \text{but } \omega = 2\pi(60) = 377$$

$$L = \frac{25.3}{377} = 67.2 milihenrys$$

$$|I| = \frac{48.8}{50+56} = \frac{48.8}{106} = 0.46 Amps\ rms$$

$$\text{So Power } = I^2 R_L$$

$$P = (0.46)^2(50) = 10.6 watts$$

CIRCUITS PROB. 2.3:

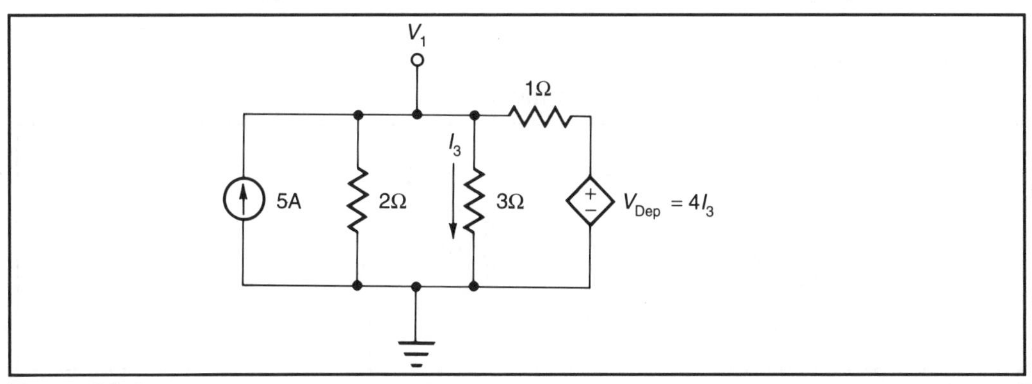

Figure P2.3

Find the current through the 3 ohm resistor for the above circuit with a dependent voltage source. Also find the power taken from the source. Check the powers being dissipated in the resistors; do they equal the power taken from the source?

SOLUTION

Sum the currents at voltage node 1:

$$-5 + V_1/2 + V_1/3 + (V_1 - V_{dep})/1 = 0$$

x6: $-5\text{x}6 + 3V_1 + 2V_1 + 6V_1 - 6(4V_1/3)$

$30 = 3V_1$; $V_1 = 10$ V; $I_\alpha = 10/3 = 3.33$ A

Power taken from current source:

$$P_{CS} = VI = 10x5 = 50 \text{ watts}$$

Power dissipated in resistors:

$P_1 = I_1^2 R_1$, where:

$$I_1 = (V_1 - V_{dep})/1 = 10\text{-}4x3.33)/1 = -3.33 \text{ A}$$

$$P_1 = (-3.33)^2 x1 = 11.09 \text{ watts}$$

$$P_2 = V_1^2/R_2 = 10^2/2 = 50 \text{ watts}$$

$$P_3 = V_1^2/R_3 = 10^2/3 = 33.3 \text{ watts}$$

$$P_{Tot} = 11.09 + 50 + 33.3 = 94.4 \text{ watts}$$

Therefore the power supplied by the dependent source is

94.4 - 50 = 44.4 watts

Check:

$$P_{dep} = V_{dep} x I_1 = (4xI_3)I_1 = (4x3.33)x3.33 = 44.4 \text{ watts}$$

CIRCUITS PROB 2.4

Determine the value of i(t) as shown in the steady state.

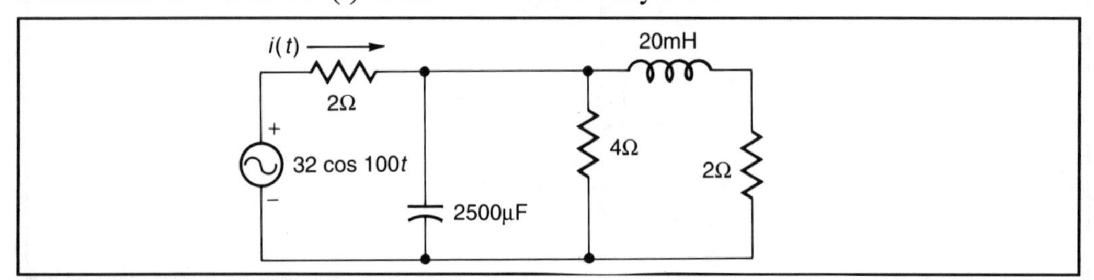

Figure P2.4

SOLUTION

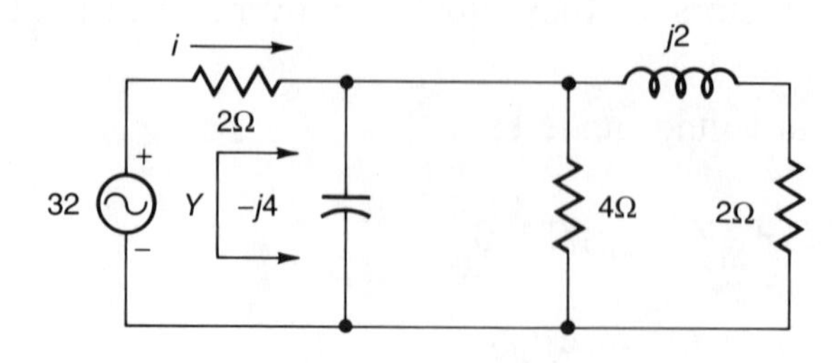

$$Y = \frac{j}{4} + \frac{1}{4} + \frac{1}{2(1+j)} = \frac{1}{4}(1+j) + \frac{1}{2}\left(\frac{1}{1+j}\right)$$

$$= \frac{1}{4}\left[\frac{(1+j)^2+2}{1+j}\right] = \frac{1}{4}\left[\frac{1+j2-1+2}{1+j}\right]$$

$$= \frac{1}{2}\left[\frac{1+j}{1+j}\right] = \frac{1}{2}$$

$$Z = \frac{1}{Y} = 2\Omega$$

$$I = \frac{32}{Z+2} = \frac{32}{4} = 8\text{ A}$$

$$i(t) = 8\cos 100t\text{ A}$$

CIRCUITS PROB 2.5

Find the phasor current I and the phasor voltage drop across each element in polar form using the source as the reference for angles.

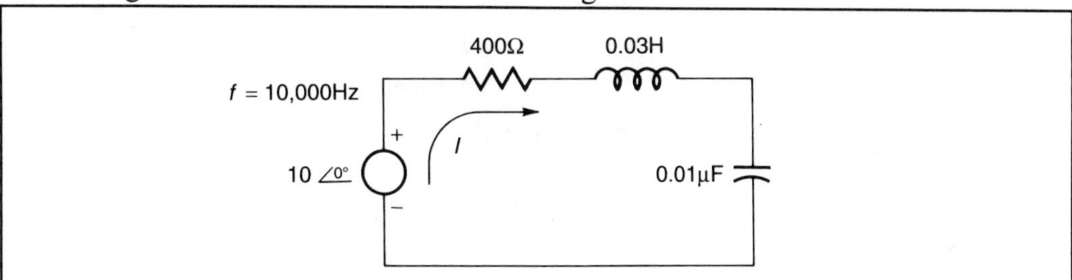

Figure P2.5

SOLUTION

$$\boldsymbol{Z_R} = 400\Omega$$

$$\boldsymbol{Z_L} = \boldsymbol{j}(20{,}000\pi)(.03) = j1885.0\Omega$$

$$\boldsymbol{Z_C} = \frac{-\boldsymbol{j}}{(20{,}000\pi)(0.01\times 10^{-6})} = -j1591.5\Omega$$

$$\boldsymbol{KVL:}\quad -10 + 400\boldsymbol{I} + j1885\boldsymbol{I} - j1591.5\boldsymbol{I} = 0$$

$$\boldsymbol{I} = \frac{10\angle 0^\circ}{400 + j293.5} = 0.0202\angle -36.3^\circ\text{ A}$$

$$\boldsymbol{V_R} = 400\boldsymbol{I} = 8.08\angle -36.3^\circ\text{ V}$$

$$\boldsymbol{V_L} = j1885\,\boldsymbol{I}$$

$$\boldsymbol{V_L} = 37.99\angle 53.7^\circ$$

$$\boldsymbol{V_C} = -j1591.5\text{ I} = 32.08\angle -126.3^\circ\text{ V}$$

$$\boldsymbol{V_{Source}} = 10\angle 0\text{ V}$$

CIRCUITS PROB. 2.6

The frequency of each source is 1000 HZ with the phase relations shown on the diagram. Replace this circuit by its Norton Phasor equivalent.

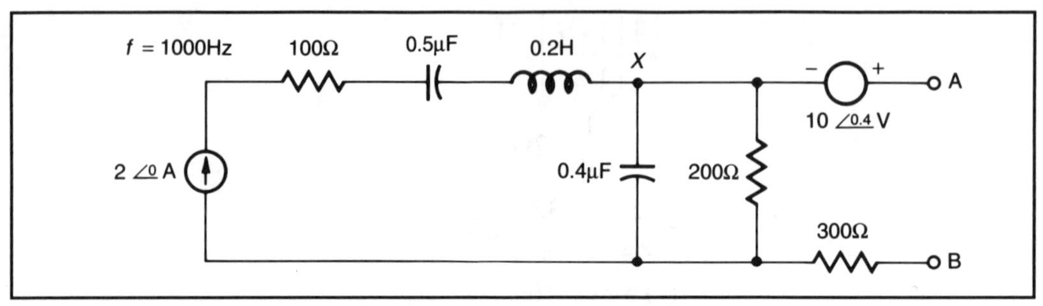

Figure P2.6

SOLUTION

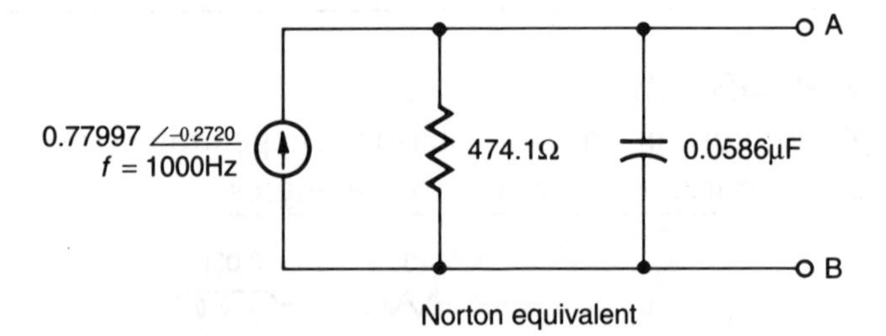

$$Y_{0.4\mu F} = j(2000\pi)(0.4\times10^{-6}) = j0.0025133$$

$$KCL \text{ Node x: } \quad -2 + j0.0025133V_{x0} + \frac{1}{200}V_{x0} + \frac{1}{300}(V_{x0} + 10\angle 0.4) = 0$$

$$V_{x0} = \frac{1.96934\angle -0.0065913}{0.008704\angle 0.29292} = 226.257\angle -0.29951$$

$$\text{Short Circuit } I_{AB} = \frac{1}{300}\left[(216.184 - j66.7576) - 10\angle 0.4\right]$$

$$I_{AB} = 0.77997\angle -0.27197 \text{ A}$$

Replacing each source by its internal impedance:

$$Z_{AB} = \frac{1}{0.005 + j0.0025133} + 300 = 459.66 - j80.254$$

$$Z_{AB} = 466.61\angle -0.17285\ \Omega$$

$$Y_{AB} = \frac{1}{Z_{AB}} = 0.0021431\angle 0.17285 = 0.0021093 + j0.0003686$$

$$R = \frac{1}{0.002193} = 474.1\Omega \ ; \ C = \frac{0.003686}{2000\pi} = 0.0586\mu F$$

TRANSIENTS PROB. 2.7

A 2.0 μf capacitor is charged so that it has 100 volts across its terminals. The terminals are suddenly connected through a negligible resistance to the terminals of a 4.0 μf capacitor having no initial charge.

(a). What are the final steady state voltages across each capacitor?
(b). What are the initial stored energies of each capacitor?
(c). What are the final steady state stored energies of each capacitor?

A 100 ohm resistor is placed in series with the 4.0μf capacitor so that the charging current will flow through the resistor. The same 2.0μf capacitor charged to 100 volts is connected to the combination of the 4.0μf capacitor and 100 ohm resistor in series with the 4.0μf capacitor having no initial charge.

(d) What is the time constant of the circuit?
(e) What are the final steady state voltages across each capacitor?
(f) What are the final steady state stored energies of each capacitor?
(g) How do you explain or account for the results of the energies calculated in (c) and (f) above in light of the different resistor power losses?

SOLUTION

a). $Q=CE=2 \times 10^{-6} \times 100 = 200 \times 10^{-6}$ coulomb

Capacitors are suddenly connected together, assuming no circuit resistance and the 4μfd capacitor has $Q_2 = 0$.

The capacitors are in parallel across V so electrons will flow until equilibrium is reached.

The total charge $Q=200 \times 10^{-6}$ coulomb remains in the system.

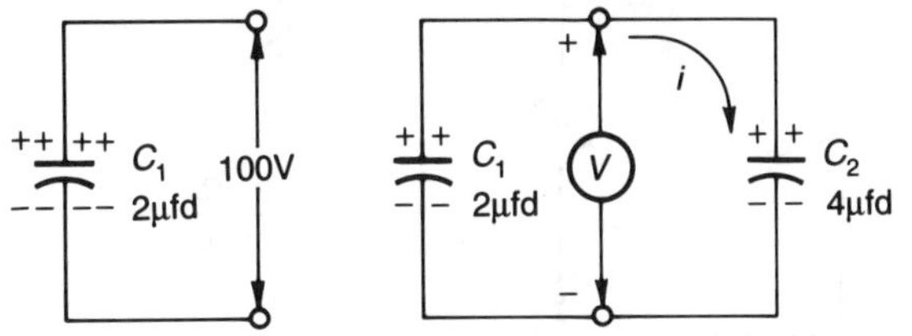

++ ++ C_1 2μfd 100V -- --

+ + C_1 2μfd - - + V - i + + C_2 4μfd - -

$$\text{So } V = \frac{Q}{C_1 + C_2} = \frac{200 \times 10^{-6}}{(2+4) \times 10^{-6}} = 33.3 \text{ Volts}$$

The voltage across each capacitor is 33.3 volts.

b) $$W = \int \frac{1}{C} dq = \frac{1}{2}\frac{Q^2}{C} \text{ and initially 0}$$

$Q_1 = 200 \times 10^{-6}$ coulomb so:

$$W_1 = \frac{1}{2}\frac{\left(200 \times 10^{-6}\right)^2}{2 \times 10^{-6}} = \frac{\left(2 \times 10^{-4}\right)^2}{4 \times 10^{-6}}$$

$$W_1 = \frac{1 \times 10^{-8}}{10^{-6}} = 1 \times 10^{-2} \text{ joules}$$

energy initially in C_1

Q_2 was zero initally, so $W_2 = 0$

c) Final energy

$$Q_1 = 2\times10^{-6}\times33.3 = 66.6\times10^{-6} \text{ coulomb}$$

$$W_1 = \frac{1}{2}\frac{\left(66.6\times10^{-6}\right)^2}{2\times10^{-6}} = \frac{4.45\times10^{-10}}{4\times10^{-6}}$$

$$= 1.11\times10^{-3} \text{ joules in } C_1$$

$$Q_2 = 4\times10^{-6}\times33.3 = 133.3\times10^{-6}$$

$$W_2 = \frac{1}{2}\frac{\left(133.3\times10^{-6}\right)^2}{4\times10^{-6}} = \frac{1.76\times10^{-8}}{8\times10^{-6}}$$

$$= 2.22\times10^{-3} \text{ joules in } C_2$$

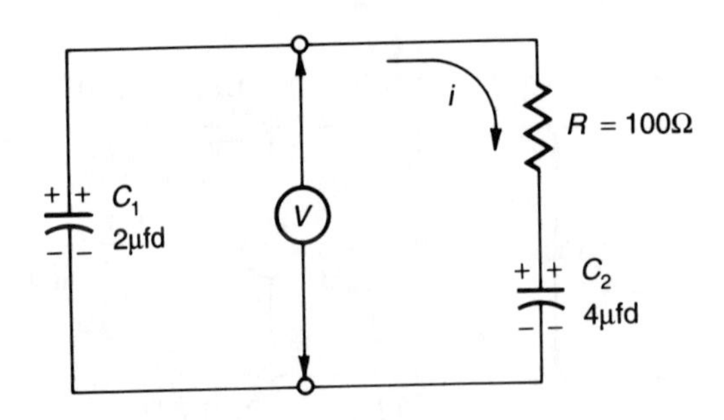

Time Constant t = RC

$$Ri + \frac{1}{C_1}\int idt + \frac{1}{C_2}\int idt = 0$$

i steady state = 0, so:

i transient is the solution of

$$R\frac{di}{dt} + \frac{1}{C_1}i + \frac{1}{C_2}i = 0 \quad R\frac{di}{dt} + \left(\frac{C_1+C_2}{C_1C_2}\right)i = 0$$

$$\text{so } i = Ke^{-\left(\frac{C_1+C_2}{C_1C_2R}\right)t} \quad \text{where } \frac{C_1+C_2}{C_1C_2R} = \frac{2+4}{2\times4R} = \frac{3}{4R}$$

$$i = \frac{VC_1}{R}e^{-\left(\frac{C_1+C_2}{C_1C_2R}\right)t} \quad \text{so } \frac{t}{RC} = \frac{t}{100\times\frac{4}{3}} = 1$$

d) Time Constant $\tau = RC = 100 \times 10^{-6}$

$= 133 \times 10^{-6}$ Seconds

e) $E_{C_1} = E_{C_2} = 33.3$ Volts, same as part (a).

f) $W_1 = \frac{1}{2}\frac{\left(66.6 \times 10^{-6}\right)^2}{2 \times 10^{-6}} = 1.11 \times 10^{-3}$ joules

$W_2 = \frac{1}{2}\frac{\left(133.3 \times 10^{-6}\right)^2}{4 \times 10^{-6}} = 2.22 \times 10^{-3}$ joules

g) The circuit resistance determines the peak discharge or charge current, so there is I^2R loss in parts a, b & c even if the resistance of the copper bar seems negligible. This accounts for the loss in energy.

TRANSIENTS PROB. 2.8

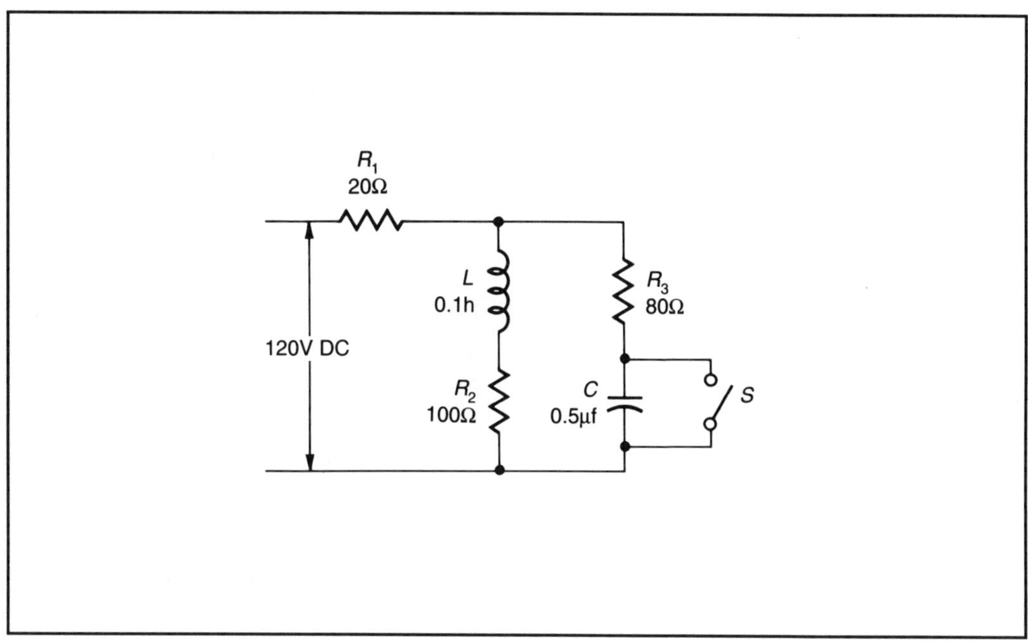

Figure P2.8

The circuit shown above is in a steady state.

REQUIRED

(a) What current is drawn from the power source in this initial steady state condition?

(b) What is the analytical expression for the current drawn from the power source after closing switch S?

SOLUTION:

a) With the switch open, and with the statement that the circuit is in steady state (to a dc source), one may make the assumption that the current through the inductor is no longer changing: The voltage across this element will then be zero. The voltage across the capacitor has built up to its steady state value and therefore no current will be flowing in this branch.

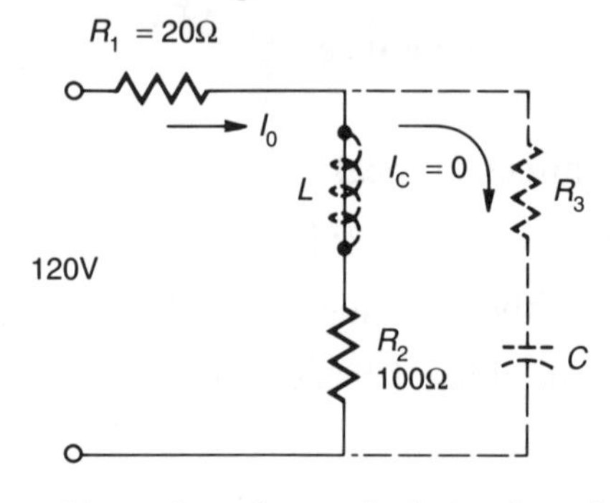

$$I_0 = \frac{E}{R_1 + R_2} = \frac{120}{120} = 1 \text{ A}$$

b) After the switch is closed,
the equivalent circuit will be as follows (with an initial inductor current as found in part a):

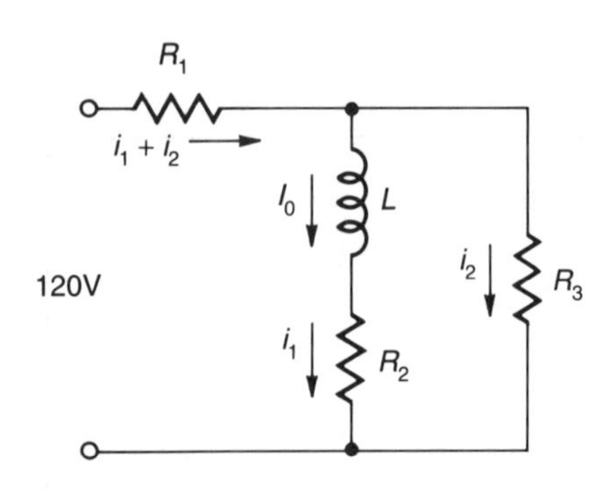

Loop (differential) equations:

1) $\text{E} = \text{R}_1(i_1 + i_2) + L\dfrac{di}{dt} + R_2 i_1$

2) $\text{E} = \text{R}_1(i_1 + i_2) + R_3 i_2$

with an initial condition of $\text{i}_1(0) = I_0$

Using Laplace transforms to solve:

1) $\dfrac{\text{E}}{\text{s}} = R_1(I_1 + I_2) + L[sI_1 - i_1(0)] + R_2 I_1$

2) $\dfrac{\text{E}}{\text{s}} = R_1(I_1 + I_2) + R_3 I_2$

Rearranging terms:

1) $\dfrac{\text{E}}{\text{s}} + Li_1(0) = [(R_1 + R_2) + Ls]I_1 + R_1 I_2$

$$\frac{120}{s} + 0.1 = (120 + 0.1s)I_1 + 20I_2$$

2) $\dfrac{120}{\text{s}} = (20)I_1 + (100)I_2$

Solving for I_1 and I_2:

$$I_1 = \frac{\begin{vmatrix} \left(\frac{120}{s}+0.1\right) & 20 \\ \left(\frac{120}{s}\right) & 100 \end{vmatrix}}{\begin{vmatrix} (120+0.1s) & 20 \\ 20 & 100 \end{vmatrix}} = \frac{\left(\frac{120}{s}+0.1\right)100-\left(\frac{120}{s}\right)20}{(120+0.1s)100-20^2}$$

$$= \frac{96+0.1s}{s(116+0.15)}$$

$$I_2 = \frac{\begin{vmatrix} (120+0.1s) & \left(\frac{120}{s}+0.1\right) \\ 20 & \left(\frac{120}{s}\right) \end{vmatrix}}{(120+0.1s)100-20^2} = \frac{120+0.1s}{s(116+0.1s)}$$

But $I_{Source} = I_1 + I_2$

$$\therefore I_s = \frac{96+0.1s+120+0.1s}{s(116+0.1s)} = \frac{216+0.2s}{s(116+0.1s)}$$

$$= \frac{216}{116}\left[\frac{1+\frac{0.2}{216}s}{s(1+\frac{0.1}{116}s)}\right] = 1.86\left[\frac{1+\tau_1 s}{s(1+\tau_2 s)}\right]$$

where $\tau_1 = 0.000927$

$\tau_2 = 0.0008625$

$$\therefore i_s(t) = \mathcal{L}^{-1}[I_s] = 1.86\left[1-(1-\frac{\tau_1}{\tau_2})e^{-\frac{t}{\tau_2}}\right]$$

$$= 1.86(1+0.072e^{-\frac{t}{\tau_2}})$$

TRANSIENTS PROB 2.9

The switch was closed sufficiently long ago such that the current i (through the source) has reached steady state. The switch is then opened at time t = t(0).

REQUIRED:

(a) Find the current i just before the switch is opened; that is, at $t = t(0^-)$
(b) Find the current i just after the switch is opened; that is, at $t = t(0^+)$
(c) Find the current i as a function of time after the switch is opened; that is, find $i(t-t_0)$, where $t_0 = t(0)$.

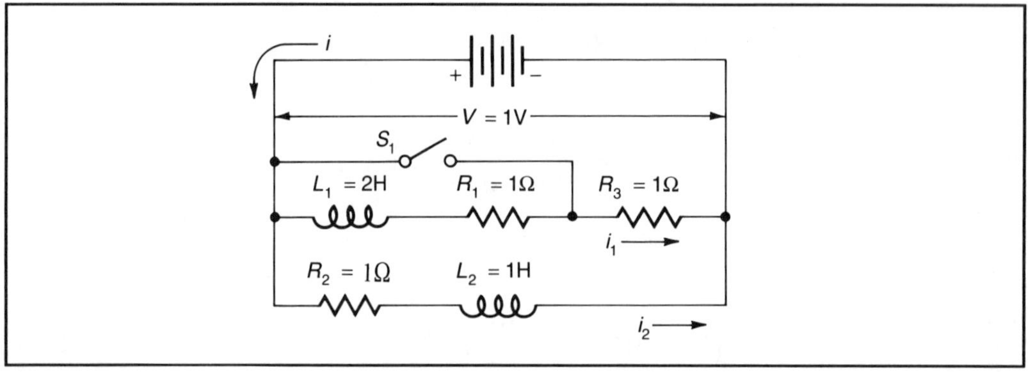

Figure P2.9

SOLUTION:

a) At $t = t(0^-)$ only R_3 and R_2 determine the magnitude of i

$$R_{total} = \frac{1 \times 1}{1+1} = 0.5 \text{ ohm}$$

$$i = \frac{E}{R_{total}} = \frac{1}{0.5} = 2 \text{ amp} \qquad \text{ANSWER}$$

b) At $t = t(0^+)$, since the current in an inductance will not change instantaneously, the current in L_1 will remain at 0. Only R_2 will determine the magnitude of i

$$i = \frac{E}{R_2} = \frac{1}{1} = 1 \text{ amp} \qquad \text{ANSWER}$$

c) At $(t - t_0)$ the R_2 and L_2 branch of the circuit is in steady state and is not considered. Therefore for the L_1, R_1 and R_3 branch:

$$i = i_{transient} + i_{R_2}$$

$$i_{transient} = \frac{E}{R}\left(1 - e^{-\frac{Rt}{L_1}}\right)$$

$i_{R_2} = 1$ amp (See above (b))

$R = R_1 + R_3 = 1 + 1 = 2$ ohms

Therefore

$$i = \frac{1}{2} - \frac{1}{2}e^{-t} + 1 = 1.5 - 0.5e^{-t} \qquad \text{ANSWER}$$

Proof: at $t = 0$, $e^{-t} = 1.0$ and $i = 1.5 - 0.5 \times 1.0 = 1$ amp

Note: Part (c) can be also worked by using the classic solutions of:

$$L_1 \frac{di_1}{dt} + (R_1 + R_3)i_1 = 1$$

$$L_2 \frac{di_2}{dt} + R_2 i_2 = 1$$

$i = i_1 + i_2 = 1.5 - 0.5e^{-t}$ by applying the proper boundary conditions.

TRANSIENTS PROB 2.10

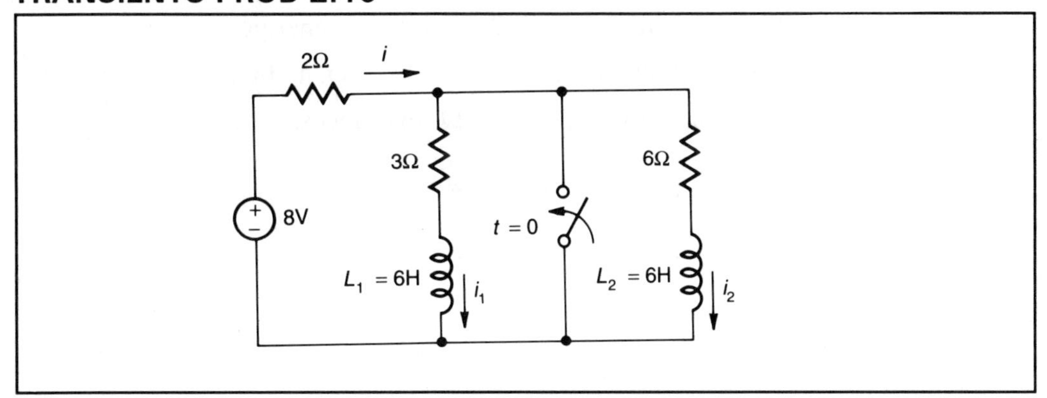

Figure P2.10

For the circuit shown determine: $i_1(0^+)$, $i_2(0^+)$, energy stored in L_1 and L_2 at $t=0$, $i_1(t)$ and $i_2(t)$ for $t>0$.

SOLUTION

The (dc steady state) circuit for $t < 0$ is shown

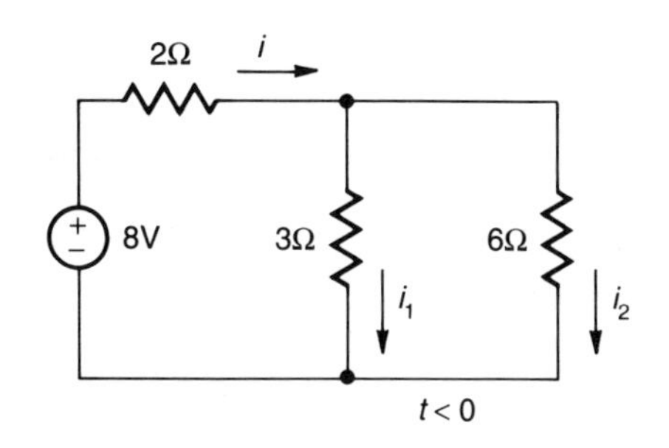

So $i(o^-) = \dfrac{8}{2 + \dfrac{3 \times 6}{3+6}} = 2$ A

By the current divider rule, $i_1(0^-) = 2 \times \dfrac{6}{9} = \dfrac{4}{3}$ A

and $i_2(0^-) = 2 \times \dfrac{3}{9} = \dfrac{2}{3}$ A

Since currents in inductors cannot change instantaneously,

$$i_1(0^-) = i_1(0^+) = \frac{4}{3} \text{ A and } i_2(0^-) = i_2(0^+) = \frac{2}{3} \text{ A.}$$

Energy stored in L_1 , at t = 0 is,

$$\frac{1}{2} L_1 i_1^2(0^+) = \frac{1}{2} \times 6 \times \left(\frac{4}{3}\right)^2 = \frac{16}{3} J.$$

Similarly energy stored in L_2 at t = 0 is,

$$\frac{1}{2} L_2 i_2^2(0^+) = \frac{1}{2} \times 6 \times \left(\frac{2}{3}\right)^2 = \frac{4}{3} J.$$

At t = 0 the switch is closed. The stored energies in L_1 and L_2 flow through the closed switch and get dissipated in the 3Ω and 6Ω resistances.

$$\text{for } t > 0\text{: } i_1(t) = i_1(0^+) e^{-\frac{3\Omega}{6H}t} = \frac{4}{3} e^{-\frac{t}{2}} \text{ A.}$$

$$\text{and } \quad i_2(t) = i_2(0^+) e^{-\frac{6\Omega}{6H}t} = \frac{2}{3} e^{-t} \text{ A.}$$

TRANSIENTS PROB 2.11

For the circuit shown below find and sketch $i_L(t)$.

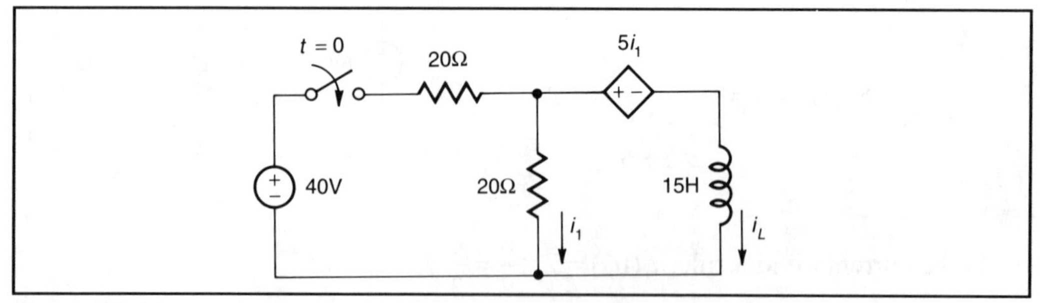

Figure P2.11

SOLUTION:

One may replace the circuit to left of inductor with its Thevenin equivalent for $t \geq 0$

$$V_{oc} = V_{Thev} = \frac{40 \times 20}{20+20} - 5 \times \frac{40}{20+20} = 15 \text{ V (open circuit voltage)}$$

To find R_{eq} "kill" 40 V source and replace inductor with 1A source.

Then

$$R_{eq} = \frac{V_0}{1A} = 1A \times \frac{20 \times 20}{20+20} - 5\frac{20}{20+20} = 7.5\Omega$$

The circuit then becomes:

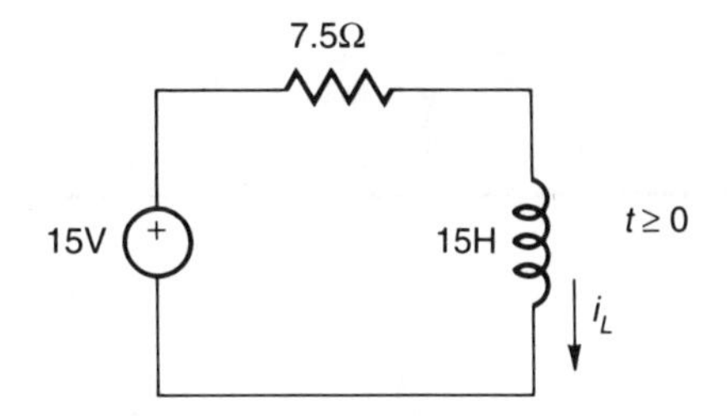

$$\text{Forced } i_L = i_{Lf} = \frac{15}{7.5} = 2 \text{ A}$$

$$\text{Natural } i_L = i_{Ln} = Ae^{-\frac{t}{L/R}} = Ae^{-\frac{t}{2}}$$

$$i_L = Ae^{-\frac{t}{2}} + 2$$

$$i_L(0) = 0 = A + 2 \ ; \ A = -2$$

$$\therefore i_L = 2\left(1 - e^{-\frac{t}{2}}\right) \text{ amps } t \geq 0$$

$$= 0 \quad t < 0$$

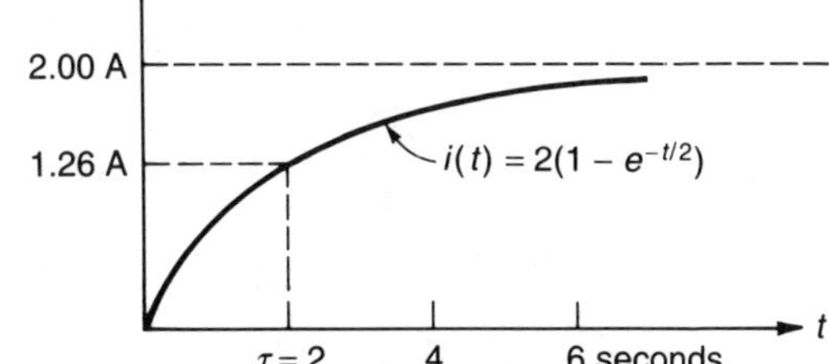

TRANSIENT PROB 2.12

Find the voltage, v, as shown. The circuit is initally unenergized.

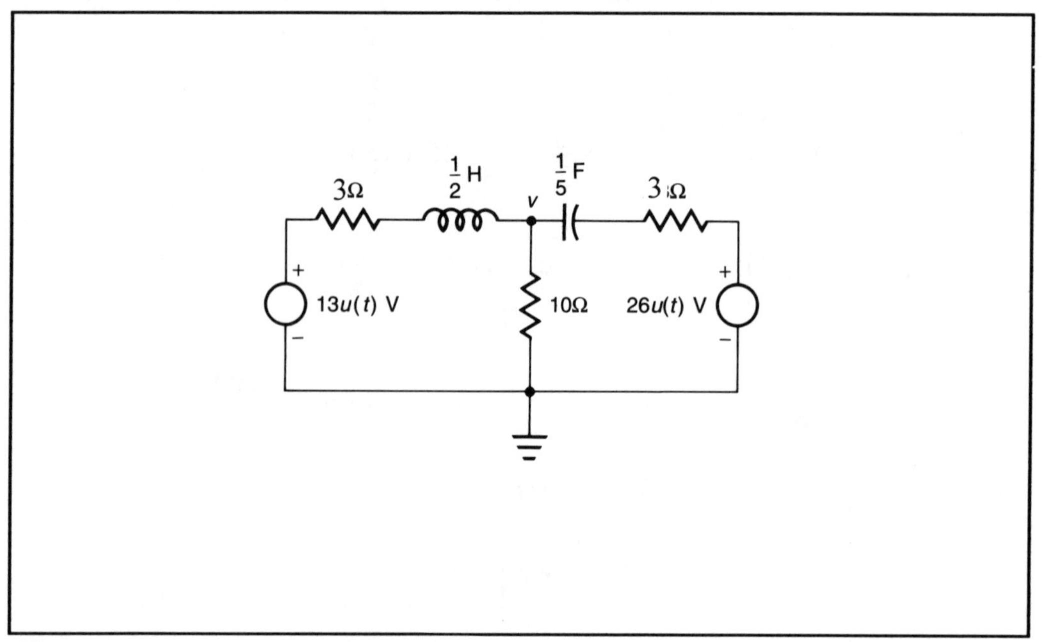

Figure P2.12

SOLUTION:

For $t \geq 0$

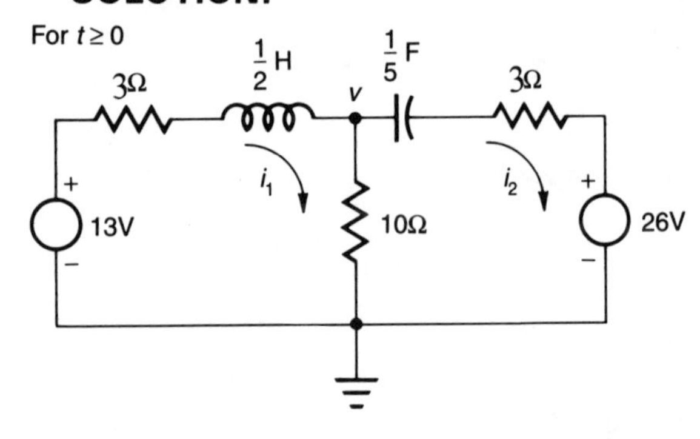

$$(3+10)i_1 + \frac{1}{2}\frac{di_1}{dt} - 10i_2 = 13$$

$$-10i_1 + (3+10)i_2 + 5\int i_2 dt = -26$$

$$10i_1 = 13i_2 + 5\int i_2 dt + 26$$

$$i_1 = 1.3i_2 + 5\int i_2 dt + 2.6$$

$$13i_1 = 16.9i_2 + 6.5\int i_2 dt + 33.8$$

$$\frac{di_1}{dt} = 1.3\frac{di_2}{dt} + 0.5i_2$$

$$\frac{1}{2}\frac{di_1}{dt} = 0.65\frac{di_2}{dt} + 0.25i_2$$

$$16.9i_2 + 6.5\int i_2 dt + 33.8 + 0.65\frac{di_2}{dt} + 0.25i_2 - 10i_2 = 13$$

$$0.65\frac{di_2}{dt} + 7.15i_2 + 6.5\int i_2 dt = -20.8$$

$$\frac{di_2}{dt} + 11i_2 + 10\int i_2 dt = 32$$

$$\frac{d^2 i_2}{dt^2} + 11\frac{di_2}{dt} + 10i = 0$$

The Laplace transform equivalent is:

$$s^2 + 11s + 10 = (s+10)(s+1) = 0$$

$$i_2(t) = Ae^{-10t} + Be^{-t}$$

$$i_1(t) = 1.3Ae^{-10t} + 1.3Be^{-t} + .5A\int e^{-10t}dt + .5B\int e^{-t}dt + 2.6$$

$$1.25Ae^{-10t} + 0.8Be^{-t} + C + 2.6$$

$$\lim_{t\to 0} i_1(t) = 1 \qquad \therefore\ \text{C} = \text{-}1.6$$

$$i_1(t) = 1.25Ae^{-10t} + 0.8Be^{-t} + 1$$

$$i_1(0) = 0 = 1.25A + 0.8B + 1$$

$$i_2(0) = -2 = A + B$$

$$A + B = -2 \qquad 1.25A + 0.8B = -1$$

$$1.25A + 1.25B = -2.5$$

$$-1.25A - 0.8B = +1.0 \qquad \therefore B = -\frac{10}{3} \quad A = \frac{4}{3}$$

$$i_1(t) = \frac{5}{3}e^{-10t} - \frac{8}{3}e^{-t} + 1$$

$$i_1(t) = \frac{4}{3}e^{-10t} + \frac{10}{3}e^{-t}$$

$$v(t) = 10(i_1 - i_2) = 10\left(\frac{5}{3}e^{-10t} - \frac{8}{3}e^{-t} + 1 - \frac{4}{3}e^{-10t} + \frac{10}{3}e^{-t}\right)$$

$$= 10\left(\frac{1}{3}e^{-10t} + \frac{2}{3}e^{-t} + 1\right)$$

$$= \frac{10}{3}e^{-10t} + \frac{20}{3}e^{-t} + 10$$

IMPEDANCE MATCHING PROB. 2.13

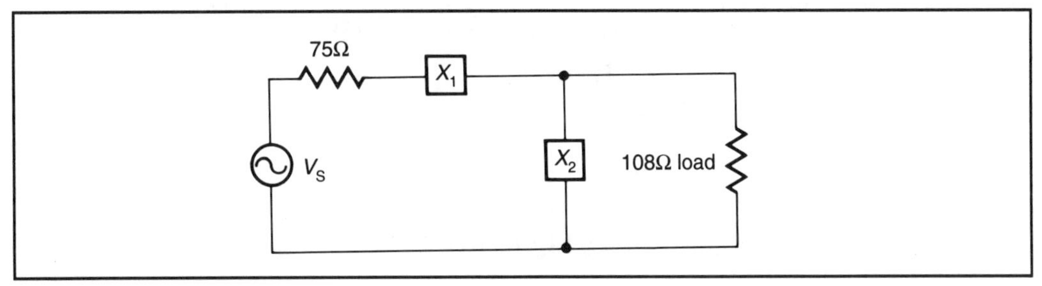

Figure P2.13

V_S is a 60 Hz source. It is desired to maximize the power delivered to the 108Ω resistor. The 75Ω resistance is internal to the source. X_1 and X_2 are reactive elements (capacitors or inductors) to be added. Find appropriate values for these elements so that the power to the load is maximum.

SOLUTION

Assume that X_1 is part of the source impedance and X_2 will be part of the load impedance. Then maximum power will be delivered to the load when:

$Z_s = Z_s^*$ (the load is the complex conjugate of the source impedance.)

Assume X_1 to be inductive so

$$Z_s = 75 + jX_1$$

Then the parallel load impedance is capacitive,

$$\frac{108(-jX_2)}{108 - jX_2} = \frac{(108)\left(\frac{1}{j\omega C}\right)}{108 + \frac{1}{j\omega C}}\left(\frac{108 - \frac{1}{j\omega C}}{108 - \frac{1}{j\omega C}}\right)$$

$$= \frac{108}{1 + \omega^2(108)^2 C^2} - \frac{j\omega(108)^2 C}{1 + \omega^2(108)^2 C^2}$$

For the conjugate match, the real part of the load must be equal to the imaginary part, thus $75 = \frac{(108)}{1 + \omega^2(108)^2 C^2}$

$$\omega = 2\pi 60 = 377$$

$$75(1 + 1.66 ' \ 10^9 C^2) = 108$$

$$C^2 = \frac{0.440}{1.66 ' \ 10^9} = 265 ' \ 10^{-12}$$

$$C = 16.3 \mu f$$

Then for complete matching, the reactive parts must cancel each other.

For the load we have $\frac{-j\omega(108)^2 C}{1+\omega^2(108)^2 C^2} = \frac{-j71.7}{1+0.439} = -j49.8$

Then

$$|jX_1| = |-j49.8|$$

$$\omega L = 49.8 \quad Z = Z_L^* = (75 + j49.8)\Omega$$

$$L = \frac{49.8}{377} = 0.132 \text{ henrys}$$

An equally valid solution can be found by assuming the load to be inductive and the source capacitive.

IMPEDANCE MATCHING PROB. 2.14

The current flowing in the 200Ω load in the circuit below is 165° out of phase with the 20 volt source. Find the coupling coefficient k and the current i(t).

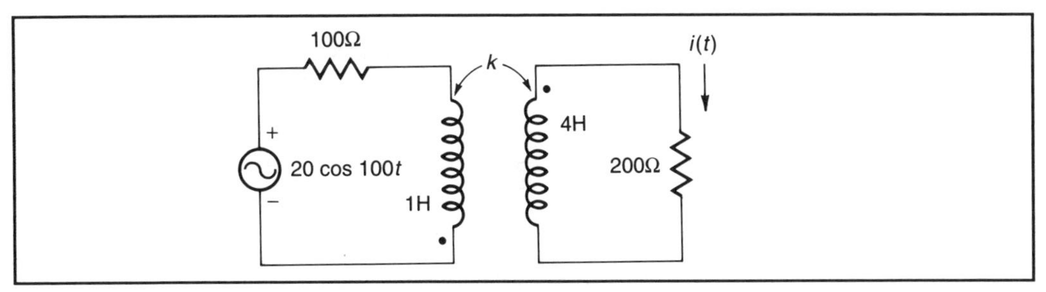

Figure P2.14

SOLUTION:

$$M = k\sqrt{L_1L_2} = k\sqrt{1\times 4} = 2k$$

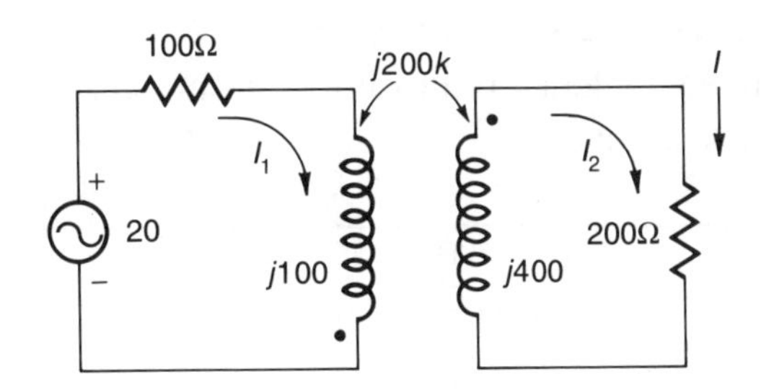

$$(100+j100)I_1 + j200kI_2 = 20$$
$$j200kI_1 + (200+j400)I_2 = 0$$
$$(1+j)I_1 + j2kI_2 = 0.2$$
$$jkI_1 + (1+j2)I_2 = 0$$

$$I_2 = \frac{\begin{vmatrix} 1+j & 0.2 \\ jk & 0 \end{vmatrix}}{\begin{vmatrix} 1+j & j2k \\ jk & 1+j2 \end{vmatrix}} = \frac{-j0.2k}{1+j+j^2-2+2k^2}$$

$$= \frac{j0.2k}{1-2k^2-j^3} = \frac{0.2k\angle 90^\circ}{\sqrt{(1-2k^2)^2+9}\angle \tan^{-1}\left(\dfrac{-3}{1-2k^2}\right)}$$

$$90^\circ - \tan^{-1}\left(\frac{-3}{1-2k^2}\right) = 165^\circ$$

$$\tan^{-1}\left(\frac{-3}{1-2k^2}\right) = -75^\circ$$

$$\frac{-3}{1-2k^2} = \tan(-75^\circ) = -(2+\sqrt{3})$$

$$\frac{3}{1-2k^2} = 2+\sqrt{3} \qquad 1-2k^2 = \frac{3}{2+\sqrt{3}}$$

$$2k^2 = 1-\frac{3}{2+\sqrt{3}} \qquad k^2 = \frac{1}{2}-\frac{\frac{3}{2}}{2+\sqrt{3}}$$

$$k = \sqrt{\frac{1}{2}-\frac{\frac{3}{2}}{2+\sqrt{3}}} = 0.313$$

$$I = I_2 = \frac{(0.2)\times(0.313)\angle 90^\circ}{\sqrt{(1-2(.313)^2} + 9\angle -75^\circ}$$

$$= 0.0202\angle 165^\circ \text{ A}$$

$$i(t) = 20.2\cos(100t+165^\circ) \text{ mA}$$

IMPEDANCE MATCHING PROB 2.15

Find the Thevenin equivalent circuit for the circuit shown below at terminals a and b, all resistor values are in ohms.

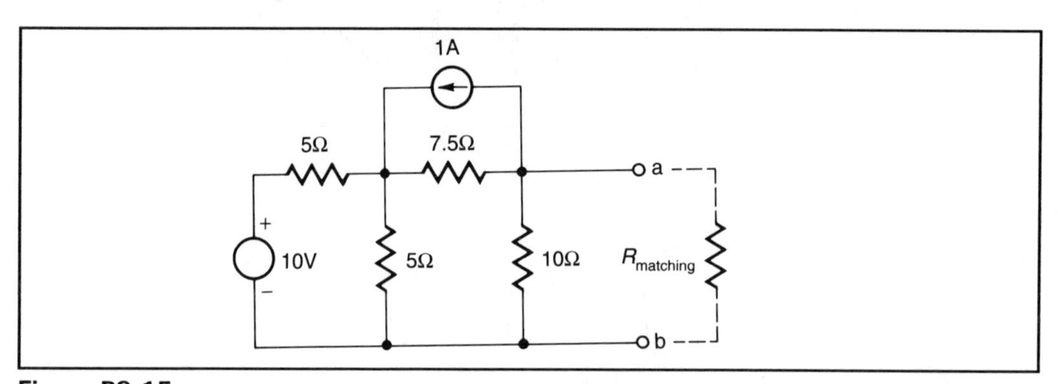

Figure P2.15

SOLUTION

Convert the current source in parallel with the 7.5Ω resistor to a voltage source in series to get:

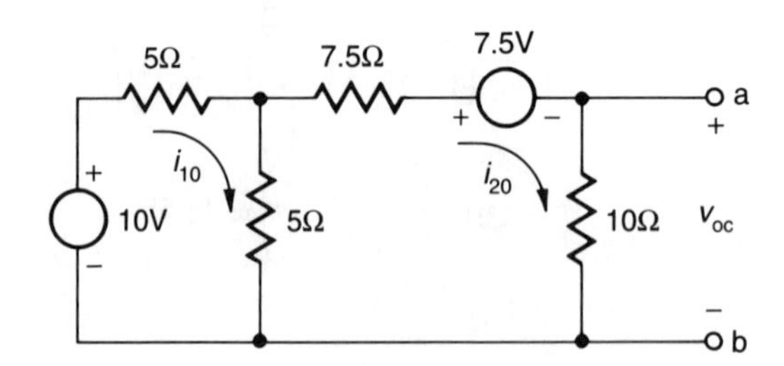

$$10i_{10} - 5i_{20} = 10$$
$$-5i_{10} + 22.5i_{20} = -7.5$$

$$i_{20} = \frac{\begin{vmatrix} 10 & 10 \\ -5 & -7.5 \end{vmatrix}}{\begin{vmatrix} 10 & -5 \\ -5 & 22.5 \end{vmatrix}} = \frac{-75+50}{225-25} = \frac{-25}{200}$$

$$i_{20} = -\frac{1}{8}\text{ A} \qquad v_{OC} = 10i_{20} = -\frac{5}{4}V = v_{Thev}$$

Short the terminals a & b to get:

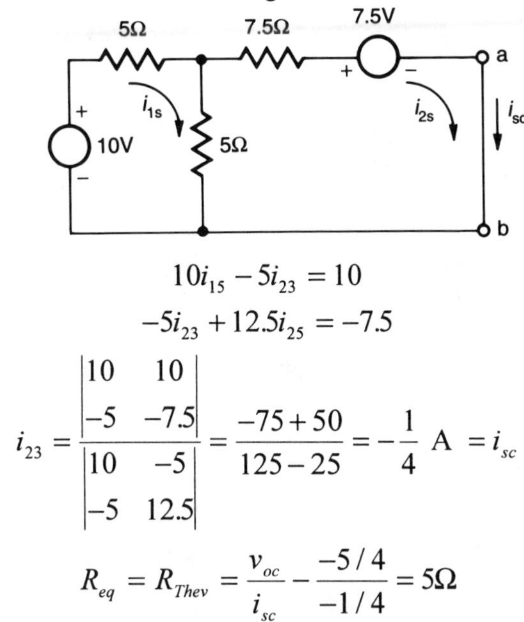

$$10i_{15} - 5i_{23} = 10$$
$$-5i_{23} + 12.5i_{25} = -7.5$$

$$i_{23} = \frac{\begin{vmatrix} 10 & 10 \\ -5 & -7.5 \end{vmatrix}}{\begin{vmatrix} 10 & -5 \\ -5 & 12.5 \end{vmatrix}} = \frac{-75+50}{125-25} = -\frac{1}{4}\text{ A} = i_{sc}$$

$$R_{eq} = R_{Thev} = \frac{v_{oc}}{i_{sc}} - \frac{-5/4}{-1/4} = 5\Omega$$

$\therefore$ The Thevenin equivalent is:

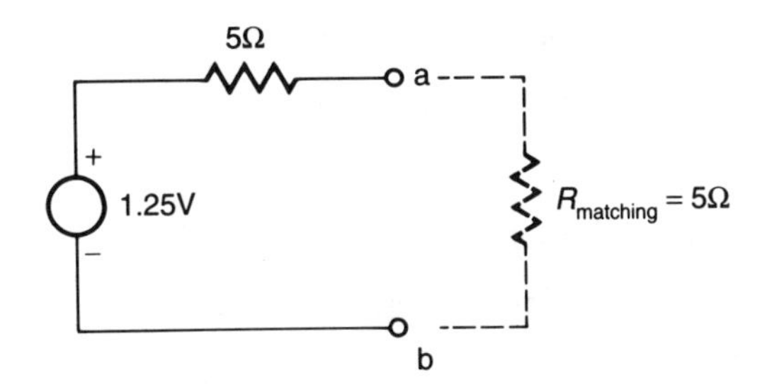

IMPEDANCE MATCHING PROB 2.16

(a) Find n for maximum power in the 2 ohm resistor.
(b) Find the power in the 2 ohm resistor if n=4.

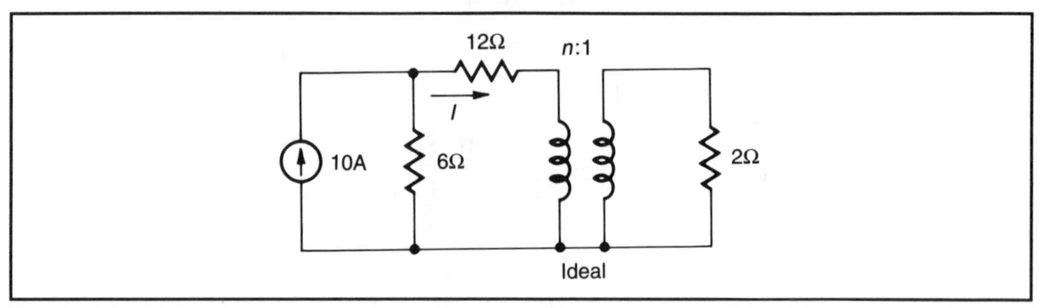

Figure P2.16

SOLUTION

a) Find the Thevenin Equivalent of the circuit to the left of the transformer.

It is:

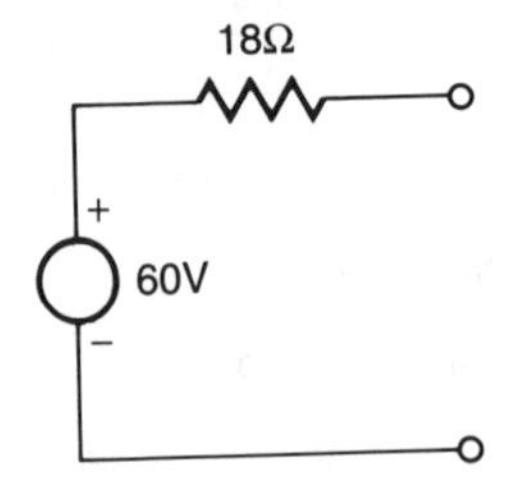

To obtain the maximum power from this circuit the load across it should be 18Ω. The impedance seen looking into the n turn side of the transformer is $4^2(2) = 32\Omega$

The impedance seen looking into the n turn side of the transformer is $n^2(2)$.

$$n^2(2) = 18$$

$$n = 3$$

For $n = 3$, Power in $2\Omega = 50$ w

b) If $n = 4$, the impedance seen looking into the transformer in 2Ω is same as power into transformer.

Using current division $I = 10\dfrac{6}{6+44} = \dfrac{6}{5}$ A

$$P = I^2(32) = 46.08 \text{ w}$$

or calculating Power in secondary

$$P = \left(\frac{6}{5}\times 4\right)^2 (2) = 46.08 \text{ w}$$

METERS & WAVEFORMS PROB 2.17

Consider the following expression for a wave:

$f(t) = 10.0 \sin \omega t + 2.0 \cos (3\omega t + 90^{o})$

(1) On the grid provided sketch the wave, f(t) to the scale given by locating ordinances for every 30° on the ωt scale.The maximum value of the fundamental component of the wave, f(t), is:

a. 0.0 b. 2.0 c. 10.0 d. none of these

(2) The d-c component of the wave, f(t), is:

a. 0.0 b. 1.0 c. 1.5 d. 2.0 e. 2.5 f. 3.0 g. 4.0 h. 5.0 i. 10.0 j. none of these

(3) The half-period average of the wave, f(t) , is:

a. 0.0 b. 1.2 c. 2.0 d. 3.28 e. 5.96 f. 6.36 g. 7.64 h. 10.0 i 12.0 j. none of these

(4) The rms or effective value of the wave, f(t), is:

a. 2.0 b. 3.18 c. 6.36 d. 7.07 e. 7.21 f. 7.63 g. 8.02 h. 9.96 i. 11.3 j. none of these

SOLUTION

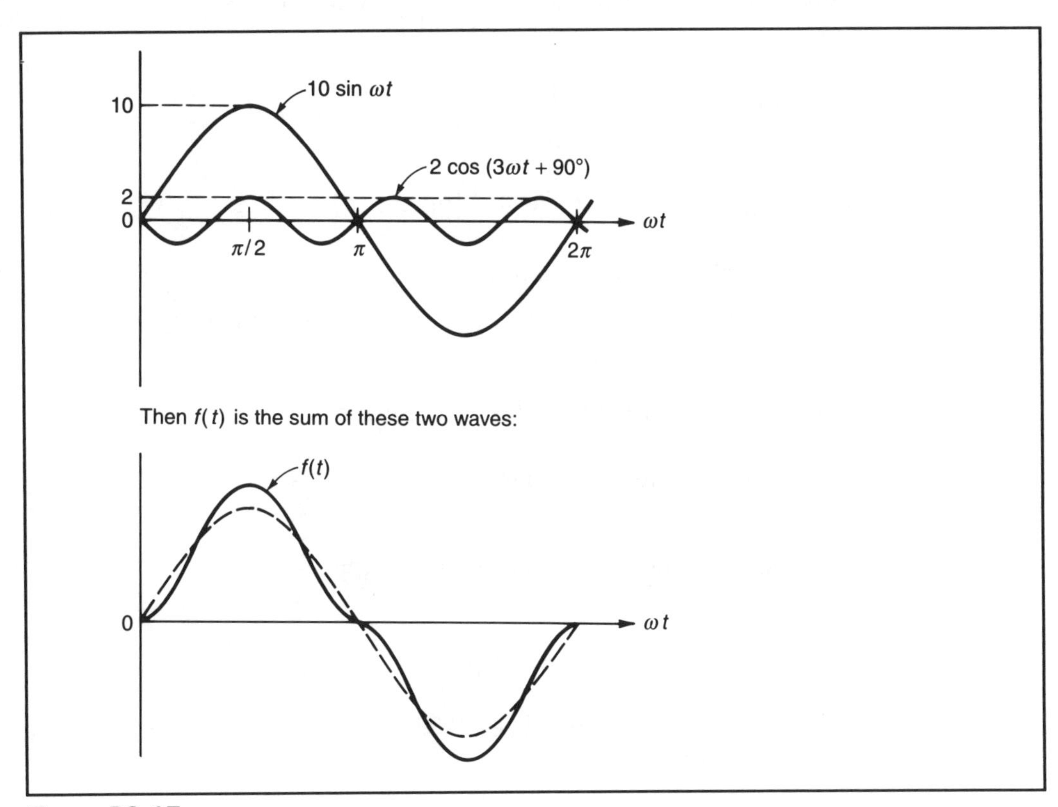

Figure P2.17

(1) $f(t) = 10.0 \sin \omega t + 2.0 \cos (3\omega t + 90^{o})$

To determine values of the second term, set up a table:

ωt	3ωt	(3ωt +90°)	Cos (3ωt +90°)
0	0	π/2	0
π/6	π/2	π	-1
π/3	π	3π/2	0
π/2	3π/2	2π	+1

(2) the d.c. component:

$$\text{d.c. component } = a_0 = \frac{1}{2\pi}\int_0^{2\pi} f(t)d(\omega t)$$

$$= \frac{1}{2\pi}\int_0^{2\pi} 10\sin\omega t \, d(\omega t) + \frac{1}{2\pi}\int_0^{2\pi} 2\cos(3\omega t + 90°)\, d(\omega t)$$

$= 0 + 0 = 0$ The answer is a. 0.0

(3) The half - period average of the wave, f(t), is:

$$\text{d.c half - period average } = \frac{a_t}{2} = \frac{1}{\pi}\int_0^{\pi} f(t)d(\omega t)$$

$$= \frac{1}{\pi}\left[\int_0^{\pi} 10\sin\omega t \, d(\omega t) + \frac{1}{3}\int_0^{\pi} 2\cos(3\omega t + 90°)\, d(\omega t)\right]$$

$$= \frac{1}{\pi}[+10 + 10 - 2/3 - 2/3] = \frac{1}{\pi}[20 - 4/3] = \frac{18.67}{\pi}$$

$= 5.95$ The answer is e. 5.96

Note that there is one more negative than positive third harmonic loop per half - cycle.

(4) The rms or effective value of the wave, f(t), is:

$$\text{effective value of fundamental} = F_1 = \frac{10}{\sqrt{2}}$$

$$\text{effective value of 3rd harmonic} = F_3 = \frac{2}{\sqrt{2}}$$

$$F_{TOT(effective)} = \sqrt{F_1^2 + F_3^2} = \sqrt{\frac{100}{2} + \frac{4}{2}} = 7.21$$

The answer is e. 7.21

METERS & WAVEFORMS PROB 2.18

Consider again the expression for the wave in Problem 2.17 as follows:

f(t) = 10.0 sin ωt + 2.0 cos (3ωt +90°)

(1) The maximum or peak value of the wave, f(t) is:

a. 2.0 b. 4.0 c. 6.0 d.8.0 e. 10.0 f. 11.5 g. 12.0
h. 14.0 i. 16.0 j. none of these

(2) If the wave, f(t), represents a current in amperes that is flowing through a resistor having a resistance of 3.0 ohms, the power loss in watts is:
a. 43 b. 55 c. 78 d. 110 e. 156 f. 156 g. 187
h. 221 i. 312 j. none of these

(3) If the wave , f(t), represent a current in amperes that is flowing through an inductor having an inductance of 0.1 henry, the rms or effective voltage in volts across the inductor is:
a. 0.00 b. 0.33 c. 0.67 d. 0.83 e. 1.00 f. 0.33 g. 0.67
h. 0.83 i 1.00 j. none of these

(4) If the wave, f(t),. represents a current in amperes having an angular velocity of 6000 radians per second which current is passing through a capacitor having negligible losses, no initial charge, and a capacitance of 0.002 farads; the rms or effective voltage in volts across the capacitor is:
a. 0.0 b. 0.21 c. 0.59 d. 0.74 e. 1.00 f. 2.10 g. 5.90
h. 7.40 i. 10.0 j. None of these

SOLUTION

(1) Note from the sketch the fundamental and the 3rd harmonic peaks occur in phase, thus $f(t)_{peak}$ is merely: $f(t)_{peak} = 10 + 2 = 12$

The answer is g. 12.0

(2) From Problem 7, part (4):

$$I_{eff.} = \sqrt{I_1^2 + I_3^2} = 7.21 \text{ amperes}$$

then:

$$P = I_{eff}^2 R = (7.21)^2 (3.00) = 156 \text{ watts}$$

The answer is f. 156

(3) the effective voltage across the inductor is:

$$E_{eff.} = \sqrt{E_1^2 + E_3^2} \quad \text{where } E_1 = I_1 X_{L1} = \left(\frac{10}{\sqrt{2}}\right)(\omega L) = \frac{1.0}{\sqrt{2}}\omega$$

$$\text{and } E_3 = I_3 X_{L3} = \left(\frac{2}{\sqrt{2}}\right)(3\omega L) = \frac{0.6}{\sqrt{2}}\omega$$

$$\therefore E_{eff.} = \sqrt{\frac{\omega^2}{2} + \frac{.36\omega^2}{2}} = \frac{\omega}{\sqrt{2}}\sqrt{1+.36} = 0.83\omega$$

The answer is h. 0.83ω

(4) for $\omega = 6000$ radians / second:

$$E_1 = X_{C1}I_1 = \left(\frac{1}{\omega C}\right)\left(\frac{10}{\sqrt{2}}\right) = \frac{10}{(6\times10^3)(2\times10^{-3})\sqrt{2}} = \frac{10}{12\sqrt{2}}$$

$$E_3 = X_{C3}I_3 = \left(\frac{1}{3\omega C}\right)\left(\frac{2}{\sqrt{2}}\right) = \frac{2}{3(6\times10^3)(2\times10^{-3})} = \frac{2}{36\sqrt{2}}$$

$$\therefore E_{eff.} = \sqrt{\left(\frac{10}{12\sqrt{2}}\right)^2 + \left(\frac{2}{36\sqrt{2}}\right)^2} = 0.59$$

The answer is c. 0.59

METERS & WAVEFORMS PROB. 2.19

An alternating current voltmeter consists of a series connection of an ideal half-wave diode and a D'Arsonval meter. The meter is calibrated to read the rms value of an applied voltage. When the waveform sketched below is applied, the meter reads 80 volts. What is the peak value of the applied waveform?

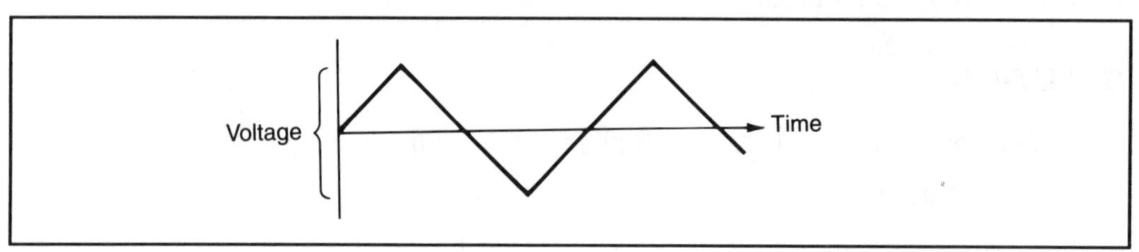

Figure P2.19

SOLUTION

The voltage as seen by the D'Arsonval meter(assuming a lossless diode rectifier) from a pure sine wave: e $= E_{\max} \sin\omega t$

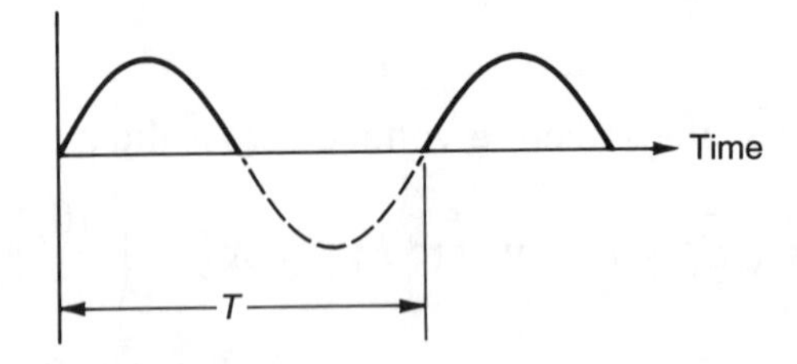

$$\text{Then: } E_{avg} = \frac{1}{T}\int_0^{\frac{1}{2}T} E_{\max}\sin\omega t\, d(\omega t) + \frac{1}{T}\int_{\frac{1}{2}T}^{T} 0\, d(\omega t)$$

Actual voltage read would be $E_{max}\pi$ but meter is calibrated to read

$\frac{E_{max}}{\sqrt{2}}$ (for an rms value).

For the wave shape given:

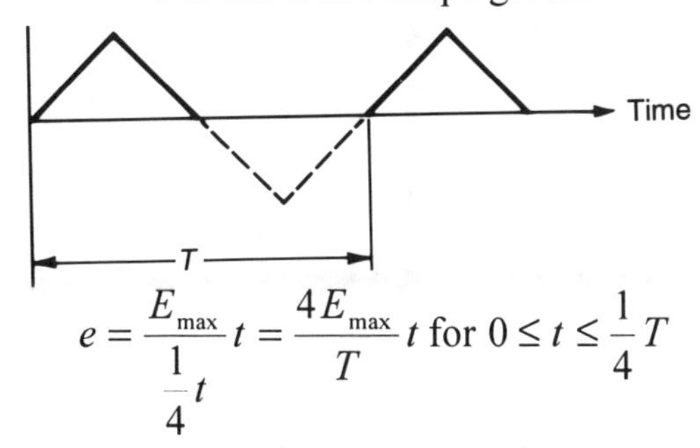

$$e = \frac{E_{max}}{\frac{1}{4}t}t = \frac{4E_{max}}{T}t \text{ for } 0 \le t \le \frac{1}{4}T$$

$$\text{Then: } E_{avg} = \frac{2}{T}\int_0^{\frac{1}{4}T} \frac{4E_{max}}{T} t\,dt = \frac{1}{4}E_{max}$$

$$\text{But meter reads (for a pure sine wave): } E_{meter} = \frac{\pi}{\sqrt{2}}E_{avg}$$

Then, for the saw - tooth wave, as given:

$$80x \text{ volts } = \frac{\pi}{\sqrt{2}}E_{avg} = \frac{\pi}{\sqrt{2}}\left(\frac{1}{4}E_{max}\right)$$

$$E_{max} = \left(\frac{\sqrt{2}}{\pi}\right)(4)(80) = 143.7 \text{ volts ANSWER}$$

x_{Note}: 80 volts is not the true rms voltage of the given wave form.

METERS & WAVEFORMS PROB. 2.20

A 50 micro-ampere meter movement with a 2k ohm internal resistance is to be used with a shunt arrangement so as to have full scale ranges of 10mA, 100mA, 500mA, and 10 amps. The maximum voltage drop across the input terminals for a full scale deflection should not exceed 250 millivolts.

Determine the resistor sizes needed and state any assumptions made.

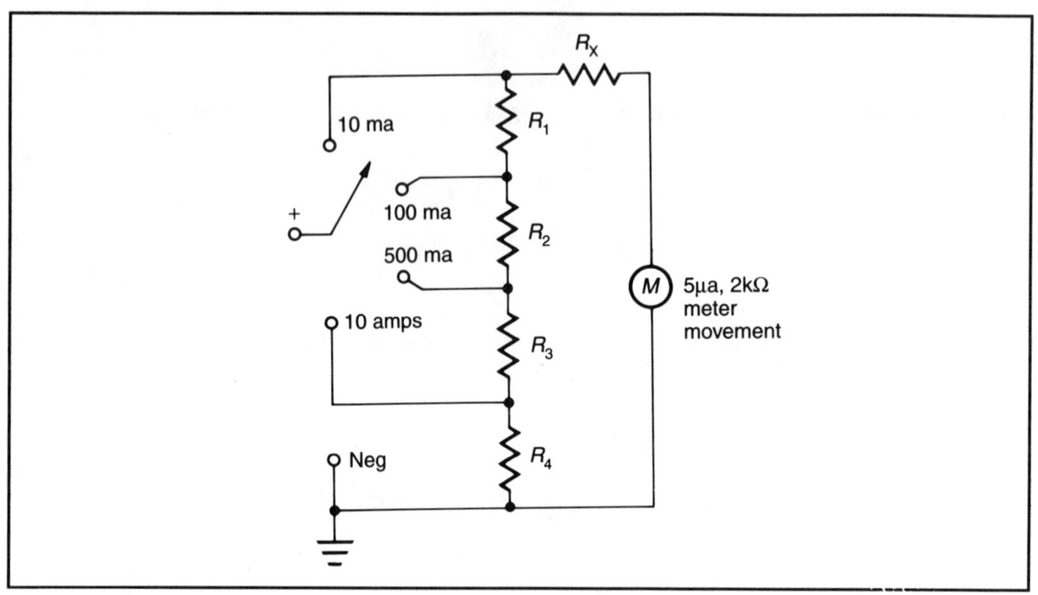

Figure P2.20

Solution: R_x can be determined from the voltage limit. At 0.250 volts, the current through the meter must be $50\mu a$ when using the 10ma setting.

$$\text{Then} \quad \frac{0.25}{R_x + 2k} = 50 \times 10^{-6}, \qquad R_x = 3k\Omega$$

Now the current through the shunt string of $R_1 + R_2 + R_3 + R_4$ must be 10ma - $50\mu A$ when the voltage is 0.250 volts.

$$10 \times 10^{-3} - 50 \times 10^{-6} = 9.95mA$$

This is only 0.5% less than 10mA.

The accuracy of our computations will not be seriously affected by ignoring this small amount when calculating the value of $R_1 + R_2 + R_3 + R_4$.

$$\text{Then } R_1 + R_2 + R_3 + R_4 = \frac{0.25}{10mA} = 25\Omega$$

Now with the switch in 100ma position, the equivalent circuit looks like this:

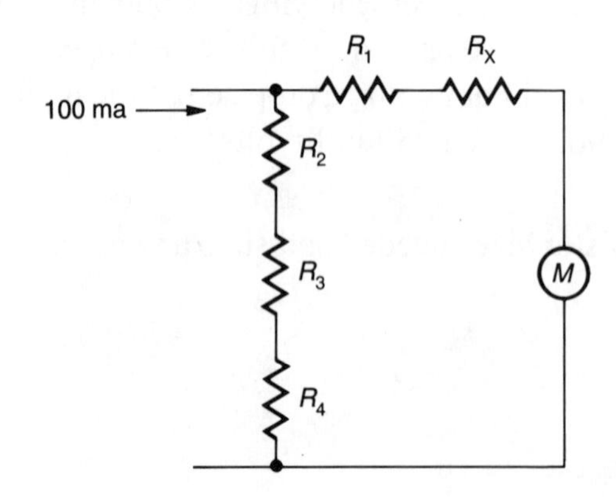

Assume $R_1 \langle\langle R_2 + 2k$ since $R_1 + R_2 + R_3 + R_4 = 25$

and $R_x + 2k = 5k$ this is at worst a 0.5% error if $R_2 + R_3 + R_4 = 0$.

$$\text{Then} \quad R_2 + R_3 + R_4 \cong \frac{0.25}{100 \times 10^{-3}} = 2.5\Omega$$

$$\text{So} \quad R_1 = 25 - 2.5 = 22.5\Omega$$

Now with the switch in 500 mA position:

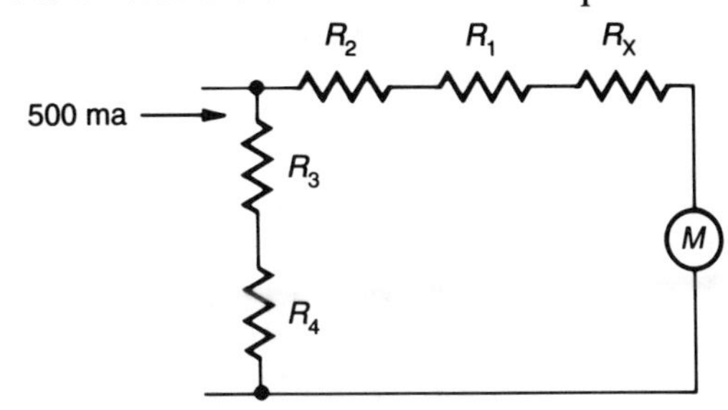

Again, the error of neglecting $R_2 + R_1 \langle\langle R_x + 2K$ is less than 0.5% so

$$R_3 + R_4 \cong \frac{0.25}{500 \times 10^{-3}} = 0.5\Omega$$

$$\text{So} \quad R_2 = 2.5 - 0.5 = 2\Omega$$

Finally, the switch at the position for 10 amps:

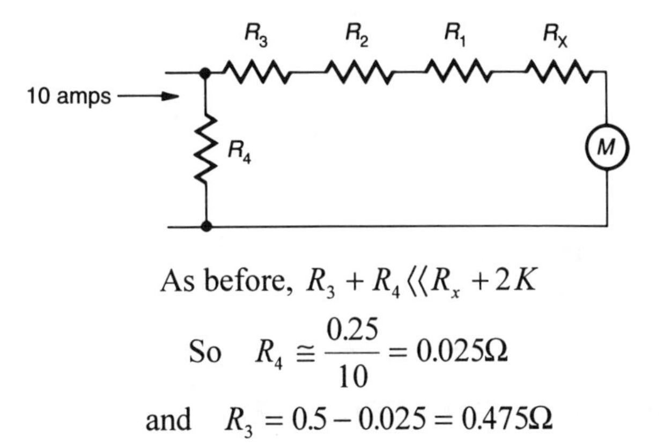

As before, $R_3 + R_4 \langle\langle R_x + 2K$

$$\text{So} \quad R_4 \cong \frac{0.25}{10} = 0.025\Omega$$

$$\text{and} \quad R_3 = 0.5 - 0.025 = 0.475\Omega$$

The absolute value of these resistors could be calculated without these simplifying assumptions by writing simultaneous equations for both parallel branches of the circuit is each case. Since resistors are seldom available in better than 1% tolerances, or at best $1/2\%$ the added labor for this additional accuracy is not warranted.

RESONANCE & BANDWIDTH PROB 2.21

An amplifier of voltage gain equal to 70dB is to be built by cascading transistor stages that have a gain-bandwidth product of 3×10^8 radians/sec. Find the maximum bandwidth that can be achieved using these stages, and the total number of stages required.

SOLUTION

A voltage gain of 70dB,

$$20\log\frac{v_0}{v_i} = 70; \qquad \log\frac{v_0}{v_i} = 3.5$$

$$\frac{v_0}{v_i} = 3162$$

for optimum bandwidth, the individual stage gain should be 1.65 (see Modern Electronic Circuit Design by David Comer, Addison - Wesley, 1976).

$$\text{Then} \quad 1.65^n = 3162$$

$$n\log 1.65 = \log 3162$$

$$0.22n = 3.5$$

$$n = 16.09$$

Since an integral number of stages is required, we select 17. This leads to an individual stage bandwidth of

$$BW = \frac{3\times 10^8}{1.65} = 182\times 10^6\ Rad/\sec$$

Bandwidth shrinkage due to cascading:

$$BW_{overall} = (\text{Bandwidth of single stage})\left(\sqrt{2^{\frac{1}{n}} - 1}\right)$$

$$= 182\times 10^6\left(\sqrt{2^{\frac{1}{17}} - 1}\right)$$

$$BW = 37.1\times 10^6\ Rad/\sec \Rightarrow \frac{37.1\times 10^6}{2\pi} = 5.91\times 10^6\ Hz$$

RESONANCE & BANDWIDTH PROB 2.22

Refer to the circuit below. If the inductance L is 10.0 microhenries and the frequency of the applied voltage is 9.55 MHz, determine the value of the reactance of C so that the circuit will be series resonant. What is the impedance looking into the circuit under these conditions?

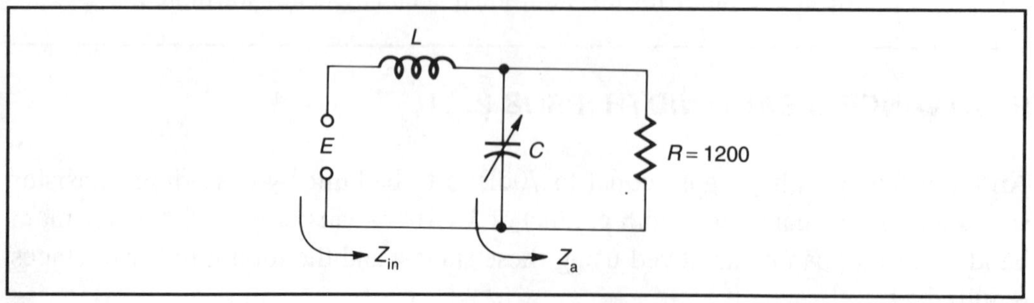

Figure P2.22

SOLUTION

(a) Solve circuit by the admittance method:

$$Y_a = \frac{1}{Z_a} = \frac{1}{1200} + jY_{ca} = 0.833 \times 10^{-3} + jY_{ca}$$

$$Z_a = \frac{1}{0.833 \times 10^{-3} + jY_{ca}} = \frac{0.833 \times 10^{-3}}{D} - \frac{jY_{ca}}{D} = R_a - jX_{ca}$$

where denominator = D after rationalizing the fraction with the

conjugate: $D = 0.693 \times 10^{-6} + Y_{ca}^2$

using: $(a-b)(a+b) = a^2 - b^2$.

New equivalent circuit:

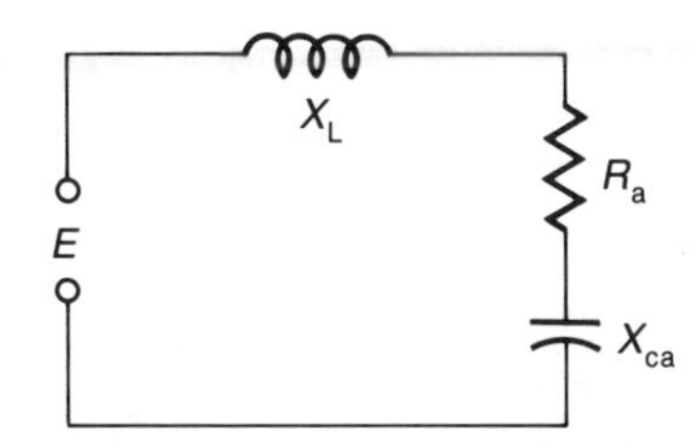

At Resonance: $X_{ca} = X_L$

$$X_L = 2\pi fL = (6.28)(9.55 \times 10^6)(10 \times 10^{-6}) = 599.7 \text{ ohms}$$

Thus $\frac{Y_{ca}}{D} = 599.7 = X_{ca}$ based on the principles of resonance.

If $\frac{Y_{ca}}{D} = 599.7$, then $Y_{ca} = 599.7(0.693 \times 10^{-6} + Y_{ca}^2)$

$$599.7Y_{ca}^2 - Y_{ca} + 4.156 \times 10^{-4} = 0$$

Solving the second degree equation by using the standard formula we obtain

$$Y_{ca} = \frac{1 \pm \sqrt{1 - 4(4.156 \times 10^{-4})(599.7)}}{2 \times 599.7}$$

$$= \frac{1 \pm \sqrt{1 - 0.9979}}{2 \times 599.7} = \frac{1}{1199.4} = 0.833 \times 10^{-3}$$

where the radical was approximated to be zero.

Therefore the reactance requested is:

$$X_{ca} = \frac{1}{Y_{ca}} = 1{,}199.4 \text{ ohms} \qquad \text{ANSWER}$$

(b) $Z_{in} = R_a$, as in resonance the only effective part of the impedance is the real part of the complex expression.

$$Z_{in} = R_a = \frac{0.833 \times 10^{-3}}{D} = \frac{0.833 \times 10^{-3}}{0.693 \times 10^{-6} + Y_{ca}^2}$$

$$= \frac{0.833 \times 10^{-3}}{0.693 \times 10^{-6} + \left(0.833 \times 10^{-3}\right)^2}$$

$$Z_{in} = 0.6 \times 10^3 = 600 \text{ ohms} \qquad \text{ANSWER}$$

RESONANCE & BANDWIDTH PROB 2.23

For the circuit below find $H(s)=V_2(s)/Vs$ and all its critical frequencies. Then find an expression for $|H(\omega)|$ and determine $|H(\omega)|_{max}$.

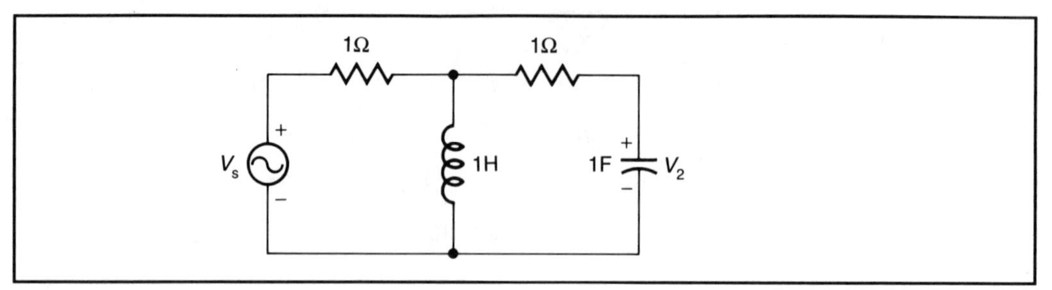

Figure P2.23

SOLUTION

$$V_2(s) = \frac{I_2(s)}{s}$$

$$(s+1)I_1 - sI_2 = V_s$$

$$-sI_1 + (s+1+\frac{1}{s})I_2 = 0$$

$$-s^2 I_1 + (s^2+s+1)I_2 = 0$$

$$I_2 = \frac{\begin{vmatrix} s+1 & V_s \\ -s^2 & 0 \end{vmatrix}}{\begin{vmatrix} s+1 & -s \\ -s^2 & s^2+s+1 \end{vmatrix}} = \frac{s^2 V_s}{s^3+2s^2+2s+1-s^3}$$

$$I_2 = \frac{s^2 V_s}{2s^2+2s+1}$$

$$V_2(s) = \frac{sV_s}{2s^2+2s+1}$$

$$H(s) = \frac{s}{2s^2+2s+1}$$

The critical frequencies are:

Zeros: $s = 0, \; s \to \infty$

Poles: $2s^2+2s+1=0$

$$s = \frac{-2 \pm \sqrt{4-8}}{4} = \frac{-2 \pm j2}{4}$$

$$s = -0.5 \pm j0.5$$

$$H(\omega) = \frac{j\omega}{-2\omega^2 + j2\omega + 1} = \frac{j\omega}{1 - 2\omega^2 + j2\omega}$$

$$|H(\omega)| = \frac{\omega}{\sqrt{(1-2\omega^2)^2 + 4\omega^2}} = \frac{\omega}{\sqrt{1+4\omega^4}}$$

$|H(\omega)|_{\max}$ occurs when H(ω) is real.

H(ω) is real when $1 - 2\omega^2 = 0$,

$$\omega = \frac{1}{\sqrt{2}}$$

$$|H(\omega)|_{\max} = \frac{j\left(\frac{1}{\sqrt{2}}\right)}{j^2\left(\frac{1}{\sqrt{2}}\right)} = \frac{1}{2} = H\left(\frac{1}{\sqrt{2}}\right)$$

RESONANCE & BANDWIDTH PROB 2.24

Determine the transimpedance $V_o(s)/I_1(s)$ for the network show.

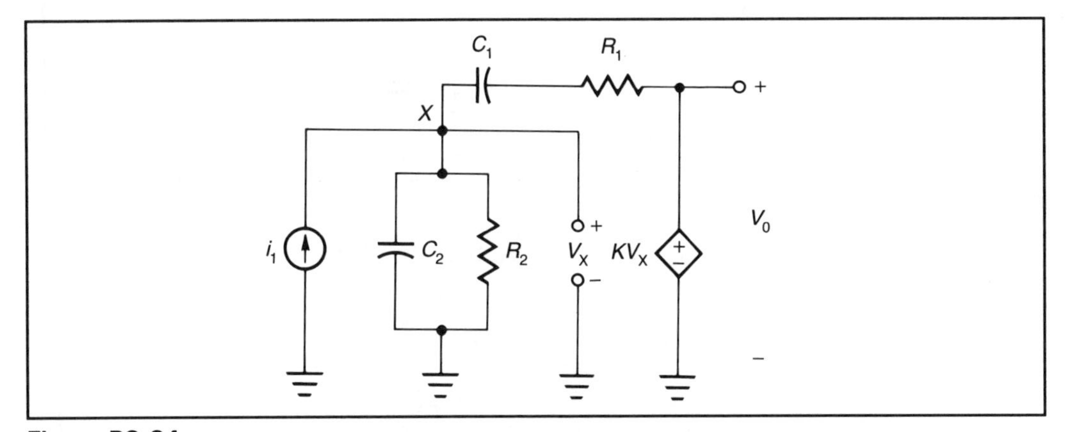

Figure P2.24

Node Equation at X:

SOLUTION

$$-I_1 + \frac{V_x(s)}{\dfrac{R_2}{sC_2R_2 + 1}} + \frac{V_x(s) - KV_x(s)}{\dfrac{1}{sC_1} + R_1} = 0$$

$$V_x = I_1 \frac{R_2(sC_1R_1 + 1)}{s^2C_1C_2R_1R_2 + s(C_1R_1 + C_2R_2 + (1-K)C_1R_2) + 1}$$

$$V_0(s) = KV_x(s)$$

$$\frac{V_0}{I_1} = \frac{K}{C_2}\left[\frac{s + \dfrac{1}{R_1C_1}}{s^2 + s\dfrac{C_1R_1 + C_2R_2 + (1-K)C_1R_2}{C_1C_2R_1R_2} + \dfrac{1}{C_1C_2R_1R_2}}\right]$$

RESONANCE & BANDWIDTH PROB 2.25

The network shown represents an oscilloscope probe connected to an oscilloscope. The components C_2 and R_2 represent the input circuitry of the oscilloscope and C_1 and R_1 represent the probe.

a) Find the transfer function $V_o/V_1(s)$.
b) Find a relationship among the components that makes the natural response equal to zero for all time.
c) Suppose the excitation, $v_1(t)$, is a unit step of voltage. Sketch $V_o(t)$ if

1) $\frac{C_1}{C_1+C_2} = \frac{R_2}{R_1+R_2}$ 2) $\frac{C_1}{C_1+C_2} > \frac{R_2}{R_1+R_2}$ 3) $\frac{C_1}{C_1+C_2} < \frac{R_2}{R_1+R_2}$

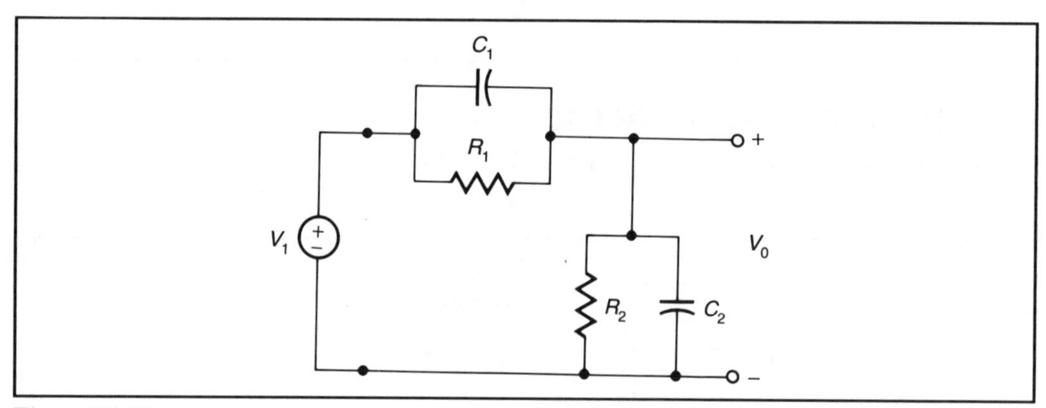

Figure P2.25

SOLUTION

a) Use a voltage divider:

$$\frac{V_0}{V_1}(s) = \frac{\dfrac{R_2}{sC_2R_2+1}}{\dfrac{R_1}{sC_1R_1+1} + \dfrac{R_2}{sC_2R_2+1}}$$

$$\frac{V_0}{V_1} = \frac{C_1}{C_1+C_2}\left[\frac{s+\dfrac{1}{C_1R_1}}{s+\dfrac{R_1+R_2}{R_1R_2(C_1+C_2)}}\right]$$

b) To make the natural response zero, eliminate the pole in $\frac{V_0}{V_1}$

by causing it to cancel with the zero.

$$\frac{1}{C_1R_1} = \frac{R_1+R_2}{R_1R_2(C_1+C_2)}$$

Thus $\frac{C_1+C_2}{C_1} = \frac{R_1+R_2}{R_2}$ or $1+\frac{C_2}{C_1} = 1+\frac{R_1}{R_2}$ or $\frac{C_2}{C_1} = \frac{R_1}{R_2}$

c) If $V_1(t) =$ Unit Step, $V_1(s) = \frac{1}{s}$

$$V_0(s) = \frac{C_1}{C_1 + C_2} \times \left[\frac{s + \frac{1}{C_1 R_1}}{s\left[s + \frac{R_1 + R_2}{R_1 R_2 (C_1 + C_2)}\right]} \right] = \frac{K_1}{s} + \frac{K_2}{s + \frac{R_1 + R_2}{R_1 R_2 (C_1 + C_2)}}$$

$$K_1 = \frac{R_2}{R_1 + R_2} \qquad \text{and } K_2 = \frac{C_1}{C_1 + C_2} - \frac{R_2}{R_1 + R_2}$$

$$v_0(t) = \frac{R_2}{R_1 + R_2} + \left[\frac{C_1}{C_1 + C_2} - \frac{R_2}{R_1 + R_2} \right] e^{\frac{-t}{\tau}} \;, \quad t\rangle 0$$

$$\text{Where} \quad \tau = \frac{R_1 R_2 (C_1 + C_2)}{R_1 + R_2}$$

$$1) \text{ if } \frac{C_1}{C_1 + C_2} = \frac{R_2}{R_1 + R_2} \quad \text{then } v_0(t) = \frac{R_2}{R_1 + R_2} = \frac{C_1}{C_1 + C_2}$$

2) and 3)

$$v_0(t = 0^+) = \frac{C_1}{C_1 + C_2}$$

$$v_0(t \to \infty) = \frac{R_2}{R_1 + R_2}$$

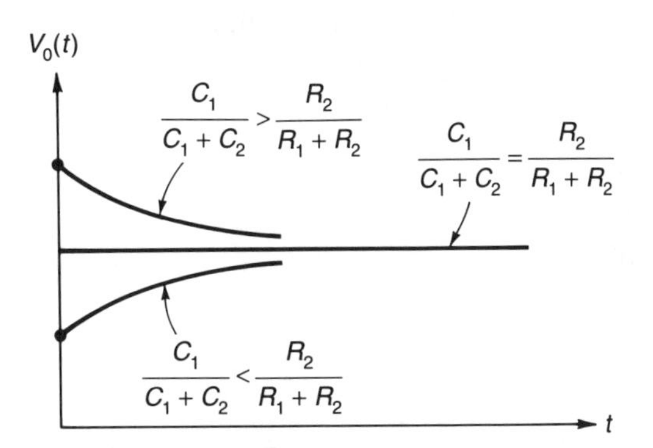

ELECTROSTATICS PROB 2.26

The following practice problems have relatively short solutions. When comparing answers, choose the nearest value to your calculated one. If your solution is a graphical one, remember that voltage and current do not have to be plotted to the same scale; and, of course, use a reasonably large plot.

Problem 2.26.1

For a parallel plate capacitor separated by an air gap of 1 cm and with an applied dc voltage across the plates of 500 volts, determine the force on an electron mass of 18.2×10^{-31} kg inserted in the space. The mass of an electron is 9.1×10^{-31} kg.

(a) 3.2×10^{-14} N
(b) 1.6×10^{-14} N
(c) 9.1×10^{-31} N
(d) 1.6×10^{-19} N
(e) 5.0×10^{-19} N

SOLUTION

When solving problems, always glance at the answer selections to see the form of the answer. Are the answers separated widely (if so, usually only 2 or 3 digits may be sufficient)? If only the magnitude is wanted (here, one may carry along any vector notation only as far as it is needed)? Is it easier to solve graphically?

The "mass" of 2 electrons has a charge $Q = 3.2 \times 10^{-19}$ C, thus the electric field is

$$E = \frac{500}{0.01} = 50 \times 10^3 \text{ V/m.}$$

The force is then:

$$F = QE = (3.2 \times 10^{-19}) \times (50 \times 10^3) = 1.6 \times 10^{-14} \text{ N.}$$

The answer is (b).

Problem 2.26.2

Assume a point charge of 0.3×10^{-3} *C* at an origin. What is the magnitude of the electric field intensity at a point located 2 meters in the X-direction, 3 meters in the Y-direction, and 4 meters in the Z- direction away from the origin?

(a) 500 kV/m
(b) 5 kV/m
(c) 93 kV/m
(d) 9.3 MV/m
(e) 1.2 MV/m

SOLUTION

The magnitude of the length of the resultant vector, R, in the x, y, z plane is

$$R = \sqrt{2^2 + 3^2 + 4^2} = \sqrt{29}$$

The magnitude of the electric field, E is :

$$E = \frac{Q}{4\pi\varepsilon R^2} = \frac{(0.3 \times 10^{-3})}{4\pi(8.85 \times 10^{-12})29}$$

$= 93{,}000$ V/m. The answer is (c).

Problem 2.26.3

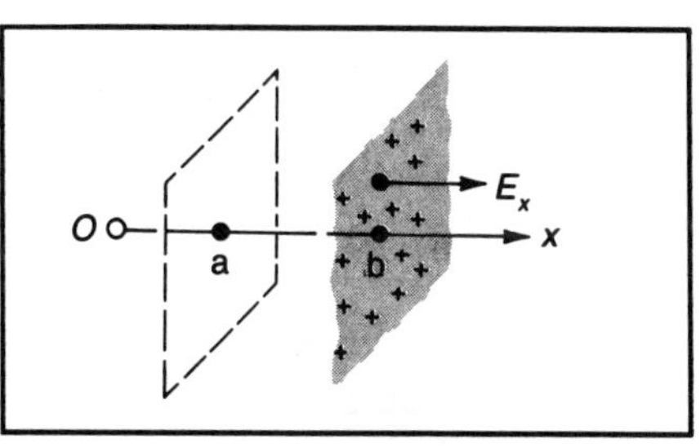

Figure P2.26.3 Charge Placement

An infinite sheet of charge, with a positive charge density, σ, has an electric field of

E = $\sigma/(2\epsilon_o)_{ax}$ for x>b.

If a second sheet of charge with a charge density of -σ is then placed at **a** (see Fig. P2.26.3), what is the electric field for a<x<b?

(a) $\sigma/\epsilon_{o\ ax}$
(b) 0
(c) $-\sigma/\epsilon_{o\ ax}$
(d) $\sigma/2\epsilon_{o\ ax}$
(e) $-\sigma/2\epsilon_{o\ ax}$

SOLUTION

On the b plane (for that plane alone), $E^+ = (-\sigma/2\varepsilon))_{ax}$, and for the negatively charged plane at the a plane (agian for that plane alone, but acting to the right of a), is the same as before, Therefore:

$$E = E^+ + E^- = \left(\frac{-\sigma}{\varepsilon_0}\right)_{ax} \quad \text{V/m}$$

The answer is (c).

Problem 2.26.4

Two equal charges of 10 micro-Columbs are located one meter apart on a horizontal line, and another charge of 5 micro-Coulomb is placed one meter below the first charge (forming a right triangle). What is the magnitude of the force on the 5 micro-Coulomb charge?

(a) 0.09 x 10^6 N
(b) 12.6 x 10^4 N
(c) 6.39 x 104 N
(d) 14.3 x 10^{-6} N
(e) 63 x 10^{-2} N

SOLUTION

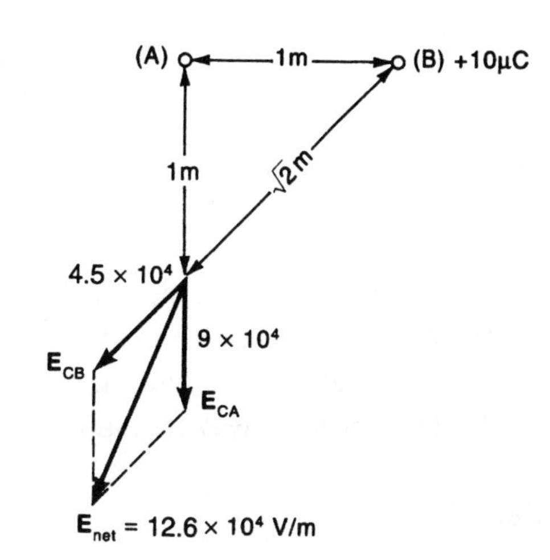

From a sketch of the vectors in Fig. 2 - 26.4, the value of E_{net} is 12.6×10^4 V / m.

F is them found to be:

$$E = \frac{10 \times 10^{-6}}{4\pi(8.85 \times 10^{-12})r^2}$$

$$F = Q_c E_{net} = 5 \times 10^{-6} \times 12.6 \times 10^4 = 63 \times 10^{-2} \text{ N}$$

The answer is (e)

ELECTROMAGNETICS PROB 2.27

The following practice problems have relatively short solutions. When comparing answers, choose the nearest value to your calculated one.

Problem 2.27.1

For a coil of 100 turns wound around a toroidal core of iron with a relative permeability of 1,000, find the current necessary to produce a magnetic flux density of 0.5 Tesla in the core. The dimensions of the core are given in Fig. 2.27-1

(a) 390 A
(b) 39 A
(c) 1.2 A
(d) 12.2 A
(e) 5π A

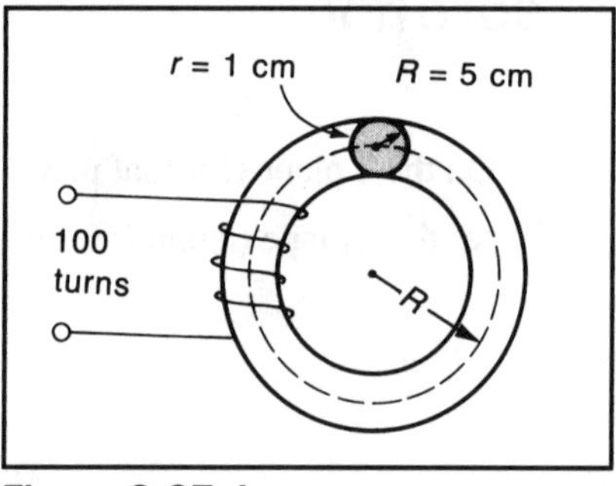

Figure 2.27-1

Solution:

The crosss - sectional area of the iron core is $\pi r^2 = \pi \times 10^{-4}$ m^2.

The path length is: $\ell = 2\pi R = 2\pi(5 \times 10^{-2}) = 0.1\pi$ m.

The permeability is $\mu = \mu_0 \mu_r = 4\pi(10^{-7}) \times 10^3 = 4\pi \times 10^{-4}$

Hence, $H = \frac{B}{\mu} = \frac{0.5}{4\pi \times 10^{-4}} = 390$ A t / m. and since $H = mmf \, / \, lenth$,

$mmf = H \times length = 390 \times 0.1\pi = 122.5$ A - t.

Thus, $I = mmf \, / \, turns = 122.5 / 100 = 1.225$ A.

The answer is (c).

Problem 2.27.2

Two long straight wires, bundled together, have a magnetic flux density around them. One wire carries a current of 5 amperes and the other carries a current of 1 ampere in the opposite direction, determine the magnitude of the flux density measured at a point 0.2 meters away (i.e., normal to the wires).

(a) $2\pi \text{x} 10^{-6}$ T

(b) $4\pi x10^{-6}$ T
(c) $4x10^{-6}$ T
(d) $16\pi x10^{-6}$ T
(e) $6x10^{-6}$ T

SOLUTION

Assume the two wires are bundled close together (with respect to the 0.2 meter position); then the net current to the right is : $I_{net} = 5 - 1 = 4$ A. The flux density is given by,

$$B = \frac{(\mu I)}{2\pi r} = \frac{(4\pi \times 10^{-7})(4)}{(2\pi \times 0.2)} = 4 \times 10^{-6} \text{ Tesla.}$$

The answer is (c).

Problem 2.27.3

Given a cast steel core of toroid with an air gap of 1 mm cut into it, determine the current necessary in a 600 turn coil wound around the core to produce a flux across the air gap of 0.6 mWb. Assume it is known from a magnetization curve that cast steel requires an H of 400 A-t/m for a flux density of 0.6 Tesla. Refer to Figure 2.27-3 and neglect any fringing in the air gap.

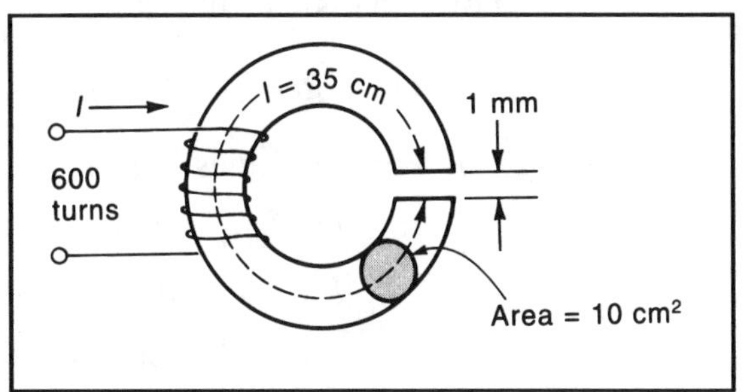

Figure 2.27-3

(a) 618 A
(b) 0.6 A
(c) 1.4 A
(d) 0.4 A
(e) 1.0 A

SOLUTION

The flux through the air gap and the core must be the same. Since there is no fringing, the effective cross - sectional gap area is the same as the core. Thus the B's are the same. For $B_{caststeel} = 0.6$ Tesla, H = 400 A • t / m = (mmf)(length), so

$$\text{mmf}_{\text{caststeel}} = 400 \times 0.35 = 140 \text{ A} \bullet \text{ t;}$$

$$\text{Since } H_{airgap} = \frac{B}{\mu_0} = \frac{0.6}{(4\pi \times 10^{-7})} = 0.48 \times 10^6$$

$$\text{and } mmf_{airgap} = H \times length = 0.48 \times 10^3.$$

$$\text{The total } mmf = mmf_{caststeel} + mmf_{airgap}$$

$$= 140 + 480 = 620 \text{ A} \bullet \text{ t}, \quad I = mmf / t = \frac{620}{600} = 1.03 \text{ A.}$$

The answer is (e).

MAGNETIC CIRCUITS PROB 2.28

A circular lifting magnet for a crane is to be designed so that with a flux density of 30,000 lines per square inch in each air gap, the length of each air gap is 0.5 inch. Leakage and saturation effects are such that the magnetomotive force for the air gaps is 0.85 of the magnetomotive force for the complete magnetic circuit. The mean length per turn of winding is 24 inches and the magnet is to operate with an applied terminal voltage of 70 volts with the winding at a temperature of 60°C.

(1) What is the pull or tractive force in pounds per square inch for the magnet?
(2) What size wire should be used for the winding?

SOLUTION

Assume a d.c. magnet:

$$\text{Force in dynes:} \quad F = \frac{B^2 A}{8\pi} \quad \text{Maxwell's Equation}$$

$$\text{Force in lbs / sq. in.} \quad = \frac{B^2}{72 \times 10^6} = \frac{(30{,}000)^2}{72 \times 10^6}$$

$$\text{(1) Force} = 12.5 \text{ lbs / sq. in. pull}$$

$$\text{(2)} \quad B = 30{,}000 \text{ Lines / sq. in.}, \quad \text{L} = 0.5 \text{ in air gap.}$$

$$mmf_{airgap} = 0.85 \text{ total NI}$$

$$L_{mean} \text{ of coil} = 24 \text{ inches}$$

$$V = 70 \text{ volts on coil}$$

$$H_{airgap} = 0.313 \text{ B NI per inch for air gap}$$

$$NI_{for\ \text{airgap}} = 0.313 \times 30{,}000 \times 0.5 \times 2 = 9{,}400 \text{ amp - turns for air gap}$$

$$= \frac{9400}{0.85} = 11{,}050 \text{ ampere turns for air gap}$$

$$NI_{total} = \frac{9{,}400}{0.85} = 11{,}050 \text{ ampere turns for air gap \& magnetic ckt.}$$

Since the coil dimensions are not given, assume 400 square inches of radiating surface and allow 0.7 watts per square inch dissipation at 60° C.

$$Watts = 400 \times 0.7 = 280 \text{ watts dissipated}$$

$$\text{and } I = \frac{280 watts}{70 volts} = 4 \text{ Amperes coil current}$$

$$N = \frac{NI}{I} = \frac{1050}{4} = 2763 \text{ turns on coil}$$

$$\text{Wire Length} = \frac{2763 \times 24 inches}{12 inches / foot} = 5526 feet$$

$$R_{wire} = \frac{70volts}{4amperes} = 17.5 \text{ ohms}$$

$$R_{wire} = \frac{\rho \; length}{area} \quad \text{where } \rho = 12 \text{ at } 60^\circ \text{C}$$

$$\text{A}_{\text{cir. mills}} = \frac{12 \times 5526}{17.5} = 3880 \text{ cir. mills}$$

No. 14 AWG Magnet wire has 4,107 cir. mills and 2.525 ohms / 1000'.

No. 15 AWG Magnet wire has 3,257 cir. mills and 3.184 ohms / 1000'.

Choose No. 14 AWG Magnet wire. ANSWER.

Machines□ 3-1

Note: Although several of the following problems are much longer than might be expected on the examination, it is suggested that the reader follow them through where appropriate.

Machines: Problem 3.1

SITUATION:

An older dc shunt motor is intended to be used to drive a load whose output power requirement varies between 5 and 15 hp but will only tolerate a small speed variation. The name plate rating of the machine is given as 15 hp, 230 volts, 57.1 A., 1,400 rpm. It is known that the field circuit resistance is 115 Ω and its armature resistance is 0.13 Ω. No data or test results are available on its no-load characteristics.

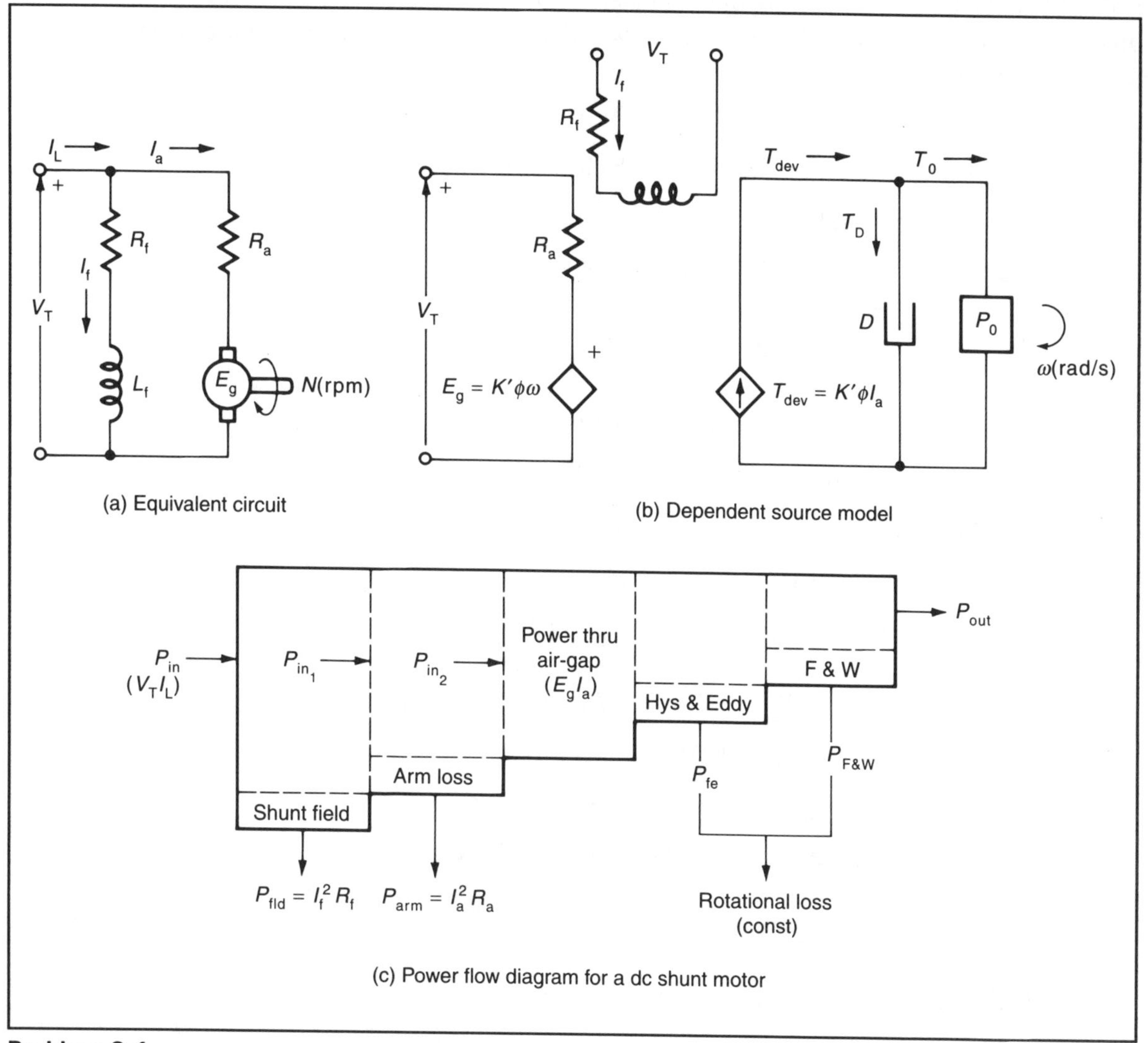

Problem 3.1

REQUIREMENTS:

The machine needs to be analyzed as to its suitability for the speed requirements (i.e., speed regulation) and its efficiency over the various load requirements. It has already been decided that the accuracy of the analysis does not require taking into account any effect that armature reaction might produce. The data needed for making a judgment on using this machine are:

1) No-load and 5 hp line currents.
2) No-load and 5 hp speeds.
3) Efficiency at both 5 and 15 hp.

SOLUTION

For the conditions stated (disregard armature reaction since data is unavailable on no-load characteristics) one can make a simplified equivalent circuit (or, for those more comfortable with using a dependent source type model, it is also shown). And, for keeping one's thought process ordered, it is recommended that a power flow diagram for known full-load parameters be made. Refer to Problem figure.

The following calculations are obvious for full load conditions using the equivalent circuit of diagram (a):

The shunt field loss (that will be constant):

$$I_f = \frac{V_T}{R_f} = \frac{230}{115} = 2 \text{ A}$$

$$P_{fld} = I_f^2 R_f = (2)^2 \times 115 = 460 \text{ watts}$$

The armature current loss (that is a function of the load):

$$I_a = I_L - I_f = 57.1 - 2.0 = 55.1 \text{ A}$$

$$P_{arm} = I_a^2 R_a = (55.1)^2 \times 0.13 = 395 \text{ watts}$$

The power transfer across the air gap ($E_g I_a$):

$$P_{in2} = P_{in} - (P_{fld} + P_{arm}) = 230 \times 57.1 - (460 + 395) = 12{,}278 \text{ watts}$$

The rotational losses (F &W plus iron loss) :

$$P_{rot} = P_{in2} - P_0 = 12{,}278 - (15 \times 746) = 1{,}088 \text{ watts}$$

The speed and machine - flux - constant relationship:

$$N = \frac{V_r - I_a R_a}{K\Theta} = 1{,}400 \text{ rpm} = \frac{230 - 55.1 \times 0.13}{K\Theta} \; ; \; K\Theta = \frac{222.8}{1{,}400} = 0.159 \text{ V / rpm.}$$

Calculations needed for no - load conditions:

$$P_{in1} = P_{arm} + P_{rot} + P_0 \;\; ; \;\; 230I_a = I_a^2 \times 0.13 + 1{,}088 + 0$$

The no - load armature current equation and solution:

$$I_a^2 - \left(\frac{230}{0.13}\right) I_a + \left(\frac{1{,}088}{0.13}\right) = 0 \;\; ; \;\; \text{from } I_a = -\frac{b}{2} \pm \sqrt{b^2 - 4ac}$$

$$I_a = -\left(\frac{-1769}{2}\right) \pm \sqrt{(-1769)^2 - 4 \times 8{,}369} = 1{,}764 \text{ A or } 4.7 \text{ A} \quad \leftarrow \text{ Choose}$$

(Alternative method - assumes $P_{rot} >> P_{arm} \approx \frac{P_{rot}}{V_a} = \frac{1088}{230} = 4.73$ A.)

The no - load speed:

$$N_{NL} = \frac{V_T - I_a R_a}{K\Theta} = \frac{230 - 4.7 \times 0.13}{0.159} = 1{,}443 \text{ rpm}$$

Calculations needed for a five hp output:

The armature current equations and solutions:

$$P_{in1} = P_{arm} + P_{rot} + P_0 \;\; ; \;\; 230I_a = I_a^2 \times 0.13 + 1{,}088 + 5 \times 746$$

$$I_a^2 - \left(\frac{230}{0.13}\right) I_a + \left(\frac{4{,}818}{0.13}\right) = 0$$

$$I_a = -\left(\frac{-1{,}769}{2}\right) \pm \frac{1}{2}\sqrt{(1{,}769)^2 - 4 \times 37{,}062} \;\; ; \;\; I_a = \; 1{,}748 \text{ or } 21.2 \text{ A} \leftarrow \;\; \text{Choose}$$

The Speed:

$$N_{5hp} = \frac{V_T - I_a R_a}{K\Theta} = \frac{230 - 21.2 \times 0.13}{0.159} = 1{,}425 \text{ rpm}$$

The efficiency at 5 hp out:

$$Eff\,(at\ 5hp) = \frac{P_0}{P_{in}} = \frac{P_0}{P_0 + \text{ all losses}} = \frac{5 \times 746}{5 \times 746 + 460 + 584 + 1{,}088} = 0.699$$

The efficiency at 15 hp out:

$$Eff\,(at\ 15hp) = \frac{P_0}{P_{in}} = \frac{P_0}{P_0 + \text{ all losses}} = \frac{15 \times 746}{15 \times 746 + 460 + 395 + 1{,}088} = 0.852$$

Table of results (assumes rotational losses are constant):

Power Out	Speed	%Speed Regulation	Arm Current	%Efficiency
15 hp	1,400 rpm	3.2%	55.1 A	85.2%
5 hp	1,425 rpm	1.8%	21.2 A	69.6%
0 hp	1,443 rpm	0	4.7 A	0

Alternate method of solution using the dependent source model of diagram b) at full load (1,400 rpm or 146.5 rad/sec):

The speed and machine-flux-constant at 15 hp using rad/sec for speed is:

$$\omega(rad/s) = \frac{V_T - I_a R_a}{K'\Phi} = \frac{E_g}{K'\Phi} \; ; \; K'\Phi = \frac{E_g}{\omega} = \frac{222.8}{146.5} = 1.521$$

for developed torque:

$$T_{dev} = K'\Phi I_a = 1.521 \times 55.1 = 83.8 \text{ N-m}$$

Torque needed for a 15 hp output:

$$P_0 = hp \times 746 = 15 \times 746 = 11{,}190 \; ; \; T_0 = \frac{11{,}190}{146.5} = 76.38 \text{ N-m}$$

Torque needed for rotational losses:

$$T_D = T_{dev} - T_0 = 83.81 - 76.38 = 7.43 \text{ N-m}$$

Torques required for a 5 hp output:

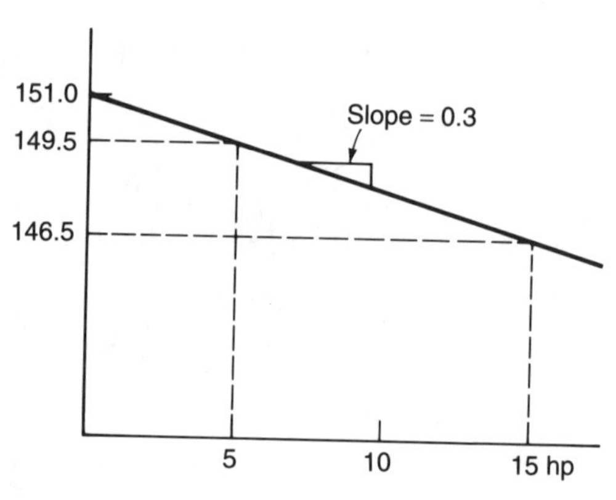

$$T_0 = 5 \times \frac{746}{\omega} = \frac{3730}{\omega} \text{ N-m}$$

Total developed torque needed:

$$T_{dev} = T_D + T_0 = 7.43 + \frac{3730}{\omega} \text{ N-m}$$

For the speed at 5hp, assume a linear relationship between no-load and full-load speed:

$$y = mx + b \; ; \; \omega = \text{ m x hp x } \omega_0 = -0.3 \times 5 + 151 = 149.5$$

$$\therefore T_{dev} = 7.43 + \frac{3730}{149.5} = 32.4 \text{ N-m}$$

Armature current needed for a 5 hp output:

$$T_{dev} = K'\Phi I_a \; ; \; I_a = \frac{T_{dev}}{K'\Phi} = 21.3 \text{ A}$$

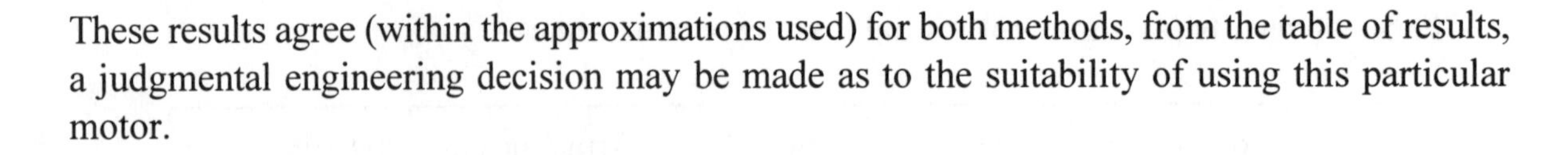

These results agree (within the approximations used) for both methods, from the table of results, a judgmental engineering decision may be made as to the suitability of using this particular motor.

MACHINES: Problem 3.2

SITUATION:

A 50kVA transformer rated at 2,300/230 volts at 60 Hz is to be tested in a laboratory so that its characteristics may be determined. The standard test requires both an open and short circuit test; the results of the tests are:

Open Circuit Test (Core Loss)			Short Circuit Test (Cu Loss)		
I	E	W	I	E	W
6.5 A	230 V	187 watts	21.7 A	115 V	570 watts

REQUIRED:

For future and present requirements, the parameters of an equivalent circuit need to be found. In addition, several values of efficiency and voltage regulation are to be determined (the assumption is that the coils are designed such that $I_1^2R_1=I_2^2R_2$):

1) Determine the coil resistance of the high windings side.
   a) 35.4 Ω, b) 0.605 Ω, c) 0.0065 Ω, d) 1.30 Ω, e) 5.29 Ω
2) Determine the copper power loss.
   a) 187 W, b) 757 W, c) 570 W, d) 1.5 kW, e) 0.255 W.
3) Determine the core loss.
   a) 187 W, b) 757 W, c) 570 W, d) 1.5 kW, e) 0.255 W
4) Determine the efficiency of the transformer at full load (assume unity power factor).
   a) 0.985, b) 0.996, c) 0.990, d) 0.636, d) 0.328
5) Determine the efficiency of the transformer at half load (assume unity power factor).
   a) 0.985, b) 0.996, c) 0.990, d) 0.636, d) 0.328
6) Find the percent voltage regulation of the transformer for unity power factor.
   a) 98.29% b) 1.175%, c) 96.44%, d) -2.62% e) 2.62%
7) Find the percent voltage regulation of the transformer for a 0.8 lagging power factor.
   a) 98.29% b) 3.56%, c) 96.44%, d) -2.62% e) 2.62%
8) Find the percent voltage regulation of the transformer for a 0.8 leading power factor.
   a) 98.29% b) 3.56%, c) 96.44%, d) -2.62% e) 2.62%
9) Determine the no-load standby current when the transformer is connected to its rated source on the high side.
   a) 6.5 A, b) 65 A, c) 0.65 A, d) 31.6 A, e) 316 A.
10) Determine the high side the steady state current when the transformer is connected to its rated voltage source on the high side and the low side is inadvertently connected to a load of 0.0394 Ω pure resistance.
   a) 6.5 A, b) 65 A, c) 0.65 A, d) 31.6 A, e) 316 A.

SOLUTION

The open circuit test measures core loss with negligible copper loss.

The short circuit test measures the copper loss with negligible core loss.

1) The coils are designed so that:

$$I_1^2 R_1 = I_2^2 R_2$$

where I_1 and R_1 are the high voltage coil current and a.c. resistance. I_2 and R_2 are the low voltage coil current and resistance.

$$\frac{E_1}{E_2} = \frac{N_1}{N_2} = \frac{I_2}{I_1}$$

where $\frac{N_1}{N_2} = a$ (turns ratio)

Since

$$W_{culoss} = I_1^2 R_1 + I_2^2 R_2 = 570 watts$$

and

$I_1^2 R_1 = I_2^2 R_2$ for good transformer design

$$570 = 2I_1^2 R_1 = 2I_2^2 R_2$$

$$I_1 = \frac{50{,}000}{2300} = 21.7 \text{ amps (Rated)}$$

$$I_2 = \frac{50{,}000}{230} = 217 \text{ amps (Rated)}$$

$\therefore R_1 = \frac{570}{2(21.7)^2} = 0.605\Omega$ (High Side) Answer b.

$R_2 = \frac{570}{2(217)^2} = 0.00605\Omega$ (Low Side)

2) Copper loss $= I_1^2 R_1 + I_2^2 R_2 = 570 watts$ Answer c.

3) Core loss $= 187$ watts neglecting the no load. Answer a.
Exciting current copper loss which would amount to

$$(6.5)^2 (0.00605) = 0.255 \text{ watt}$$

4) Full load efficiency $= \frac{output}{(\text{output} + \text{cu loss} + \text{core loss})}$

Assume Pf = 1

$$Efficiency = \frac{50{,}000}{50{,}000 + 570 + 187} = \frac{50{,}000}{50{,}757} = 0.985 \text{ Answer a.}$$

5) Half load efficiency

Assume Pf = 1

$$Efficiency = \frac{25{,}000}{25{,}000 + \frac{570}{4} + 187} = \frac{25{,}000}{25{,}329.5} = 0.99 \text{ Answer c.}$$

6) Voltage Regulation $= \dfrac{(\text{No Load Voltage - Full Load Voltage})}{(\text{Full Load Voltage})}$

Convert the equivalent transformer circuit, referring it to the low voltage coil.

Assume the low voltage coil has constant voltage $E_2 = 230$ volts.

$$R_{Equiv.Low} = \frac{Power}{I_2^2} = \frac{570}{(217)^2} = 0.0121\ ohm$$

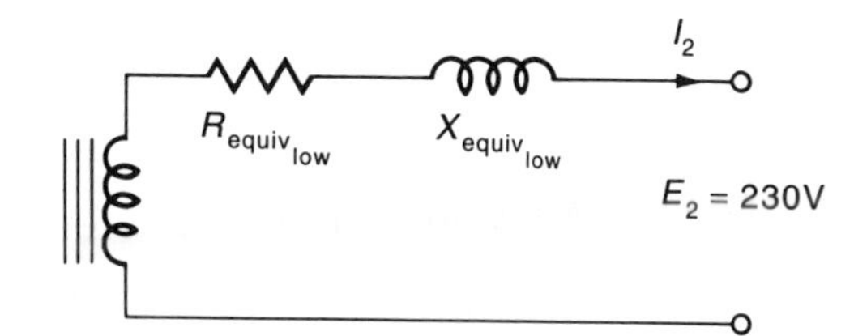

$$Z_{Equiv_{High}} = \frac{115}{21.7} = 5.3\Omega$$

$$Z_{Equiv_{Low}} = \frac{Z_{Equiv_{High}}}{a^2} = \frac{5.3}{100} = 0.053\Omega$$

$$X_{Equiv_{Low}} = \sqrt{\left(Z^2_{Equiv_{Low}} - R^2_{Equiv_{Low}}\right)} = \sqrt{(0.053)^2 - (0.0121)^2}$$

$$= 0.0516 ohms$$

Assume rated current $I_2 = 217$ amps (at unity power factor)

$$\frac{\overline{E}_1}{a} = E_2 + I_2\left(R_{Equiv_{Low}} + jX_{Equiv_{Low}}\right)$$

$$= 230 + 217(0.0121 + j0.053)$$

$$= 230 + 2.63 + j11.5$$

$$= 232.63 + j11.5$$

$$= 232.7\angle 2.84°$$

$$\text{Voltage Regulation} = \frac{232.7 - 230}{230} = \frac{2.70}{230}$$

$= 0.01175$ or 1.175 percent Answer b.

7) Find Voltage regulation for P.f.= 0.8 Lag

Assume rated current $I_2 = 217$ amps

Draw Phasor Diagram

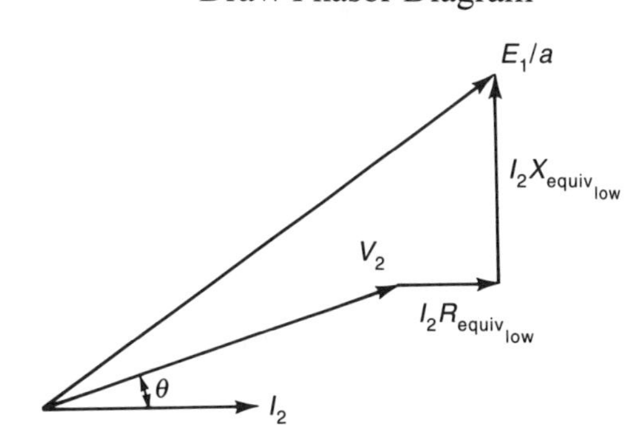

$$\frac{\overline{E}_1}{a} = V_2(\cos\theta + j\sin\theta) + I_2 R_{Equiv_{Low}} + jI_2 X_{Equiv_{Low}}$$
$$= 230(0.8 + j0.6) + 217(0.0121 + j0.053)$$
$$= 184 + j138 + 2.63 + j11.5$$
$$= 186.63 + j149.5$$
$$= 238.2\angle 38.7°$$

$$\text{Voltage Regulation } = \frac{238.2 - 230}{230} = \frac{8.2}{230} = 0.0356 \text{ or } 3.56\% \text{ Answer b.}$$

8) Voltage Regulation for P.f. = 0.8 Lead
Assume $I_2 = 217$ amperes, rated current

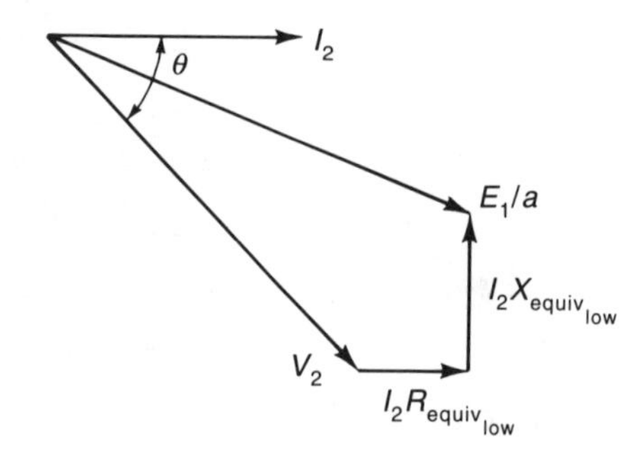

$$\frac{\overline{E}_1}{a} = V_2(\cos\theta + j\sin\theta) + I_2 R_{Equiv_{Low}} + jI_2 X_{Equiv_{Low}}$$
$$= 230(0.8 - j0.6) + 2.63 + j11.5$$
$$= 184 = j138 + 2.63 + j11.5$$
$$= 186.63 - j126.5$$
$$= 224\angle 34.2°$$

$$\text{Voltage Regulation } = \frac{224 - 230}{230} = -0.0262 \text{ or } -2.62\% \text{ Answer d.}$$

9) $$Z_{\text{core-hi-side}} = \frac{V_1}{I_1} = \frac{10V_2}{0.1I_2} = \frac{10 \times 230}{0.65} = 3{,}538\Omega$$

$$I_{\text{core-hi-side}} = \frac{V_1}{Z_1} = \frac{2300}{3538} = 0.65 \text{ A (Neglecting Cu Loss.) Answer c.}$$

10) $$Z_{\text{Copper-hi-side}} = \frac{115}{21.7} = 5.29\Omega \text{ (Neglecting core loss.)}$$

$$R_{\text{Eq-hi-side}} = \frac{W}{I_1^2} = \frac{570}{(21.7)^2} = 1.21\Omega$$

$$\cos\phi = \frac{W}{VA} = \frac{570}{115 \times 21.7} = 0.228\,, \quad \phi = \cos^{-1}(0.228) = 76.8°$$

$$X_{Eq-hi-side} = Z \sin 76.8° = 5.29 \times 0.974 = 5.15$$

$$R_{Load-hi-side} = a^2 R_L = 10^2 \times 0.0394 = 3.94\Omega$$

$$\text{new } Z_{\text{hi-side}} = 1.21 + j5.15 + 3.94 = 5.15 + j5.15$$

$$= \sqrt{2} \times 5.15\angle + 45°$$

$$I_{hi-side} = \frac{V}{Z_{new}} = \frac{2300 \text{ v}}{\sqrt{2} \times 5.15\angle + 45°} = 316 \text{ A}\angle\text{-}45° \quad \text{Answer e.}$$

(Neglecting core loss current.)

Problem 3.3

SITUATION:

For a perfectly operating dc generator the input mechanical power source may be considered to have constant speed over a range of torque requirements from no-load to full-load; however, the field current may be considered constant for a particular application. The generator is rated at 15 kW, 240 volts, 62.5 amperes, at a rated speed of 1,200 rpm. The armature resistance is known to be 0.2 Ω.

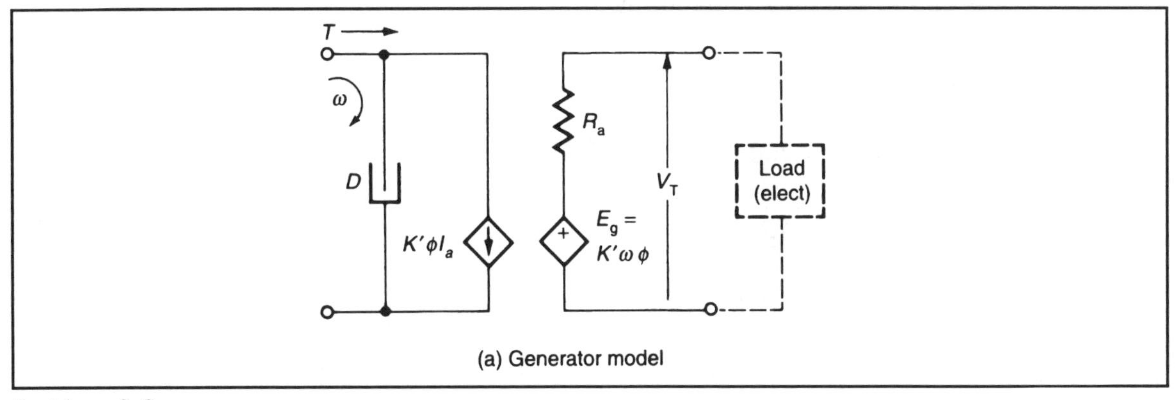

(a) Generator model

Problem 3.3

REQUIREMENT:

1. For the above stated conditions determine:

    a) The no-load terminal voltage (neglect any armature reaction effects).

    b) The percent voltage regulation and the input torque taken from the mechanical power source (neglect the rotational losses).

    c) Repeat 1b) but assume the rotational torque loss is 5.0 N-m.

2. Now assume the speed control mechanism is faulty and while the speed at no-load is still 1,200 rpm, the terminal voltage drops because the load resistance is unchanged, the speed

drops to 1,000 rpm. Determine:

a) The terminal voltage and the new power output.

b) The new input torque taken from the mechanical power source (again, neglect the rotational losses).

SOLUTION:

1a) The no-load terminal voltage (since there is no speed change) is E_g (may be given as K'Φω) remains same as, for a generator, is given as,

$$E_g = V_T + I_a = 240 + 62.5 \text{ x } 0.2 = 252.5 \text{ volts.}$$

b) and the percent voltage regulation is,

$$\%\text{V.R.} = (V_{no\text{-}load} + V_{full\text{-}load})\text{x}100/\text{V}_{\text{full-load}}$$
$$= (252.5\text{-}240)\text{x}100/240 = 5.2\ \%.$$

c) The input torque is found $T\omega = E_g I_a$= 252.5x62.5= 15,781 watts, thus,

$$T = E_g I_a/\omega = 15{,}781/[(1{,}200\text{x}2\pi)/60] = 125.6 \text{ N-m.}$$

The voltage regulation is the same (5.2%) but the new required torque is increased by the amount rotational loss torque,

$$T' = T + 5.0 = 125.6 + 5 = 130.6 \text{ N-m,}$$

and the new power input requirement is,

$$P_{in} = T\text{x}\omega = 130.6\text{x}(1{,}200\text{x}2\pi/60) = 17.93 \text{ kW}$$

2a) E_g is directly proportional to speed for a constant field,

$$E_g' = E_g(1{,}000/1{,}200) = 210.4 \text{ volts.}$$

The load current then becomes $E_g/(R_a+R_L)$, where the original load resistance was 240/62.5 =3.84 Ω,

$$I_a = 210.4/(0.2+3.84) = 52.8 \text{ A.}$$

$$V_T = E_g - I_a R_a = 210.4\text{- }52.8\text{x}.2 = 200 \text{ V.}$$

$$P_{out} = 200\text{x}52.8 = 10.56 \text{ kW.}$$

b) The new required input torque is easily found as,

$$T = P_{in}/\omega = 210.4\text{x}52.8)/(1{,}000\text{x}2\pi/60) = 11{,}109/104.6 = 106.1 \text{ N-m}$$

MACHINES: Problem 3.4

SITUATION:

Two single phase, 120 volt, 60 Hz motors are being considered for a particular application requiring one motor of 1/4 hp and the other motor a 1/2 hp one. These two motors are located at some distance from a power source so that a low line current is important such that the line voltage drop is negligible. The lower hp one will be a standard split-phase induction motor. Because of the low current constraint, The other motors will be a special type that has a slightly leading power factor (a built in capacitor in series with one of the windings). The following data has been obtained for these two motors:

MOTOR	POWER OUT (HORSEPOWER)	MOTOR EFFICIENCY	MOTOR POWER FACTOR
"A"	1/4	60%	0.7 lagging
"B"	1/2	70%	0.95 leading

REQUIRED:

To make an engineering judgment on the suitability of using these two motors, it will be necessary to find the total power needed, the combined line current and the combined power factor of these two motors operating in parallel. In addition, it will be necessary to have a complete carefully labeled phasor diagram indicating the line voltage (reference), and the individual and total line currents.

Solution 3.4

(a) Using the data given in the table above:

$$\text{Output Motor A } = VI\cos\Theta \times Eff.$$

In this case:

$$\text{HP}_{\text{OP}} = \text{Fraction of Load} \times \frac{746 \text{ watts}}{\text{HP}} = 0.25HP \times \frac{746}{HP} = 186.5watts$$

$$I_{Line} = \frac{VI\cos\Theta}{V \times Eff. \times Pf} = \frac{0.25 \times 746}{120 \times 0.7 \times 0.6} = 3.7amp$$

$$\text{Input Motor A } = \frac{Output}{Eff.} = \frac{186.5}{0.6} = 311watts$$

$$E_{Line} = 120volts$$

$$\text{Output Motor B} = VI\cos\Theta \times Eff.$$

$$= 0.5HP \times \frac{746}{HP} = 373watts$$

$$I_{Line} = \frac{0.5 \times 746}{120 \times 0.95 \times 0.7} = 4.66amp.$$

$$\text{Input Motor B} = \frac{Output}{Eff.} = \frac{373}{0.7} = 534watts$$

Motor	Pf	VA	Θ	Sin Θ	Input Power
A	0.7 lag	445	45°	0.715	311 watts
B	0.95 lead	562	18.2°	0.3123	534 watts

Total power = Input power A + Input power B
= 311+534 = 845 watts

Motor A VARS = 318 lag
Motor B VARS = -176 lead
Total VARS = 142 lag

$$\Theta' = Tan^{-1}\frac{142}{845} = 9.6°$$

$$Cos\Theta' = Cos(9.6°) = 0.986 \text{ Line Power Factor}$$

b)

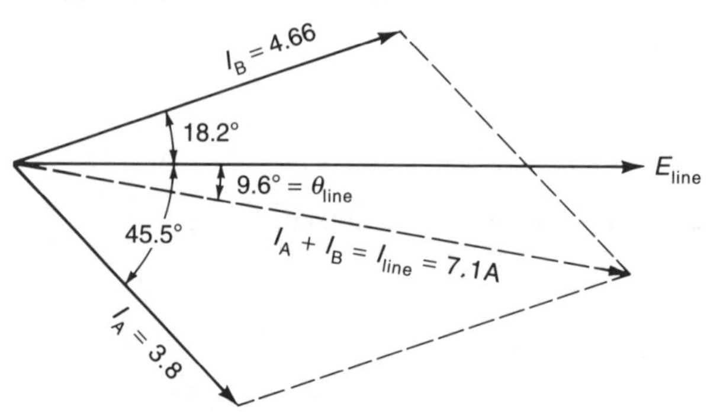

$$\bar{I}_{Line} = I_A(\cos 45.5° - j\sin 45.5°) + I_B(\cos 18.2° + j\sin 18.2°)$$

$$= 3.7(0.7 - j0.71) + 4.66(0.95 + j0.3123)$$

$$= 2.59 - j2.62 + 4.43 + j1.42 = 7.02 - j1.2$$

$$I_{Line} = 7.1 \text{ amp}$$

P_A=311 **P_B=534**

$$P_{Total} = P_A + P_B = 845 \text{ watts}$$

Machines: Problem 3.5

Situation:

In an emergency, a d.c. motor must be used as a generator. The motor is a cumulative-compound motor; its efficiency is 90%, and its positive and negative terminals are marked.
The cumulative-compound characteristic must be maintained when it is used as a generator, the rotation must be kept in the same direction and the rpm will be the same. The positive terminal must remain the positive terminal in the operation as a generator. The interpoles must aid commutation in both motor and generator mode. The machine is to deliver the same power to the line in the generator-operation as it took from the line in the motor-operation, and the looses in the machine are the same in both cases.

Required:

(a) Determine if any of the following changes are necessary, and do enough calculations to verify your answer:

1) Should the shunt-winding connections be changed?
2) Should the compound-winding connections be changed?
3) Should the interpole-winding connections be changed?
4) Should the field resistance be changed?

(b) Has the load on the mechanical clutch of the machine increased or decreased when the operation changed from motor to generator if it was running at full capacity as a motor?
(c) How much does the electrical power transferred at the line connection increase or decrease if the same mechanical torque is maintained on the shaft?

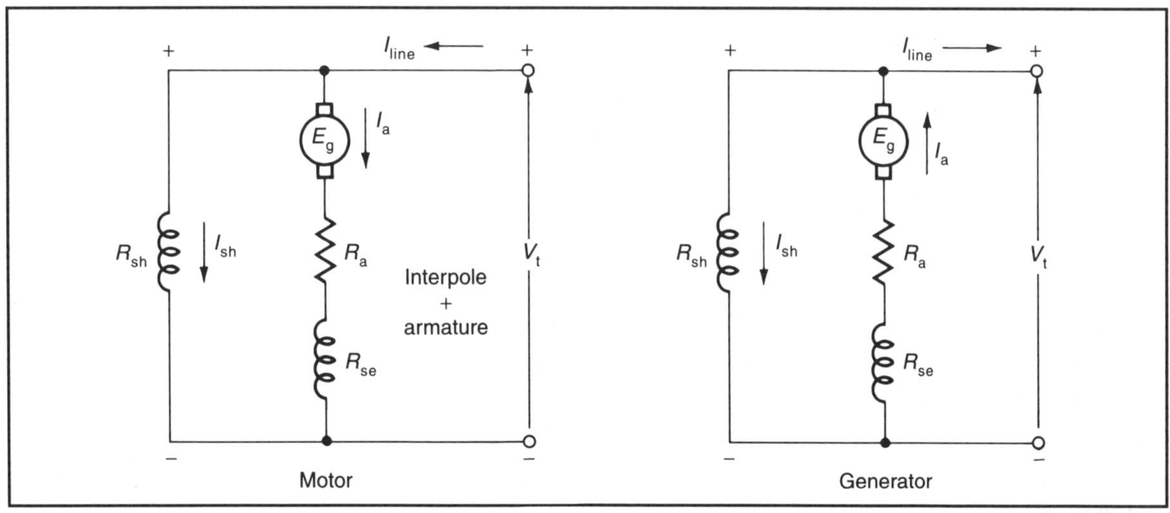

Problem 3-5.

Solution:

(a) 1. No, the shunt winding should not be changed. The problem clearly states that direction of rotation and polarity (positive terminal) stay the same. Therefore regardless of whether we have a motor or generator operation, the direction of the current stays unchanged. The only drawback is the subtractive MMF of shunt and series field winding, giving a differential compounding. This matter is brought up in the next point.

2. Yes, the compound (series) winding connections should be changed, since the armature current during generator operations is reversed and now opposes (differentially compounded) the shunt field. To still aid (cumulatively compounded) the shunt field, the compound (series) winding should be reversed.

3. No, the interpole winding connections should not be changed. Interpoles or commutating poles are narrow laminated auxiliary poles placed midway between the main poles and the plane of commutation. These interpoles are in series with the armature and are wound to oppose and nullify the armature reaction in the commutating plane. This prevents sparking that might cause flashover and also reduces iron losses in the armature teeth. Changing from motor to generator action, the polarity of the commutating pole automatically changes, with the change of the armature reaction MMF. Therefore, commutation in interpole machines is not affected by a change from motor to generator operation or a change in the direction of rotation..

4) Yes, the field (shunt) resistance should be changed. Since the generator voltage has to be larger than the terminal voltage due to the ohmic voltage drop in the armature winding, the flux has to be increased (E_G=Kϕ rpm). Increased ϕ means increased field current. Therefore, the shunt field resistance should be decreased.

(b) The load on the mechanical clutch of the machine has increased.

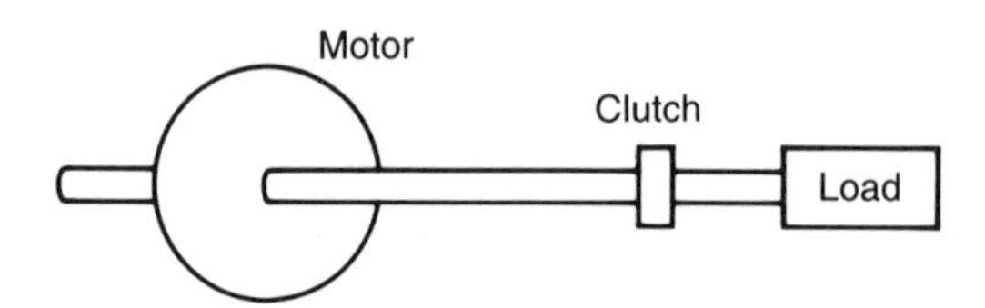

Rated Power $_{input}$= Losses + Load on clutch (for motor)
Rated Power $_{output}$ + Losses = Load on clutch (for generator)

Since losses remain the same in either mode of operation, the load on the clutch increases when machine is operated as a generator.

(c) Using the line connections as reference point, due to internal losses, the power delivered to the motor shaft is 90% of the reference or input power.
Assuming that the same mechanical torque is maintained, above shaft power is now the input (Prime mover) of the generator operation. Therefore, using the same 90% efficiency of the machine, the generator output referred to the original electrical power transferred at the line connections has decreased to:

$$0.90 \times 0.90 = 0.81 \text{ or } 81\%$$

Machines: Problem 3.6

Situation:

A 10 kVA, 240 ac voltage source sends power through two ideal transformers that are separated by short transmission line to a single phase induction motor load whose impedance is 4+j3 ohms. The first transformer is a 1:2 step-up and the second is a 2:1 step-down; the equivalent series impedance of the interconnecting transmission line is 2+j2 ohms. The operator is considering purchasing an ac capacitor to connect across the motor load to reduce some of the line loss.

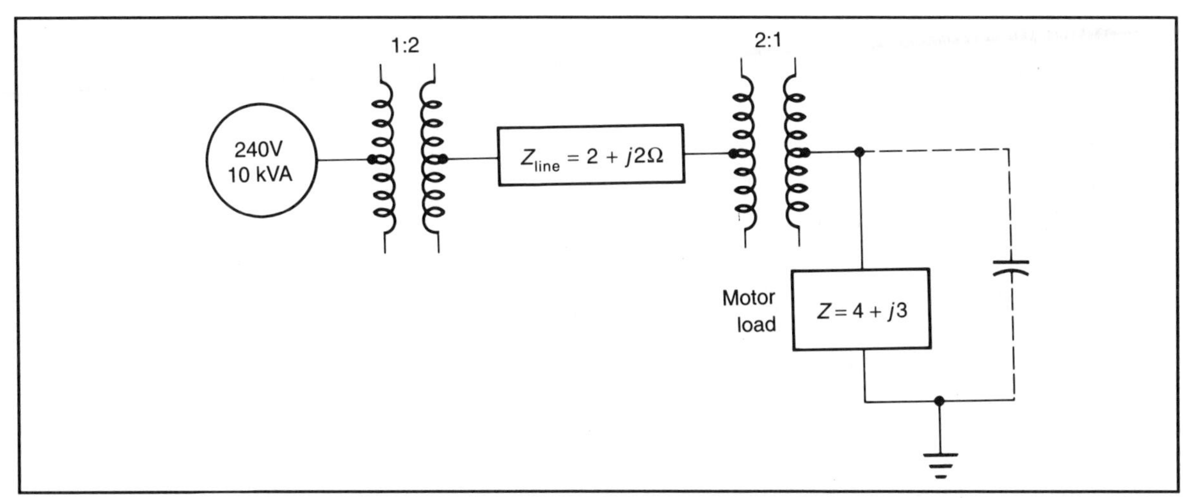

Problem 3.6

REQUIRED:
To analyze the system using per-unit values to determine the various voltages, currents, and power taken from the source both before and after the capacitor is added and also calculate the kVAR rating for the capacitor (to correct the load power factor to unity).

SOLUTION:

First the per-unit values need to be defined for each portion of the circuit (without the capacitor):

Generating Source Motor-load

V_{baseG}= 240 V --> 1 pu ---------------------> same
kVA_{baseK}= 10 kVA --> 1 pu ---------------> same
I_{baseI} = 10000/240=41.67A --> 1 pu ------> same
Z_{baseZ} = 240/41.67=5.760Ω --> 1 pu -----> same
Z_{motor}=(4+j3)/5.760= 0.6944+j0.5208 pu

For the transmission line:

V_{baseL} = 2x240=480 V --> 1 pu
kVA_{baseL} = 10 kVA --> 1 pu
I_{baseL} = 10000/480=20.83 A --> 1pu

Z_{baseL} = 480/20.83=23.04 Ω --> 1pu
Z_{line} =(2+j2)/23.04=0.0868+j0.0868 pu

The total pu impedance as seen at the source is:

$Z_{Tot} = Z_{line} + Z_{motor}$ = 0.0868+j0.0868 + 0.6944+j0.5208
= 0.7812 + j0.6076 = 0.9897/37.9° pu
I_{source} = 1.0/0°/ 0.9897/37.9° = 1.010/-37.9° pu
= 1.010/-37.9° pu x 41.67 A/pu = 42.10 A.
$P_{source} = VI\cos\Phi$ = 240x42.10 cos37.9° = 7974 watts.

The motor load voltage is $I_{motor}Z_{motor}$= 1.010x0.8680= 0.8767 pu
V_{motor} =0.8767pu x 240 V/pu = 210.4 V.

Check:
P_{motor}= I^2R =42.10^2x4 = 7089 watts
$P_{line} = (I_{source}/2)^2R_{line}$= $(21.05)^2$x2= 886 watts

Choose capacitor such that $Z_{motor}||Z_{capacitor}$ has only a real part,

Y_{motor} =1/5/36.9° = 0.16-j0.12,
Y_{cap} = +j0.12, $Z_{cap} = 1/Y_{cap}$ =1/j0.12 = -j8.33 Ω
Y_{Tot} = 0.16 + j0,
$Z_{Tot\ load}$ = 1/0.16 = 6.25 Ω --> 6.25/5.76 = 1.09 pu

The new total pu impedance as seen at the source is:

Z_{Tot} = 0.0868+j0.0868 + 1.09 = 1.177+j0.0868 pu = 1.180/4.24°
≡≡ 1.180x5.795 = 6.838/4.24° Ω
I_{source} = 1.0/1.180/4.24° = 0.8475/-4.24° pu
= 0.8475/-4.24° x 41.67 A/pu = 35.41 A
$P_{source} = VI\cos\Phi$ = 240x35.41 cos 4.24° = 8475 watts
$P_{line} = I_{line}^2xR_{line}$ =$(35.41/2)^2$x2 = 626.9 watts
$P_{motor} = I_{motor}^2xR_{motor}$= 44.3^2x4 = 7850 watts

It is interesting to compare the new load voltage with the original value (210.4 V),

$V_{motor\text{-}cap} = I(Z_{motor}||Z_{cap})$ = 35.41x6.25 = 221.3 V

Not only is motor voltage higher, for a greater motor output, but the line loss power loss is less. Also, the capacitor should have a kVAR voltage rating of 240 volts (when operating with the motor unloaded), so instead of a current of 221.3/(-j8.33)= 26.56A, it should be 240/(-j8.33)= 28.8 A, giving a kVAR of 6.9 kVAR.

Problem 3.7

Situation:

An emergency 120/208-volt, 3-phase, 4-wire, 60 Hz generator supplies an external circuit.

The load on the external circuit consists of 9000 watts of incandescent light connected between line and neutral and evenly distributed among the 3 phases, and a 10-HP, 3-phase air conditioner motor or 83% efficiency and 0.707 P.F.

Required:

(a) Show the phasor-diagram of the currents and voltages on the load-side of the generator.
(b) Determine the microfarads of the capacitor to be connected to the generator in order to reduce the generator load-current to 105% of that which would flow if the P.F.=1.
(c) Determine the size of the conduit and wire to be used as a feeder and the required fuse size to protect the feeder when it is run between the generator and its distribution board in the next room, if the total load is continuous, and the capacitor calculated in (b) is connected.

Solution:

(a) Designating: I_m =Motor current; I_{mR} = Real component of I_m; I_{mQ}= Quadrature component of Im; I_i = Incandescent lights current; I_1 = Total load current; I_1' = Corrected total current; I_c = Capacitor current,
We obtain as follows:

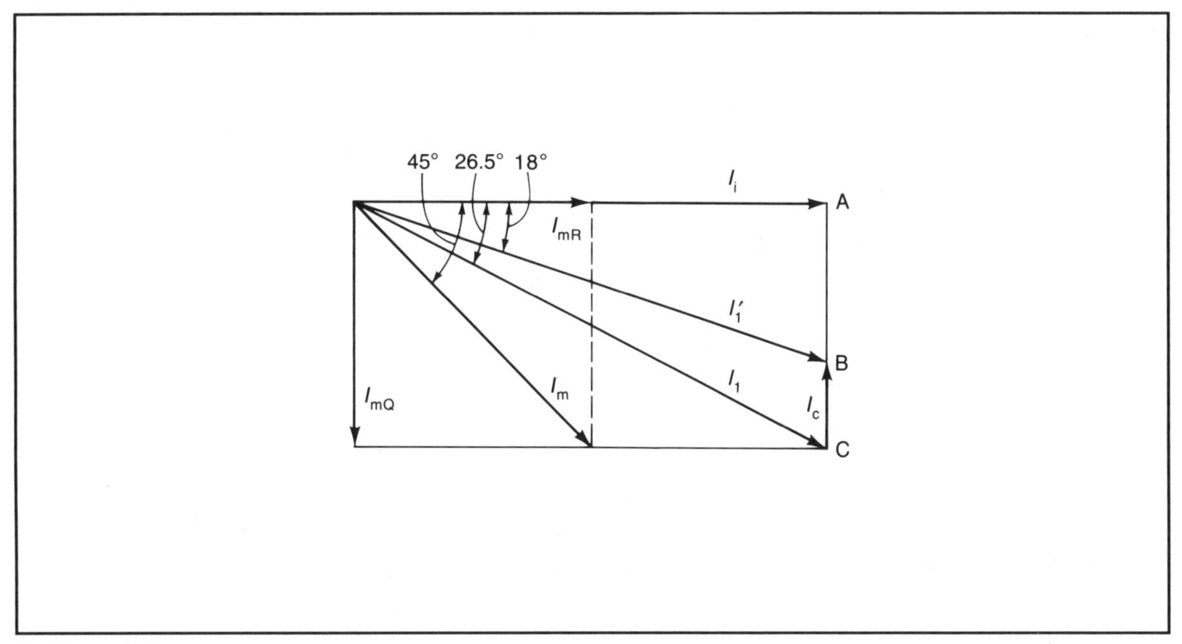

Problem 3.7

(b)

$$I_i = \frac{9{,}000}{\sqrt{3}\times 208} = 25\angle 0^\circ \text{ amps}$$

$$I_m = \frac{10\times 746}{\sqrt{3}\times 208\times 0.83\times 0.707} = 35.3\angle 45^\circ \text{ amps } = 25 + j25$$

$$\text{Thus } I_{mR} = 25\angle 0^\circ$$

$$I_{mQ} = 25\angle -90^\circ$$

$$I_1 = 25 + 25 + J25 = 50 + J25 = 56\angle 26.5^\circ \text{ amps}$$

if the p.f. $= 1$, the I_1 reduces to the real component of above current i.e. $50\angle 0^\circ$. Therefore, we have to reduce $I_1 = 56\angle 26.5^\circ$ to 105% of 50 amps $= I_1' = 52.50$ amps at an angle to be determined.

From above vector diagram I_1' (52.50) ends in point B.

$$AC = 25$$

$$AB = \sqrt{52.5^2 - 50.0^2} = 16.5$$

$$BC = 25 - 16.5 = 8.5 \text{ amps } = I_c$$

$$\text{Then: } X_{c/phase} = \frac{E}{\sqrt{3}\times I_c} = \frac{208}{\sqrt{3}\times 8.5} = 14.1 \text{ ohms}$$

$$\text{If } X_c = \frac{1}{2\pi fC}$$

$$\text{Then } C = \frac{1}{377\times 14.1} = 18.7\mu F / phase.$$

$$C_{total} = 3\times 18.7 = 56.1\mu F \qquad \text{ANSWER}$$

$$\text{(c) } I_1' = (25+25) + j16.5 = 52.5\angle 18^\circ \text{ amps}$$

$$I_{m_{NEW}} = 25 + j16.5 = 30\angle 34^\circ \text{ amps}$$

Rating of wire size:

$$125\%\ I_{m\,NEW} + I_1 = 1.25\times 30 + 25 = 62.5 \text{ amps}$$

From copper wire tables (NEC, Article 310, Table 310-16) for 70 amps continuous current we obtain Size No. 4. ANSWER.
From over current protection table at 30 amps
(no 125% factor is needed) +25 =55 amps we obtain a fuse size of 120 amps. ANSWER.

(no 125% factor is needed) +25 =55 amps we obtain a fuse size of 120 amps. ANSWER.

From conduit table for 4 wires and 55 amps we obtain a conduit size of 1 1/4" Answer.

*NEC refers to the "National Electrical Code. Obtainable from the National Fire Protection Association, Inc. Quincy, MA 02269.

Problem 3.8

Situation

Two Y connected, induction motors are fed by a 4160 V, line-to-line, 3-phase 60 Hz motor-control center 20 feet away. Motor #1 drives a 600 HP compressor. The efficiency of the motor is 90%, and its power factor is 0.5. Instruments of motor #2 indicate 1730 kW, 277 amps.

Required:

(a) Show the phasor-diagram of the loads, kW and kVA.
(b) Determine the capacity in microfarads of a wye-connected capacitor bank that is required to correct the power factor of the total load to 0.966.
(c) If a synchronous motor is installed in place of motor #2 and used instead of the capacitor bank to achieve the same overall power factor (0.966), what must its power factor be?

Solution:

$$(a)\ \text{Motor \#1} = \frac{600}{0.90} \times 0.746 = 497\text{kW} \cong 500\text{kW}$$

500 kW
$\theta = 60°$
866 kVAR
1000 kVA

$$\text{pf} = 0.5,\ \text{thus } \theta = 60°$$

$$\text{kVA} = \frac{500\text{kW}}{\cos 60°} = \frac{500}{0.5} = 1000$$

$$\text{kVAR} = 1000 \times \sin 60° = 1000 \times 0.866 = 866$$

$$\text{kVA} = \sqrt{3} \times 4{,}160 \times 277 = 1994 \cong 2{,}000$$

$$pf = \cos\theta = \frac{1730}{2000} = 0.866\ ;\ \ \theta = 30°$$

$$\text{kVAR} = 2{,}000 \times \sin 30° = 2000 \times 0.5 = 1000$$

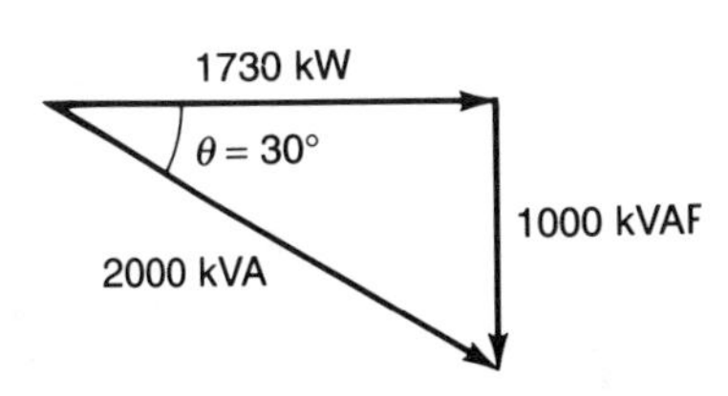

(b) Total load of motore #1 and #2

$$\text{kW} = 500 + 1{,}730 = 2{,}230$$

$$\text{kVAR} = 866 + 1000 + 1{,}866$$

$$\text{kVA} = \sqrt{2{,}230^2 + 1{,}866^2} = \sqrt{8.45 \times 10^6} = 2{,}907$$

$$\text{Actual combined pf} = \cos\theta = \frac{2{,}230}{2{,}907} = 0.767 \text{ lag;} \quad \theta = 40°$$

$$\text{Desired combined pf} = 0.966 \text{ lag;} \quad \theta = 15°$$

$$\text{kVA new} = \frac{2{,}230}{0.966} = 2{,}308$$

$$\text{BC} = \text{Required leading kVAR} = 1{,}866 - 2{,}230 \tan 15°$$

$$= 1{,}866 - 2{,}230 \times 0.268 = 1{,}866 - 598 = 1{,}268$$

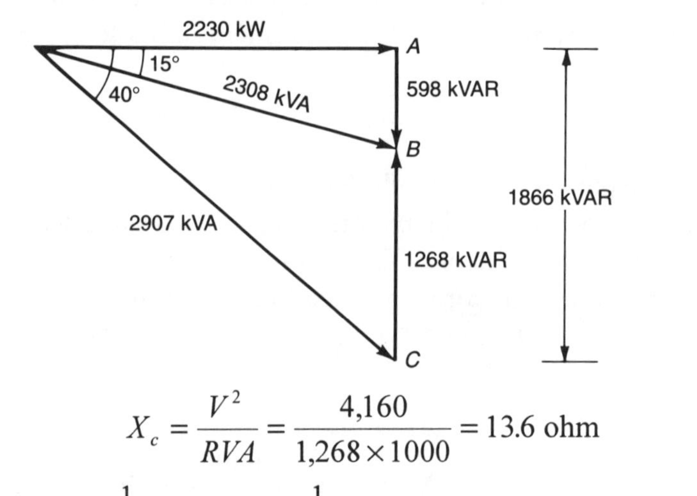

$$X_c = \frac{V^2}{RVA} = \frac{4{,}160}{1{,}268 \times 1000} = 13.6 \text{ ohm}$$

$$\text{C} = \frac{1}{2\pi f \times X_c} = \frac{1}{6.28 \times 60 \times 13.6} = 195\mu\text{F} \qquad \text{ANSWER.}$$

(c) Assuming that synchronous motor has same efficiency as motor #2 which it replaces,

$$\text{kVAR syn. motor} = \text{kVAR motor \#1} - \text{kVAR desired} = \text{AC} - \text{AB}$$

$$= 866 - 598 = 268$$

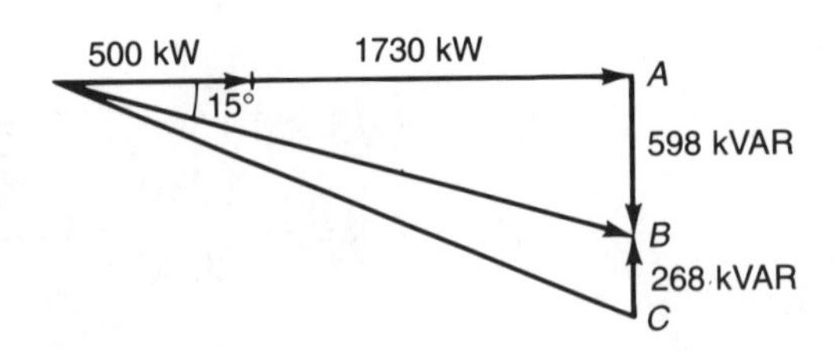

The synchronous motor alone:

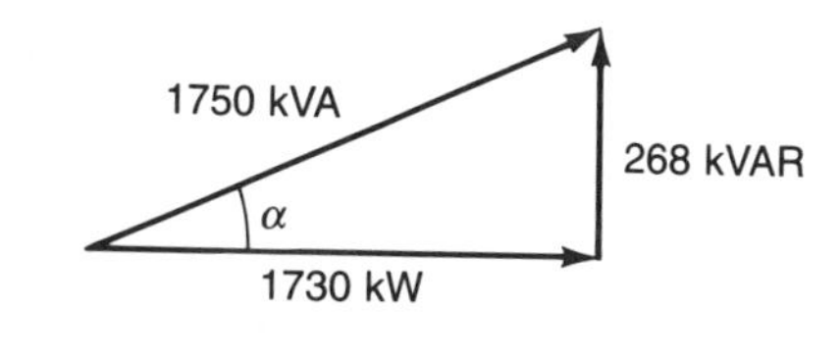

$$\tan\alpha = \frac{268}{1{,}730} = 0.155\ ;\ \ \alpha = 9^\circ$$

$$\text{kVA} = \frac{1730}{\cos 9^\circ} = \frac{1{,}730}{0.988} = 1{,}750$$

$pf = \cos 9^\circ = 0.988$ ANSWER.

Problem 3.9

Situation:

A 12,500 k VA, 6600 volt, 3600 r.p.m., 60 Hz, three phase, Y connected alternator has magnetization and short circuit characteristics curves as shown on the next page.

Required:

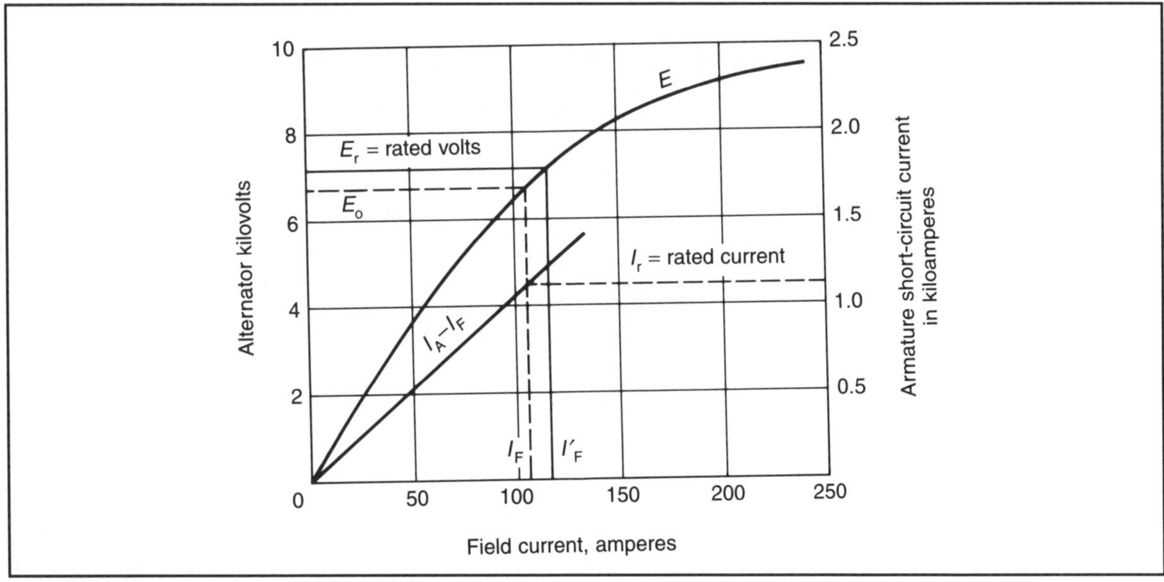

Problem 3.9

Determine the percentage voltage regulation for a 0.707 lagging power factor. Let the a.c. armature resistance be 0.5 ohms and make (and state) any reasonable assumption necessary for your solution.

Solution:

$$\text{Rated I} = \frac{12{,}500{,}000}{6600\sqrt{3}} = 1090 \text{ A}$$

Field current necessary to give short circuit line current, from graph:

$$I_f(\text{dc}) = 105 \text{ A}$$

Terminal voltage (per phase) from graph:

$$\frac{6300}{\sqrt{3}} = 3630 \text{ volts}$$

Terminal rated voltage (per phase)

$$\frac{6600}{\sqrt{3}} = 3800 \text{ volts}$$

(Here, one could find the synchronous reactance or impedance by taking these operating values and defining

$$X_S \approx \frac{3630 \text{ V}}{1090 \text{ A}},$$

However, since only one condition is asked for, the synchronous voltage drop is the 3630 volts.)

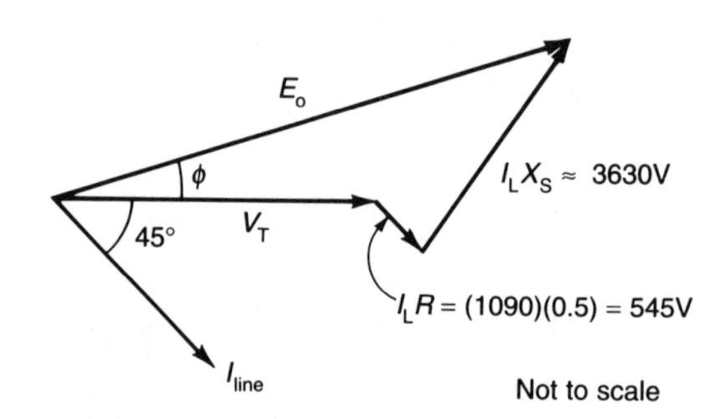

$$\mathbf{E}_0 = \mathbf{V}_T + R_A\mathbf{I}_L + jX_S\mathbf{I}_L$$

$$= 3800 + 545(\cos 45° - j\sin 45°) + 3630(\cos 45° + j\sin 45°)$$

$$= 6752.4 + j2181.6 = 7096\angle\phi \text{ volts}$$

$$\therefore \ \% \text{ V.R.} = \frac{E_0 - V_T}{V_T}(100) = \frac{7096 - 3800}{3800}(100)$$

$$= 86.7\%$$

$$\text{Note that the calculated no - load voltage is}$$
$$7096\sqrt{3} = 12{,}290 \text{ volts}$$

which is well beyond the range of available field current; thus the saturated limit of no - load voltage would be between 9,000 and 10,000 volts. The the saturated value of synchronous reactance would be less than the value used here.

This page intentially left blank.

POWER DISTRIBUTION: Problem 4.1

SITUATION:

A three-phase, 60 Hz, 173 kV, 50 mile long transmission power line delivers complex power to known load. The line has been tested for it's characteristic parameters. These parameters (on a per-phase basis) are found to be:

Resistance = 0.1 ohm/mile
Inductance = 2.0 millihenry/mile
Capacitance = 1.0×10^{-2} μF/mile

REQUIRED:

For a known load at the end of the line of 75+j30 (three-phase, total megavolt-amperes), determine the input (sending) current and the input power necessary to support the complex load.

Solution

(See Elgerd: "Basic Electric Power Engineering. Addison-Wesley, any edition.)
The equivalent circuit may be represented as:

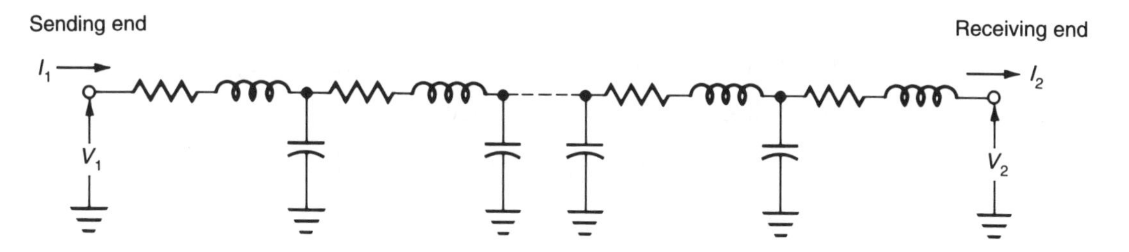

And a simpler approximation (on a per phase basis) is:

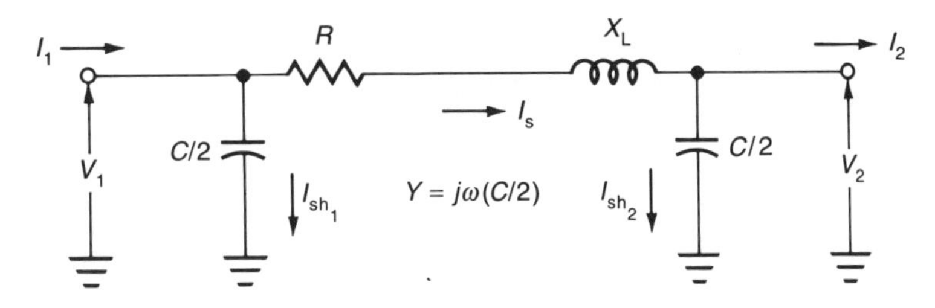

The complex power, S_{pp}, (which may also be expressed as the phasor voltage times the complex conjugate of the phasor current) on a per phase basis is:

$$\mathbf{S}_{pp} = \frac{1}{3}(75 + j30) = 25 + j10 \text{ mega volt - amps.}$$

The current at the receiving end, then, is:

$$\mathbf{I}_2 = \frac{(25 + j10) \times 10^6}{\left(\frac{173}{\sqrt{3}}\right) \times 10^3} = (0.25 + j0.1) \times 10^3 \text{ A / phase}$$

The shunt and series currents are then found

$$\mathbf{I}_{Sh_2} = \mathbf{V}_2 \mathbf{Y}_2 \text{ where } \mathbf{Y}_2 = j\frac{1}{2}(50)(377)(1 \times 10^{-8})$$

$$= j94.25 \times 10^{-6}$$

$$= \left(\frac{173}{\sqrt{3}}\right) \times 10^3 (j94.25) \times 10^{-6}$$

$$= j9.425 \text{ A / phase}$$

$$\mathbf{I}_{Ser} = \mathbf{I}_{Sh_2} + \mathbf{I}_2 = j9.425 + 250 + j100$$

$$= 250 + j109.4 \text{ A / phase}$$

And the voltage drop across the series branch may be found by first finding the series impedance:

$$\mathbf{Z} = R + jX_L = (50)(0.1) + j(50)(377)(2 \times 10^{-3})$$

$$= 5 + j37.7 \text{ ohms / phase}$$

$$\mathbf{V}_{drop} = \mathbf{I}_{Ser} \mathbf{Z} = (250 + j109.4)(5 + j37.7)$$

$$= -2.874 + j10.07 \text{ kV / phase}$$

Therefore the input voltage is

$$\mathbf{V}_1 = \mathbf{V}_{drop} + \mathbf{V}_2 = -2.874 + j10.07 + \frac{173}{\sqrt{3}}$$

$$= 97.13 + j10.07 \text{ kV}$$

And the sending end shunt current is

$$\mathbf{I}_{Sh1} = \mathbf{V}_1 \mathbf{Y} = (97.13 + j10.07)(j94.25 \times 10^{-6})$$

$$= 0.949 + j9.1545 \text{ A / phase}$$

Therefore the sending end current is

$$\mathbf{I}_1 = \mathbf{I}_{Sh_1} + \mathbf{I}_{Ser} = 0.949 + j9.154 + 250 + j109.4$$
$$= 250.9 + j118.55 \text{ A / phase}$$

And then one may find the sending end complex power (again complex voltage times the complex conjugate of the current) as

$$\text{Complex Power } = (97.13 + j10.07) \times 10^3 (250.9 - j118.55)$$
$$= 25.56 + j1.38 \text{ mega volt - amps / phase}$$

And, of course, the total complex power is three times this value:

$$76.68 + j4.14 \text{ mega volt - amps}$$

Problem 4.2

SITUATION

A 345 kV power transmission line has two bundled conductors per phase, spaced 18 inches horizontally. The conductor used in the bundle has a self GMD of 0.0403 feet and the phases are spaced horizontally 15 1/2 feet apart.

REQUIRED

Determine the following:
(a) The self GMD of the bundled conductors.
(b) The mutual GMD of the line.
(c) The inductive reactance per phase per mile.

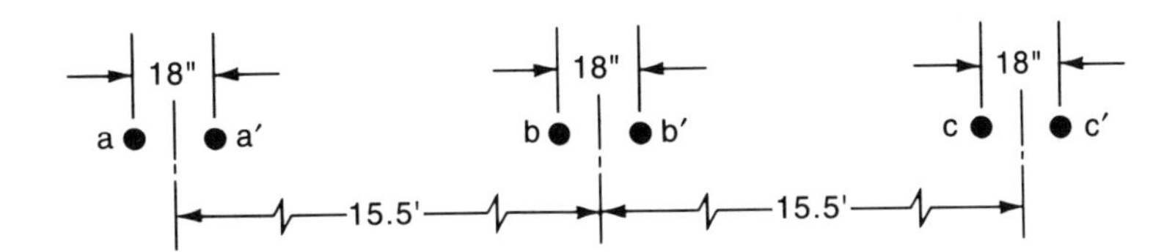

SOLUTION

(a)

Often self GMD of a bundled or composite conductor is called "geometric mean radius", or GMR. Self GMD may be denoted as D_s. This term includes the distances of a strand or conductor from all other strands within the same bundle plus the "distance of the strand from himself" or the self GMR of the strand.

In above line configuration we have two strands per bundle, thus we have four distances: $D_{aa'}$, $D_{a'a}$, D_{aa}, and $D_{a'a'}$. The self GMR of a single strand is less than the actual physical radius (R x 0.7788). This reduced radius is the above given self GMD of of 0.0403 ft. and is available from tables. Converting all distances into feet, we obtain for one bundle:

$$D_s = D_{sa} = \sqrt[4]{D_{aa'} \times D_{a'a} \times D_{aa} \times D_{a'a'}} = D_{sb} = D_{sc}$$

Note: We extract the fourth root as we have hour distances under the radical

$$D_s = \sqrt[4]{(0.0403)^2 \times \left(\frac{18}{12}\right)^2} = \sqrt[4]{0.00366}$$

$= 0.246$ ft. ANSWER

(b) The mutual GMD of the line or D_{eq} is the geometric mean of all mutual GMD values outside the bundles, i.e. between the three phases.

$$D_{ab} = D_{bc} = \sqrt[4]{(15.5)^2 \times 17.0 \times 14.0}\,, \quad \text{where ab} = 15.5' \text{ and a'b'} = 5.5'$$

$$ab' = 15.5 + 2\frac{18}{12} \times \frac{1}{2} = 17.0'$$

$$a'b = 15.5 - 2\frac{18}{12} \times \frac{1}{2} = 14.0'$$

$= \sqrt[4]{57{,}180} = 15.5$ ft.

$$D_{ac} = \sqrt[4]{(15.5 \times 2)^2 \times 32.5 \times 29.5} = \sqrt[4]{921{,}400} = 31 \text{ ft.}$$

$$D_{eq} = \sqrt[3]{D_{ab} \times D_{bc} \times D_{ac}} = \sqrt[3]{15.5 \times 15.5 \times 31.0} = 19.5 \text{ ft.}$$

(c) The inductive reactance in ohm / mile or $X_L = 2\pi f \times 10^{-3} \times 0.7411 \log \dfrac{D_{eq}}{D_s}$.

The 10^{-3} factor is needed to convert the inductance L, obtained in mh / mile to h / mile to finally yield ohm / mile.

$$X_L = 0.377 \times 0.7411 \log \frac{19.5}{0.246} = 0.530 \text{ ohm / mile} \quad \text{ANSWER.}$$

Reference: William D. Stevenson, Jr. "Elements of Power System Analysis", McGraw - Hill. p. 30 - 38.

Problem 4.3

SITUATION

Power for a remote building on an industrial site is supplied through an existing buried cable from a fixed voltage 60 Hz supply.

The load in the remote building consists of lighting and induction motors. During periods of peak demand, when the cable is carrying approximately its rated current, the resulting steady-state load voltage is well below the desired value because of the characteristics of the load. A small amount of additional constant-speed motor load is anticipated in the near future.

REQUIRED

What equipment can be installed at the building to improve the present situation and to permit the additional load? Explain how the equipment you recommend will improve the situation; the use of phasor diagrams is suggested.

SOLUTION

The crux of the whole problem is the large portion of induction motors. The power factor of induction motors at rated load are typically from 0.70 to 0.90 with some groupings of motors resulting in even lower power factors.

For a power factor of 0.8, a motor drawing 225 kVA of power will utilize only 180 kW.

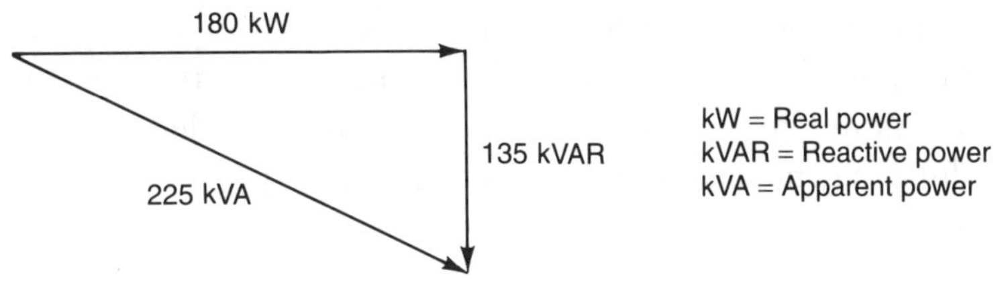

If the kVA drawn in this case were equal to the real power required (kW) a 20% reduction in current would result. The reduction in current with present load would reduce the voltage drop, thus improve the voltage at the load.

Fluorescent lighting with capacitors usually has power factors from 0.95 to 0.97, therefore are not practical to try to improve the power factor any higher.

To improve the power factor with the existing loads, capacitors should be applied. They have the characteristic of a leading power factor whereas induction motors have a lagging power factor. By adding capacitors their leading kVAR cancels out the equivalent amount of lagging kVAR, i.e.:

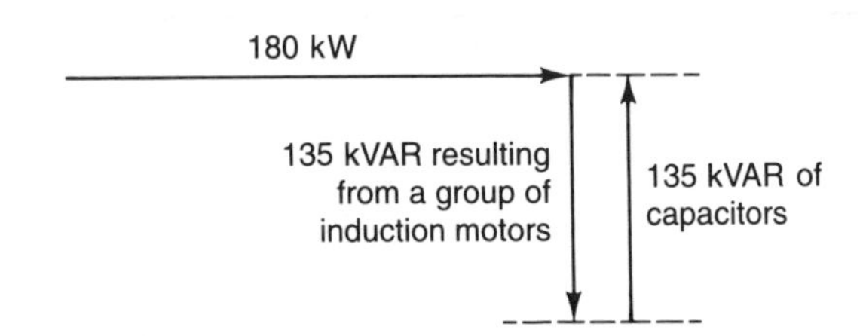

Above power factor correction results in 180kVA=180 kW or power factor=1.0.

Another method of improving power factor is to add synchronous motors for the additional motor requirements. The synchronous motor acts like a capacitor producing a leading power factor and leading kVAR.

Normally capacitors are the most effective in reducing system costs when located near the devices with low power factor. Here we are primarily concerned with the feeder to the building, but if there are any large induction motors or a grouping of motors it would minimize local branch circuit voltage drop as well as the feeder, if the capacitors were located near the source of the low power factor (pf).

An economic study of the situation should be made. Data from recording pf meters and kW meters should be gathered from as many places as feasible on feeder and branch circuits. Then a comparison of how much and where the capacitors should be installed. The installation of the synchronous motors vs. induction motors with capacitors should be evaluated.

With the additional load the voltage drop in the existing feeder may be too much even with unity pf. Then consideration should be given to using boost transformer. If may be well that a combination of boosting and capacitor and synchronous motor will be the most economical solution. Boost transformers are much less in cost than regular transformers as they are just an auto-transformer. Improving the power factor beyond a certain point increases the cost disproportionately to the gain obtained, thus all alternatives should be weighed in making the ultimate decision. Using a boost transformer alone may mean that at light load an over voltage may result which could be undesirable.

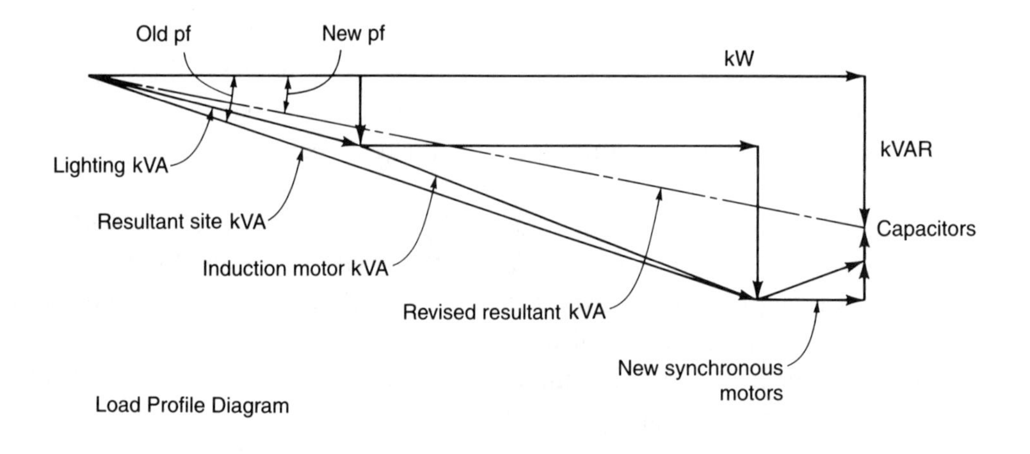

Load Profile Diagram

Problem 4.4

SITUATION

The given transmission system involving power distribution develops a fault in one of its phases.

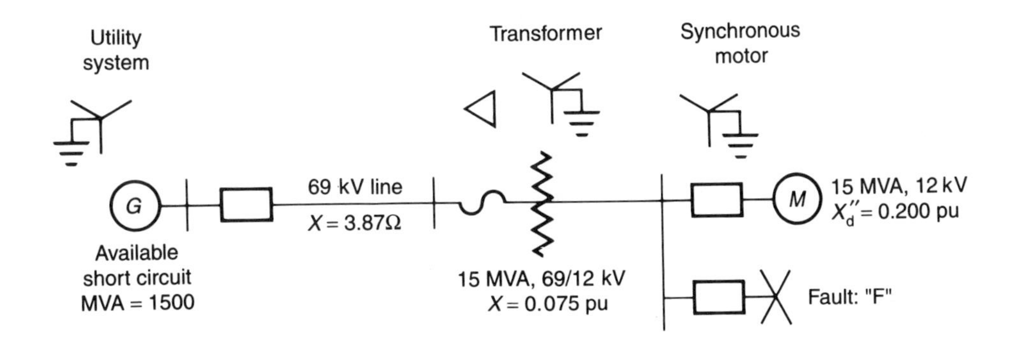

REQUIRED

Determine the fault currents at point "F" for the following conditions:

(a) Assume the fault is a three-phase one.

(b) Assume the fault is a single line to ground.

SOLUTION

kVA_{base} = 150 MVA (This value was selected to be a practical base between the two given MVA values)

$\mathrm{kV}_{\mathrm{base}}$ = 69 and 12 respectively

$$Z_{\text{base 69}} = \frac{(kV)^2 \times 1000}{kVA_{base}} = \frac{(69)^2 \times 1000}{150{,}000} = 31.8 \text{ ohms}$$

$$Z_{\text{base 12}} = \qquad = \frac{(12)^2 \times 1000}{150{,}000} = 0.96 \text{ ohms}$$

$$I_{\text{base 69}} = \frac{kVA_{base}}{\sqrt{3}kV_{base}} = \frac{150{,}000}{\sqrt{3} \times 69} = 1{,}250 \text{ amps}$$

$$I_{\text{base 12}} = \frac{kVA_{base}}{\sqrt{3}kV_{base}} = \frac{150{,}000}{\sqrt{3} \times 12} = 7{,}230 \text{ amps}$$

$$Z_{\text{line p.u.}} = \frac{Z_{\text{rated ohms}}}{Z_{base}} = j\frac{3.87}{31.8} = j0.121 \text{ p.u.}$$

$$Z_{\text{trans p.u.}} = Z_{\text{rated p.u.}} \frac{kVA_{base}}{kVA_{rated}} = j0.075\frac{150{,}000}{15{,}000} = j0.750 \text{ p.u.}$$

$$Z_{\text{utility}} = 1.0\frac{kVA_{base}}{kVA_{sh.ckr.}} = j \times 1.0\frac{150{,}000}{1{,}500{,}000} = j0.100 \text{ p.u.}$$

$$Z_{\text{motor d1}''} = j0.200\frac{150{,}000}{15{,}000} = j2.00 \text{ p.u.}$$

(a) Three phase fault:

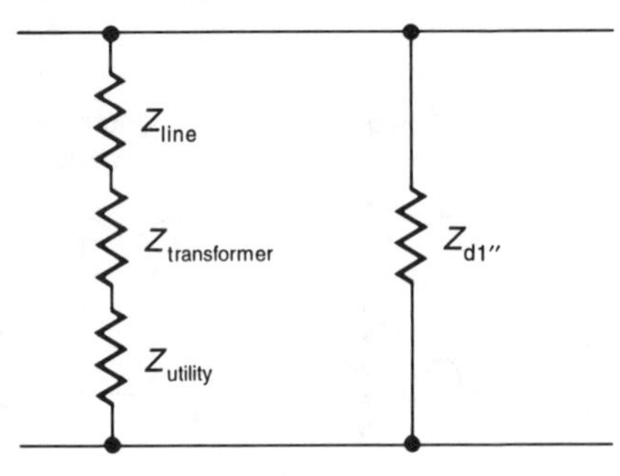

$$Z_A = Z_{line} + Z_{transf} + Z_{ut} = j0.121 + j0.75 + j0.100 = j0.971$$

$$Z_B = Z_{d1''} = j0.200$$

$$Z_{eq} = \frac{1}{j0.971} + \frac{1}{j0.200} = \frac{j0.971 \times j0.200}{j0.971 + j0.200} = j0.655$$

$$I_{fault} = \frac{E}{Z_{eq}} = \frac{1.0}{j0.655} = -j1.525 \text{ amps}$$

$$I_{fault} \text{ 3 phase at 12kV } = I_{fault\text{ p.u.}} \times I_{base\ 12}$$

$$= j1.525 \times 7{,}230 = -j11{,}000 \text{ amps ANSWER}$$

(b) Single Phase Fault:

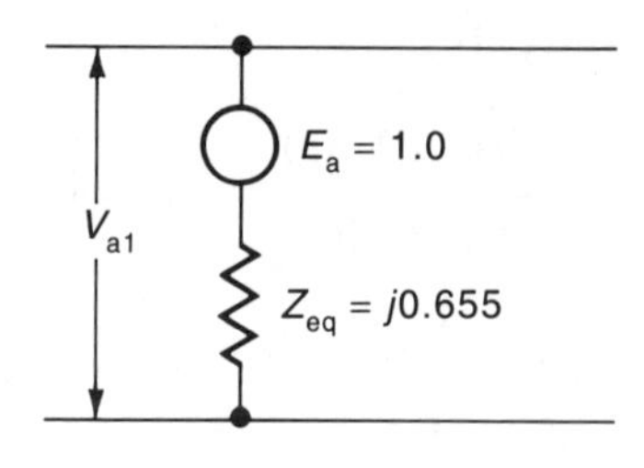

Negative sequence impedance diagram:

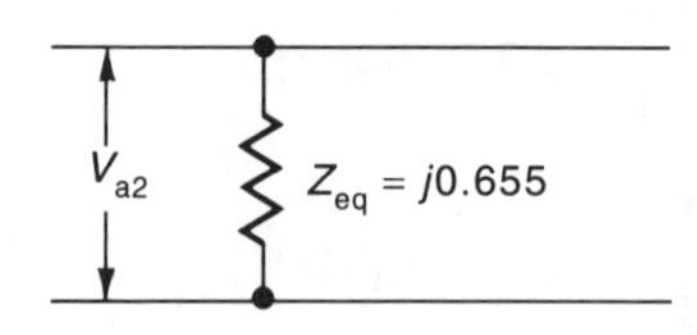

Zero Sequence Impedance diagram:

69 kV system zero sequence fault currents are isolated from fault "F" by ΔY transformer.

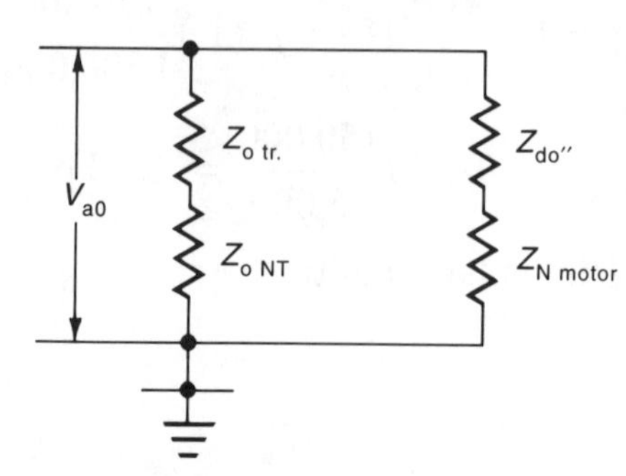

$$Z_{0\,\text{tr}} = j0.750$$

$$Z_{\text{motor do"}} = \frac{1}{2} Z_{\text{motor d1"}} (assumed) = j1.000$$

$$Z_{0\,\text{NT}} = 3Z_{\text{N motor}} = 0 \text{ ...directly connected neutrals}$$

$$Z_{0\,\text{eq}} = \frac{j0.750 \times j1.000}{j0.750 + j1.000} = j0.428$$

SEQUENCE NETWORK:

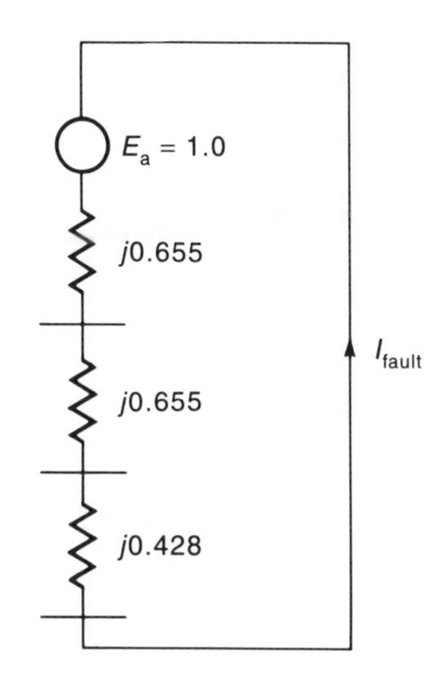

$$I_{a1} = I_{a2} = I_{a0} = \frac{E_a}{Z_1 + Z_2 + Z_0 + 3Z_N + 3Z_{fault}}$$

where $Z_N = 0$ and $Z_{fault} = 0$

thus:

$$I_{a1} = I_{a2} = I_{a0} = \frac{1.0}{j(0.655 + 0.655 + 0.428)} = -j0.580$$

$$I_{fault} = I_{\text{fault p.u.}} \times I_{\text{base at 12KV}}$$

$$= -j1.740 \times 7{,}230 = -j12{,}500 \text{ amps.} \quad \text{ANSWER.}$$

Problem 4.5

SITUATION

The voltages of an unbalanced 3-phase supply are V_a=(200+j0)V, V_b=(-j200)V and V_c=(-100+j200)V.

Connected in star across this supply are three equal impedances of (20+j10) ohms. There is no connection between the star point and the supply neutral.

REQUIRED

Evaluate the symmetrical components of the A phase current and the three line currents.

SOLUTION

The voltage components are as follows:

$$V_1 = \frac{1}{3}\left(V_a + aV_b + a^2V_c\right)$$

$$= \frac{1}{3}\left[200 + (-0.5 + j0.866)(-j200) + (-0.5 - j0.866)(-100 + j200)\right]$$

$$= \frac{1}{3}(200 + j100 + 173.2 + 50 - j100 + j86.6 + 173.2)$$

$$= \frac{1}{3}(596.4 + j86.6) = 198.8 + j28.86$$

$$V_2 = \frac{1}{3}\left(V_a + a^2V_b + aV_c\right)$$

$$= \frac{1}{3}\left[200 + (-0.5 - j0.866)(-j200) + (-0.5 + j0.866)(-100 + j200)\right]$$

$$= \frac{1}{3}(200 + j100 - 173.2 + 50 - j100 - j86.6 - 173.2)$$

$$= (-96.4 - j86.6) = -32.13 - j28.86$$

$$V_0 = \frac{1}{3}\left(V_a + V_b + V_c\right) = \frac{1}{3}(200 + j0 - j200 - 100 + j200)$$

$$= \frac{1}{3}(100) = 33.33$$

$$V_{a1} = 198.8 + j28.86 = (20 + j10)I_{a1}$$

$$V_{a2} = -32.13 - j28.86 = (20 + j10)I_{a2}$$

$V_{a0} = 33.33 = \infty$. I_{a0}, since neutral is not connected, i.e., there is no connection between the star point and the supply neutral.

$$I_{a1} = \frac{198.8 + j28.86}{20 + j10}$$ ANSWER.

$$I_{a2} = \frac{-32.13 - j28.86}{20 + j10}$$ ANSWER.

$$I_{a0} = \frac{33.33}{\infty} = 0$$ ANSWER.

$$I_a = I_{a1} + I_{a2} + I_{a0} = \frac{1}{20 + j10}(198.8 + j28.86 - 32.13 - j28.86)$$

$$= \frac{1}{20+j10}(166.67) = \frac{20-j10}{500} \times 166.67 = I_a = 6.67 - j3.33 \text{ amps in line 3.}$$

ANSWER.

To obtain the other (b,c) line currents:

$$I_b = a^2 I_{a1} + aI_{a2} + I_{a0}$$

$$= \left[(-.5 - j0.866)(198.8 + j28.86) + (-.5 + j0.866)(-32.13 - j28.86)\right]\frac{1}{20+j10}$$

$$= (-99.4 - j172.16 - j14.43 + 24.99 + 16.65 - j27.82 + j14.43 + 24.99)\frac{20-j10}{500}$$

$$= (-33 - j200)\frac{20-j10}{500} = -5.33 - j7.33 \text{ amps in line b. ANSWER.}$$

$$I_c = aI_{a1} + a^2 I_{a2} + I_{a0}$$

$$= \left[(-.5 + j0.866)(198.8 + j28.86) + (-.5 - j0.866)(-32.13 - j28.86)\right]\frac{1}{20+j10}$$

$$= (-99.4 + j172.16 + j14.43 - 24.99 + 16.65 + j27.82 - j14.43 - 24.99)\frac{20-j10}{500}$$

$$= (-133.33 + j200)\frac{20-j10}{500} = -1.33 + j10.67 \text{ amps in line c. ANSWER.}$$

as a check: $0 = I_a + I_b + I_c$ or

$$0 = 6.67 - j3.33 - 5.33 - j7.33 - 1.33 + j10.67$$

Reference: "Elements of Power Systems Analysis" by William D. Stevenson, Jr. Chapter 13.

Problem 4.6

SITUATION:

The following one-line diagram is for a single-phase system involving a voltage source of 240 volt 10 kVA generator that supplies a load through a 1:2 step-up transformer, a relatively short transmission line with an impedance of 1+j4 ohms, and a 4:1 step-down transformer. The transformers are assumed to be ideal.

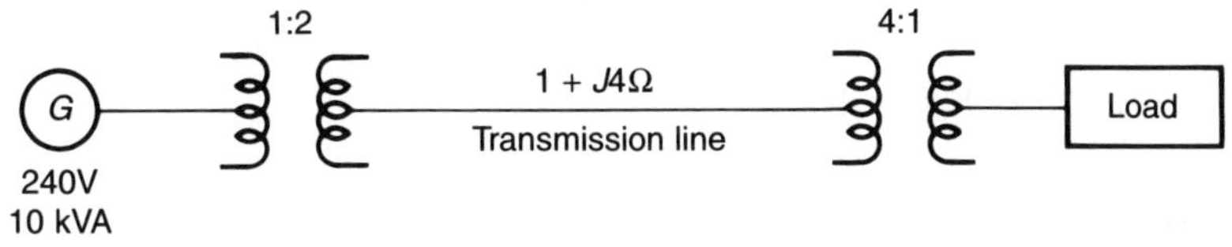

REQUIREMENT(S):

1. Assume the load is known to be 1+j1 Ω, what equivalent load (in ohms) does the generator see?

2. Assume the load is known to be 1+j1 Ω, determine the per unit values of the various system parameters along the system and determine the per unit current through the load.

3. (Optional for practice.) Repeat part 2 except Z_L=2+j2 Ω.

SOLUTION

1. The load as seen at the left of the last transformer is:

$$\mathbf{Z} = 4^2 x1 + j4^2 x1 = 16 + j16\ \Omega.$$

The impedance of the load and the transmission line as seen on the right side of the first transformer is:

$$\mathbf{Z} = (1+16) + j(4+16) = 17 + j20.$$

The impedance of the load and transmission line as seen by the generator (on the left side of the first transformer) is:

$$\mathbf{Z} = (1/2)^2 x17 + j(1/2)^2 x20 = 4.25 + j5.0 = 6.56\underline{/49.6^\circ}\ \Omega$$

(The magnitude of the current through the generator is:

$$I = V/Z = 240/6.56 = 36.6\text{ A.})$$

2. Of the several solutions possible, one method is to define various base quantities along the system as:

a) At the generator,

$$V_{B\text{-}g} = 240\text{ V} = 1\text{ pu}$$

$$\text{k}VA_{B\text{-}g} = 10\text{ kVA} = 1\text{ pu}$$

$$I_{B\text{-}g} = 10{,}000/240 = 41.66\text{ pu}$$

$$Z_{B\text{-}g} = 240/41.66 = 5.761\ \Omega = 1\text{ pu.}$$

b) For the transmission line,

$$V_{B\text{-}tl} = 240/(1/2) = 480\text{ V} = 1\text{ pu}$$

$kVA_{B\text{-}tl}$ = 10 kVA = 1 pu

$I_{B\text{-}tl}$ = 10,000/480 = 20.83 A = 1 pu

$Z_{B\text{-}tl}$ = 480/20.83 = 23.04 Ω = 1 pu

[$\mathbf{Z}_{tl}$ =(1+j4)/23.04 = 0.0434+j0.1736 pu]

c) At the load,

$V_{B\text{-}L}$ = 480/4 = 120 V = 1 pu

$kVA_{B\text{-}L}$ = 10kVA = 1 pu

$I_{B\text{-}L}$ = 10,000/120 = 83.33 A = 1 pu

$Z_{B\text{-}L}$ = 120/83.33 = 1.44 = 1 pu

[$\mathbf{Z}_L$ = (1+j1)/1.44 = 0.6944+j0.6944 pu].

The total impedance (as seen to the left of the first transformer) is,

$\mathbf{Z}_{Tot} = \mathbf{Z}_{tl} + \mathbf{Z}_L$ = 0.0434+j0.1736 + 0.6944+j.6944
= 0.7378+j0.868 = 1.139 $\angle 49.6^\circ$ pu

which is equivalent to,

= 1.139x5.76$\angle 49.6^\circ$ = 6.56 Ω.

(The magnitude of the per unit current is,
I = 1/1.139 = 0.878 pu,

and the actual current is,

I = 0.878x41.66 = 36.6 A.)

3. If load is 2+j2 then, by using pu values only $\mathbf{Z}_L$ needs to be recomputed:

$\mathbf{Z}_L$ = (2 + j2)/1.44 = 1.389 + j 1.389 pu

$\mathbf{Z}_{Tot} = \mathbf{Z}_{tl} + \mathbf{Z}_L$ = 0.0434+j0.1736 + 1.389+j1.389 =1.432+j1.563 pu

and the actual total impedance is,

$\mathbf{Z}$ = 1.432x5.761 + j1.563x5.761 = 8.25 + j9.00 = 12.2$\angle 47.5^\circ$ Ω

Problem 4.7

SITUATION

Originally it is planned to furnish a plant load requirement of 1000 hp at 2200 volt, 3-phase by induction motors operating at 80% power factor and 90% efficiency.

REQUIRED

(a) Find the line current necessary to supply this load and generator capacity.

(b) Assume that rather than supplying the 1000 hp by induction motors, it is decided to produce 400hp of this load by a synchronous motor operating in the over-excited leading mode of 85% (assume same efficiency as for an induction motor). Find the new total line current requirement and the overall power factor.

(c) If rather than installing the 400 hp synchronous motor as in Part(b), it is considered feasible to use power factor correcting capacitors for the 1000 hp motors in Part (a) to achieve the same power factor correcting as obtained in Part (b). Determine the size (kVAR) of the capacitors needed.

SOLUTION

(a) Power input to motors:

$$\frac{1000}{0.9} = 1111 \text{ hp} = 830 \text{ kW}$$

Generator Capacity

$$\frac{830 \text{ kW}}{0.8} = 1036 \text{ kVA}$$

Current Requirement

$$\frac{1036 \times 1000}{\sqrt{3} \times 2200} = 272 \text{ A}$$

(b) Power requirements:

Induct. motor input

$$\frac{600\text{hp}}{0.9} \times \frac{746}{1000} = 497 \text{ kW}$$

Induct. motor current

$$\frac{497 \times 1000}{\sqrt{3} \times 2200 \times 0.8} = 163 \text{ A}$$

Synch. motor input

$$\frac{400\text{hp}}{0.9} \times \frac{746}{1000} = 332 \text{ kW}$$

Synch. motor current

$$\frac{332\times 1000}{\sqrt{3}\times 2200\times 0.85}=102\ \ \text{A}$$

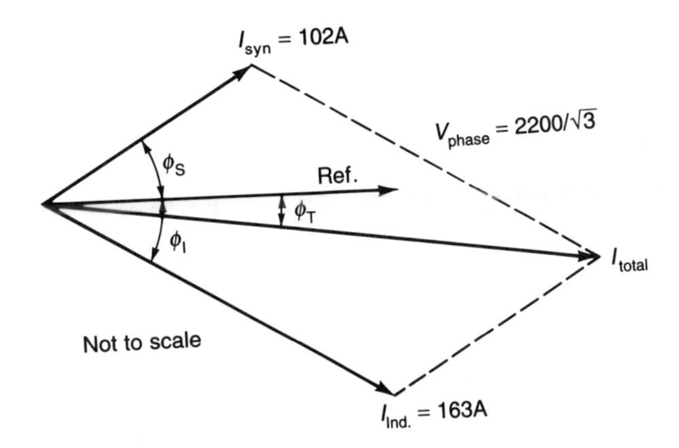

$$\mathbf{I}_{Total}=\sqrt{\left(I_I\cos\Phi_I+I_S\cos\Phi_S\right)^2+\left(I_I\sin\Phi_I+I_S\sin\Phi_S\right)^2}$$
$$=\sqrt{\left(163\times 0.8+102\times 0.85\right)^2+\left(163\times 0.6-102\times 0.53\right)^2}$$
$$=221\ \angle 11.5^\circ\ \text{A (lagging)}$$

(c)

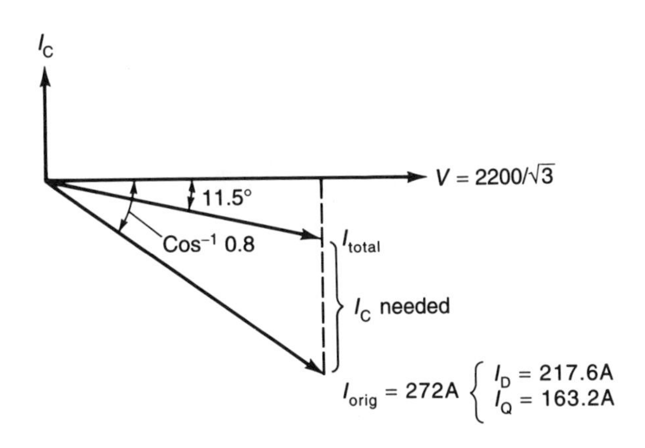

$$\text{I}_{\text{Total}}=\frac{I_D}{\cos 11.5^\circ}=\frac{217.6}{0.98}=222.06$$
$$\text{I}_{\text{Total}_Q}=222.06\sin 11.5^\circ=44.3\ \text{A}$$
$$\therefore \text{I}_\text{C}=\text{I}_{0_Q}-\text{I}_{\text{Total}\,Q}=163.2-44.3=118.9\ \text{A}$$
$$\therefore \text{kVAR}=\frac{118.9\times 2200}{1000\sqrt{3}}=151\ \text{kVAR / phase}$$

This page intentionally left blank.

Many problems on the examination are electronic in nature (see chapter 5) and are adapted and presented using the notation and concepts given in this chapter.

Electronics: Problem 5.1

SITUATION:

A single stage FET power amplifier needs to be analyzed so that parameter values and specifications may be established; in addition the FET circuit may have an interchangeable op-amp as a preamplifier for various input signal conditions.

REQUIREMENTS:

An n-channel junction FET power amplifier circuit has the drain characteristic curves as shown. The FET circuit eventually will be driven by an op-amp acting as preamplifier for various signal sources. The FET has a maximum drain voltage upper limit of 16 volts. Assume the operating quiescent current, I_D, for the load line is 22 mA and any operation is at mid-band frequencies.

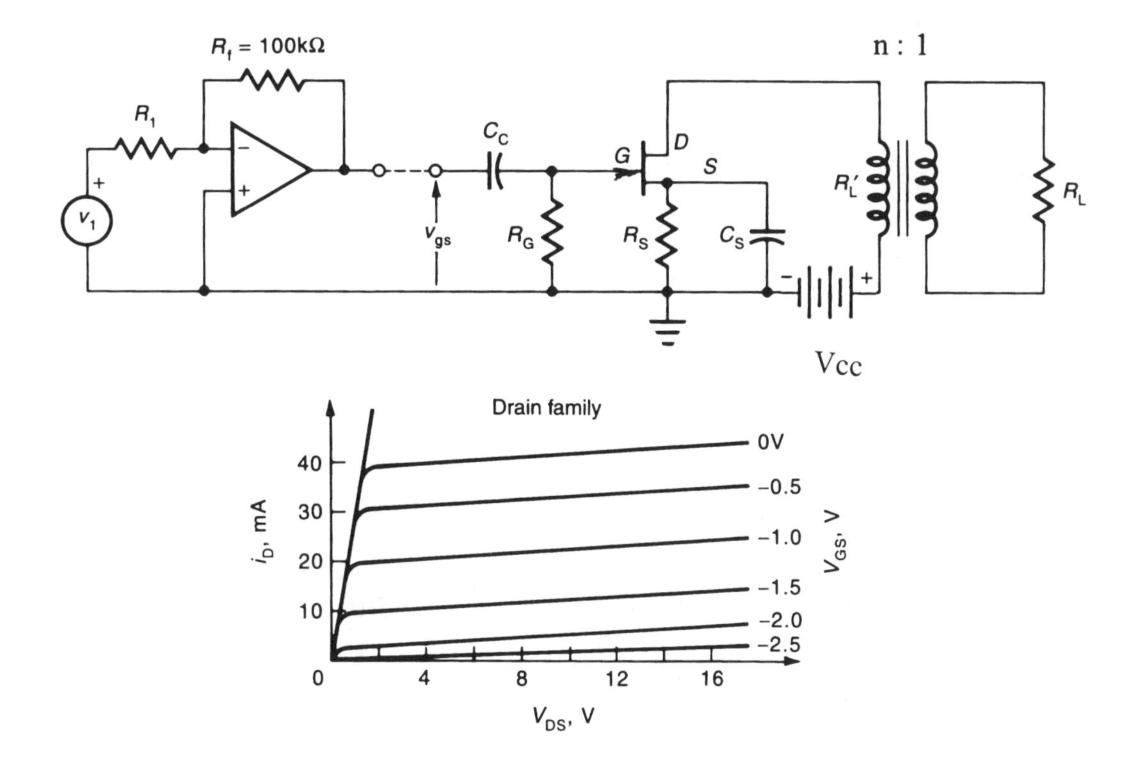

The following information is needed:

1) V_{DS} and R'_L for maximum possible output signal excursion.
2) The transformer turns ratio n for an R_L= 8 Ω for the R'_L(as found in 1).
3) Graphically determine the approximate power output when v_{gs} = 0.5 sin ωt volts.
4) The percentage of second-harmonic distortion for i_D for the same signal as in (3).
5) With the op-amp connected and with a V_I= 35 mV (rms), determine the value of R_I to produce the signal in (3).

Solution

1) $V_{DS} = \dfrac{V_{D(Upper)}}{2} = \dfrac{16}{2} = 8 \text{ v}$

$$\text{R}'_\text{L} \text{ (as referred to "n" side of transformer)} = \frac{V_{DS(Upper} - V_{DS}}{I_D} = \frac{16-8}{22ma} = 354\ \Omega$$

2) $\text{R}'_\text{L} = n^2 R_L\ ,\quad n^2 = \dfrac{\text{R}'_\text{L}}{\text{R}_\text{L}} = \dfrac{364}{8} = 45.5\ ,\quad n = 6.7$

3) From a load line plot for a v_{gs} of 0.5V on each side of the quiescent point yields:

$v_{ds(max)}$ and $v_{ds(min)}$ of 11.1 and 4.2 volts, and $i_{d(max)}$ and $i_{d(min)}$ of 31 and 12 mA

$$P_0 = \frac{\Delta v_{gs}}{2\sqrt{2}} \cdot \frac{\Delta i_d}{2\sqrt{2}} = \frac{(11-4.2)\times(31-12)ma}{8} = 16.4 \text{ mW.}$$

4) Percent distortion $= \left| \dfrac{\dfrac{(i_{d\max} + i_{d\min})}{2} - I_D}{i_{d\max} - i_{d\min}} \right| \times 100 =$

$$= \left| \frac{\left(\dfrac{(31+12)}{2} - 22\right)}{(31-12)} \right| \times 100 = 2.6\%$$

5) $\text{V}_{1(rms)} = 35 \text{ mV} \equiv 0.05 \text{ (0-to-peak)}\ ,\qquad \text{V}_{gs} = 0.5 \text{(0-to-peak)}$

$\therefore$ Op-Amp circuit gain is 10 and $\dfrac{\text{R}_\text{f}}{R_1} = \dfrac{100\times 10^3}{R_1}\ ,\quad \text{R}_\text{i} = 10 \text{ k}\Omega$.

Electronics: Problem 5. 2*

SITUATION:

A transistor (BJT) amplifier is driven by a high input impedance MOSFET (p-chan, depletion mode) preamplifier. The tentative circuit design has already been proposed. However, a check of the circuit and performance specifications need to be completed.

REQUIREMENTS:

The component values for the circuit are given and some of the parameters of the MOSFET and BJT are known, these are as indicated.

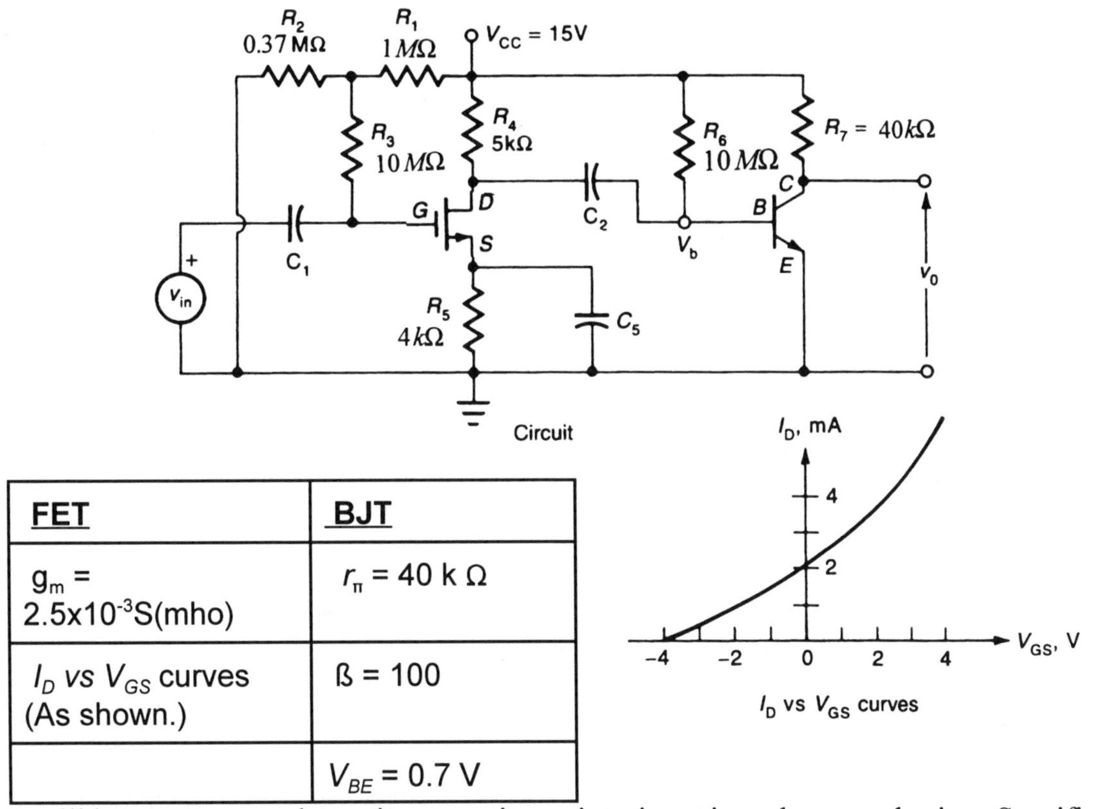

FET	BJT
g_m = 2.5x10^{-3}S(mho)	r_π = 40 k Ω
I_D *vs* V_{GS} curves (As shown.)	ß = 100
	V_{BE} = 0.7 V

It will be necessary to determine operating points, input impedance, and gains. Specifically the following items are needed:

1) The dc operating point, Q for the FET.
2) The dc operating point, Q for the BJT.
3) The small signal input impedance.
4) The voltage gain at mid-frequency.

* This problem is an adaption of Problem 3 as presented in James W. Morrison's "*Principles and Practice of Electrical Engineers Examination P&P/EE*". Published by ARCO., 1977, Pages 113-117.

SOLUTION

1) In finding the dc operating point Q for the MOSFET, the dc voltage on the gate, V_G is needed.

$V_G = [R_2/(R_1+R_2)]V_{CC} = 0.4/(1.1+0.4)]15 = 4$ V.

Then, by using the I_D vs V_{GS} characteristic curve and drawing a dcload-line on it, the voltage from gate to source, V_{GS}, may be found. Here $V_{GS} = -1$V. (The load-line is a plot of $V_G = V_{GS} + I_D R_5$.)

The summing voltages around the loop yields the voltage, V_S, as

$V_S = -V_{GS} + V_G = -(-1) + 4 = 5$ V.

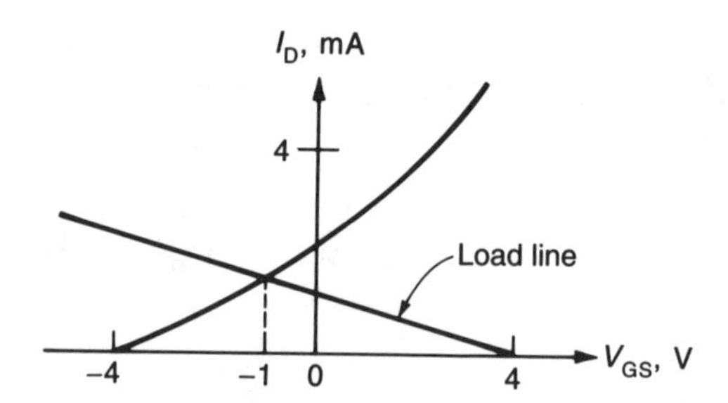

Therefore the sink current, I_S, is

$I_s = V_S/R_5 = 5/(4\text{x}10^3) = 1$ mA,

And the drain to source voltage, V_{DS}, is:

$V_{DS} = V_{CC} - I_D(R_4+R_5) = 3.75$ V.

2) In finding the operating point Q for the BJT, the base current first needs to be found,

$I_B = (V_{CC} - V_{BE})/R_6 = (15 - 0.7)/10\text{x}10^6 = 1.4$ μA.

Since the collector current, I_C is $ßi_b$, then $I_B = 140$ μA, the collector to emitter voltage, V_{CE}, is:

$V_{CE} = V_{CC} - I_C R_7 = 15 - (140\text{x}10^{-6})(40\text{x}10^3) = 9.4$ V.

3) The small signal input impedance, R_{in}, is found almost by inspection, since the MOSFET G terminal draws negligible current,

$R_{in} = (10)\text{x}10^6 = 10^7\ \Omega$

4) The voltage gain of the entire circuit (at mid-band) may be determined by drawing the equivalent small signal circuit model (where $R_A = R_4||R_6$, and $g_{m2} = ß/r_\pi$),

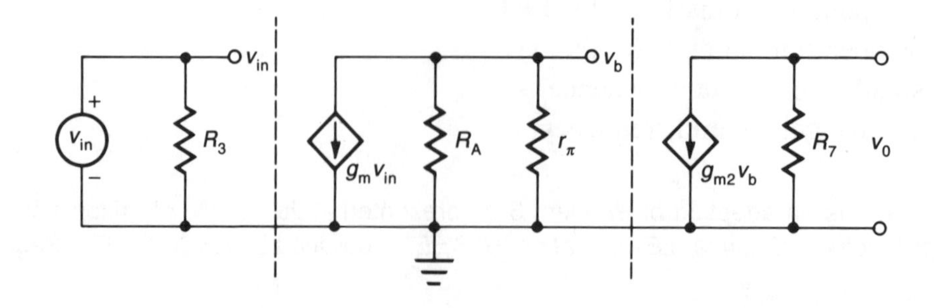

Here, R_A is essentially R_4, which was 5 kΩ, and,

$$g_{m2}=100/(40x10^3) = 2.5x10^{-3} \text{ S (mho)}.$$

The circuit model may be redrawn, making use of Thevenin's equivalent circuit as,

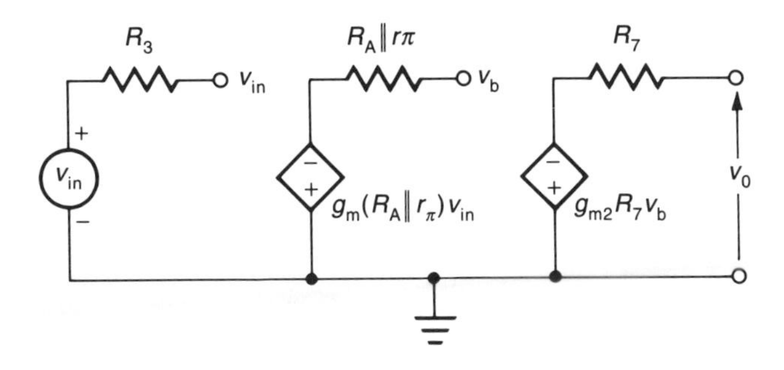

The following stage parameters may then be found as:

$$v_o = -g_{m2}R_7v_b = -(2.5x10^{-3})(40x10^3)v_b = -100v_b \text{ V}.$$
$$v_b = -g_m(R_A \| r_\pi)v_{in} = -2.5[(5x40)/(5+40)]v_{in} = -2.5 \text{ x } 4.4 \, v_{in} = 11.1 \, v_{in} \text{ V}.$$
$$\therefore v_0 = -100 \, v_b = -(100)(-11.1 \, v_{in}) \approx 1{,}100 \, v_{in} \text{ V}.$$

Therefore, the overall gain is approximately,

$$v_o / v_b = 1{,}100.$$

Electronics Problem 5. 3.

SITUATION

Assume the BJT transistor shown in the circuit has a ß = 100 and that $V_{BE} = 0.7$ volts,

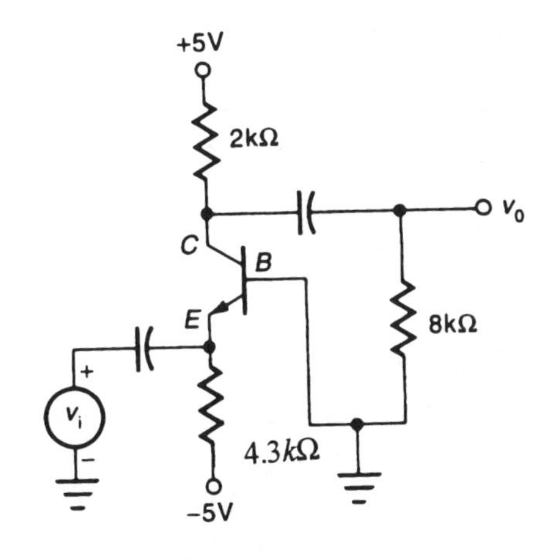

REQUIRED

Find the following quantities:

a) I_C, I_E, and V_C.
b) Find g_m, r_π, and r_e.
c) Draw the small signal equivalent circuit using a T model for the BJT and calculate the gain, v_o/v_i.

SOLUTION

a) $I_E = [-0.7-(-5)]/4.3 = 1$ mA,
$I_C \cong I_E = 1$ mA,
$V_C = 5 - I_C x2 = 3$ V.

b) $g_m = I_C/V_T = 1$ mA/0.025 V $\cong$ 40 mA/V,
$r_\pi = ß/g_m \cong 100/40 = 2.5$ kΩ,
$r_e = r_\pi/(ß+1) \cong = 2.5$ kΩ$/101 = 25$ Ω.

c) The equivalent circuit is as shown, and the values are then:

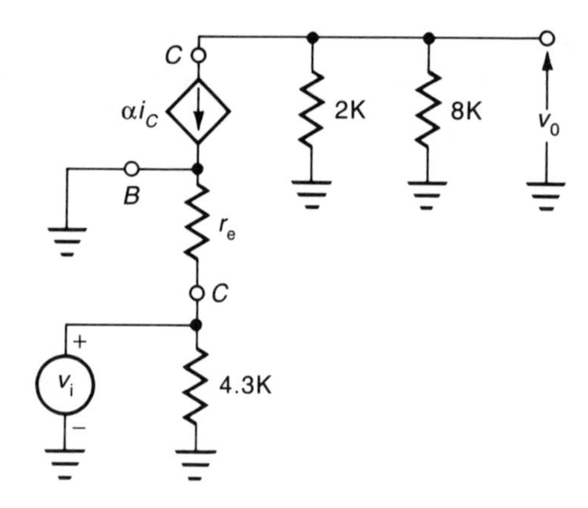

$v_o = -\alpha i_e(2k||8k)$,
$i_e = -v_i/r_e$,
$v_o = +\alpha(v_i/r_e)(2k||8k)$,
$v_o/v_i = g_m(2k||8k) \cong 40x1.6 = 64$ V/V.

Electronics: Problem 5.4

SITUATION:

A two stage transistor (BJT) amplifier uses constant current sources for the bias portions of the circuit; for the first stage the constant current source is fixed while the second is adjustable. The component values for the circuit are as given and some of the parameters of the BJTs are known to be $|V_{BE}|= 0.65$ volts, ß=100 (since high accuracy is not expected, ß is usually considered high), $V_T= 26$ mV, and all capacitors are considered large unless otherwise indicated. The circuit is as follows:

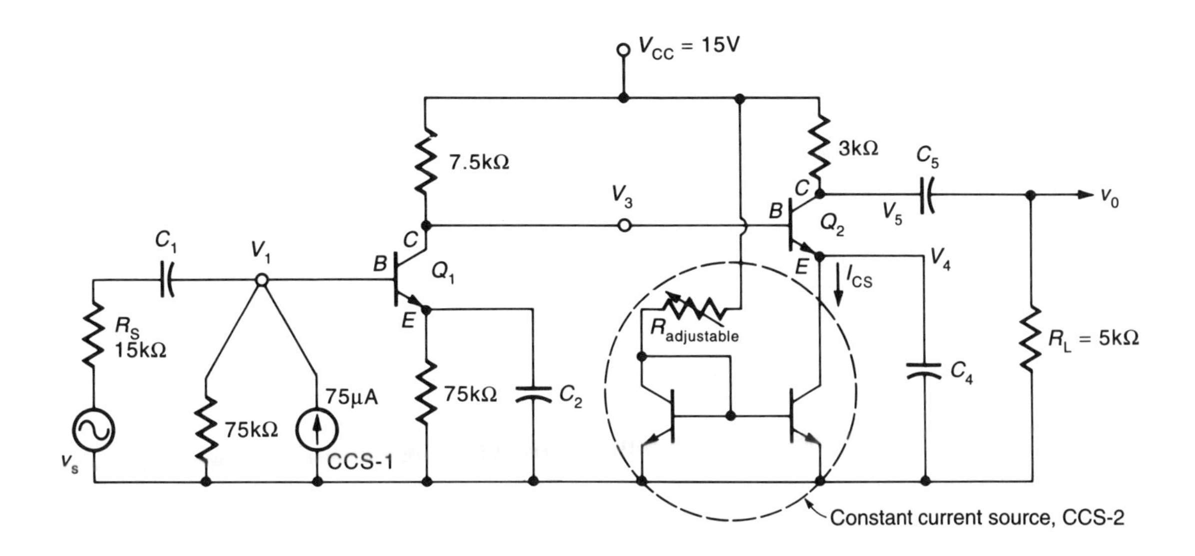

REQUIREMENTS:

The tentative circuit design has already been proposed. However, a recheck of the circuit and its performance specifications need to be completed.
Assume the second current source (at V_4) has already been adjusted and tested to produce a constant 2.0 mA.

1. Determine the dc bias voltages at V_1.
2. Determine the dc bias voltages at V_3.
3. Determine the dc bias voltages at V_5.
4. Find the input resistance R_{in}.
5. Find the ac voltage gain from V_1 to V_5 and assume R_L is disconnected.
6. Determine the transconductance of the transistors $Q_{1,2}$.
7. Determine the output resistance at V_o (include R_L).

Now assume the simplified constant current source is made up of matched transistors with very high ß's and is readjusted to give constant current source of 3.0 mA. Again, assume V_{BE}= 0.65 volts.

8. If the original setting of R_{adj} was R_x, what should the new setting be (in terms of R_x)?

9. Determine the numerical original value of $R_{adj} = R_x$ (assume V_{cc} = 15 volts).

10. Determine the dc bias voltages at V_4.

SOLUTION

1. V_1= (75x10^3)(75x10^{-6})= 5.63 volts (i_b-->0 for high ß),

2. I_C=I_E= 5.0V/7.5kΩ = 0.667 mA, V_3= 15-7.5x0.667= 10 V.
   (The 5V comes from V_1-V_{BE}-5.63-0.65 ≈5V.)

3. V_5= 15 - (3x10^3)(2x10^{-3})= 9 V.

4. To find R_{in}, it is first necessary to find r_{e1} or g_{m1}:

   r_{e1}= V_T/I_c = 26 mV/0.667 mA = 39 Ω,

   R_{in}= 75k‖r_π= 75k|(ß+1)r_{e1}= 75k‖3.94k = 3.76 kΩ

5. To find the overall voltage gain, it will also be necessary to find either r_{e2} or g_{m2}:

   r_{e2}= V_T/I_c = 26 mV/2x10^{-3} = 13 Ω
   A = {[-7.5k‖(ß+1)r_{e2}(ß/(ß+1))]/r_{e1}}{[-3k/r_{e2}][ß/(1+ß)]}
   = {[-7.5k‖(101x13)]/39}(-3k/13) = 6,470 V/V.

6. Since the r_e's have already been found,

   g_{m1} = 1/r_{e1}= 1/39= 25.6 mA/V and g_{m2}= 1/r_{e2}= 1/13= 77mA/V.

7. R_{out}= R_L‖3k= 1.87 kΩ.

For the adjustable constant current source, the transistors are matched and I_{C1}= I_{C2}, the I_{ref} is given by:

I_{ref}= I_{C1}+2I_{C1}/ß,
I_{C1}= I_{ref}/(1+2/ß) = I_{C2},
for ß>>1,
I_{C1}= I_{C2}= I_{ref}
= [(-V_{cc})-V_E(on)]/R_{adj}

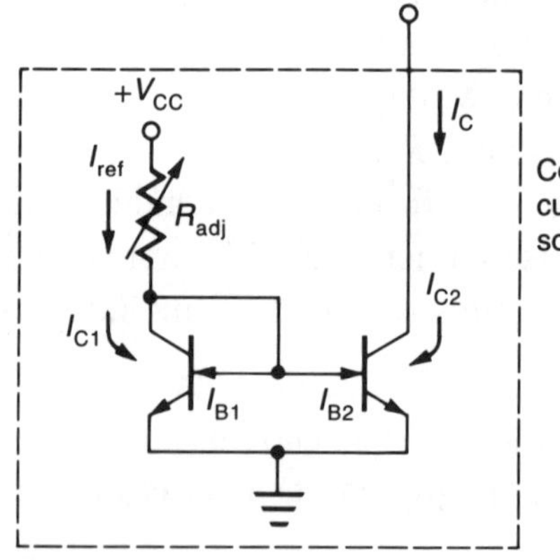

8. Changing the constant current source from 2 mA to 3 mA makes $R_{adj\text{-}new}$= (2/3)R_x.

9. I_C= 2 mA = [-(-15)-0.65]/R_x, R_x= 7.18 kΩ.

10. V_3 - V_{BE} = 10 - 0.65 = 9.35 volts.

ELECTRONICS: Problem 5.5

SITUATION:

Two different MOSFETs are being considered for use in two different circuits. Both transistors have a V_t of 2 volts; however, each unit has a different K factor (conductivity parameter).

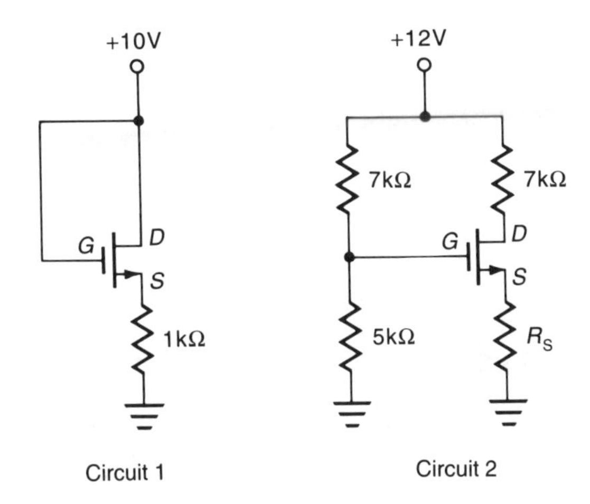

REQUIREMENT:

1. For circuit #1 the K factor is 0.25 mA/V^2 and you are to determine I_D.

2. For circuit #2 the K factor is 1.0 mA/V^2 and you are to determine R_S for a drain current of 1 mA.

SOLUTION

For circuit #1 the following relationships are obvious:

$$V_D = V_G = 10 \text{ V}; \ V_{DS,sat.} = V_{GS} - V_t; \ V_{DS} = V_{GS}; \ V_{DS} > V_{DS,sat.}$$

Then,

$$I_D = \mathrm{K}(V_{GS} - V_t)^2,$$

$$V_{GS} = V_G - V_S = 10 - 1\mathrm{kxI}_D = 10 - \mathrm{I}_D,$$

$$I_D = 0.25(10 - \mathrm{I}_D - 2)^2,$$

$$4I_D = (8 - I_D)^2 = 64 + \mathrm{I}_D^2 - 16I_D,$$

$$I_D = +10 \ +/- \sqrt{100-64} = 10 \ +/- \ 6,$$

$I_D = 4$ mA (the 16 mA is not acceptable).

For circuit #2 (with a K= 1 mA/V^2) the calculations for R_S for a drain current of 1 mA are as follows:

$$V_G = [12/(7+5)]5 = 5 \text{ V},$$

$$V_D = 12\text{-}(7\text{k}\Omega)(1\text{mA}) = 5 \text{ V},$$

$$I_D = \text{K}(V_{GS}\text{-}V_T)^2 = 1 \text{ mA} = 1(V_{GS}\text{-}2)^2,$$

$$V_{GS} = 3 \text{ V}, \;\; V_{GS} = 1 < V_t \text{ (not acceptable)},$$

$$V_{GS} = V_G - V_S,$$

$$3 = 5 - V_S \text{ ===> } V_S = 2 \text{ V},$$

$$R_S = V_S/I_D = 2 \text{ V}/1 \text{ mA} = 2 \text{ k}\Omega.$$

Electronics: Problem 5. 6.

SITUATION:
A BJT transistor is being considered for use in a single stage amplifier circuit. As part of the analysis, with only limited information on the device, you are asked to present an equivalent hybrid-π model that will "fit" into a previously designed circuit. All that is known about the BJT is that ß is 100, and that V_A is 100 volts (Early voltage).

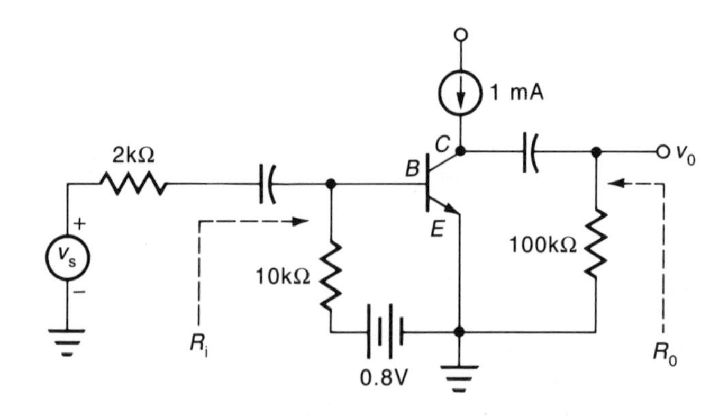

Previous Designed Amplifier Circuit

REQUIREMENTS:

1. Draw the small signal equivalent circuit using the hybrid-π model and calculate R_i and R_o.

2. Calculate the overall gain v_o/v_s.

Solution

1. The model, using the hybrid-π for the BJT, is as follows:

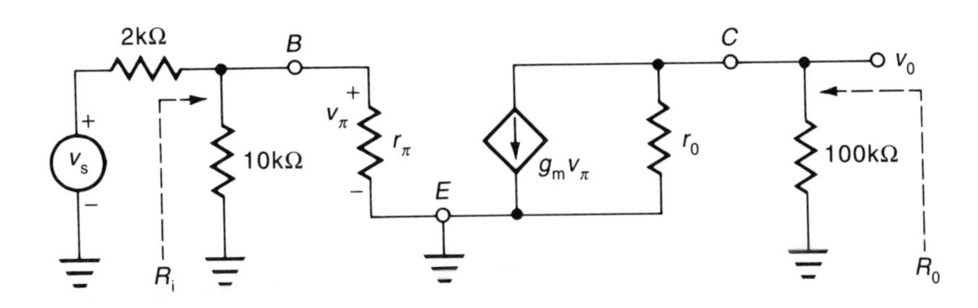

And R_i and R_o may be found by going through the following calculations:

$I_C = 1$ mA (constant current source),

$g_m = I_C/V_T = 1 \text{ mA}/0.025 = 40$ mA/V,

$r_\pi = ß/g_m = 100/(40 \text{ mA/V}) = 2.5$ kΩ,

$r_o = V_A/I_C = 100/1 \text{ mA} = 100$ kΩ,

then

$r_i = 10 \text{ k}||r_\pi = (10\text{x}2.5)/(10+2.5) = 2$ kΩ.

And for r_o set $v_s = 0$, then $v_\pi = 0$ and $g_m v_\pi = 0$, therefore $R_o = r_o||100k$,
$R_o = (100\text{x}100)/(100+100) = 50$ kΩ.

2. The circuit gain may now be found as,

$v_o = -g_m v_\pi(r_o||100k)$,

$v_\pi = [r_i/(r_i+2k\Omega)]v_s = [2/(2+2)]v_s = (½)v_s$,

then the overall gain is,

$v_o/ v_s = -(1/2)g_m(r_o||100k) = -(1/2)40(50) = -1{,}000$.

Electronics:Problem 5.7.

SITUATION:

A high gain op-amp having a unity gain bandwidth of f_t= 1 MHz is being considered in two different circuits as shown.

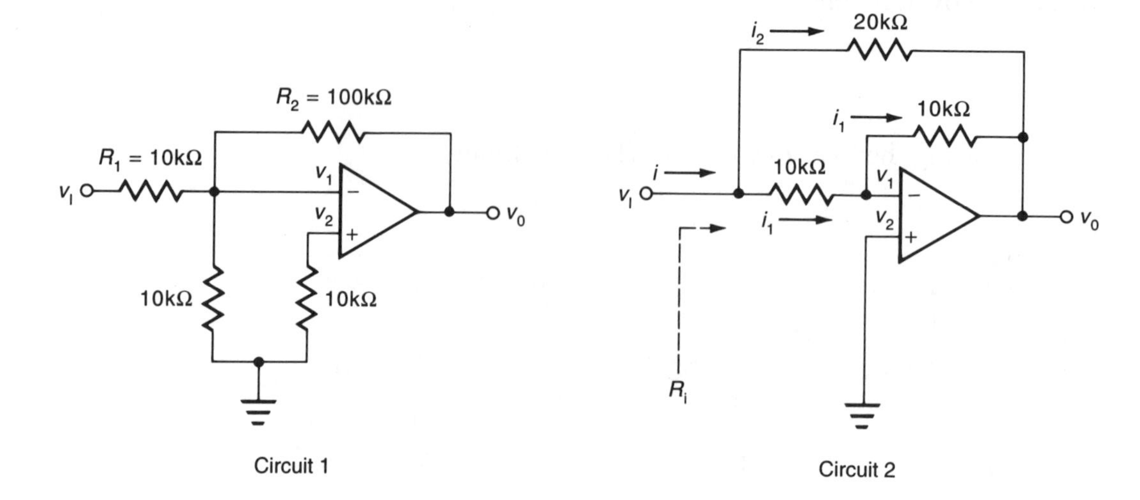

REQUIRMENT:

1. For circuit #1 you are to determine the voltage gain and the bandwidth.
2. For circuit #2 you are to determine the input resistance, R_i.

SOLUTION

1. Since the difference in the input voltages v_1 and v_2 must be very small and the input currents to the op-amp itself is negligible, then,

$$v_1 = v_2 = 0,$$

$$v_o = -(R_2/R_1)V_I$$

$$v_o/v_I = -R_2/R_1 = -100/10 = -10.$$

And, for the bandwidth,

$$\omega_t = (1+R_2/R_1)\omega'_b,$$

$$f'_b = f_{-3db} = f_t/(1+R_2/R_1) = 10^6/(1+10) = 90.9 \text{ kHz}$$

2. For circuit #2, the input resistance may be found as follows,

$$R_i = v_I/I,$$

$$v_1 = v_2 = 0,$$

$$i_1 = v_I/10\text{k},$$

$$v_o = v_1 - 10i_1 = 0 - 10\text{k}(v_I/10\text{k}) = -V_I,$$

$$i_2 = (v_I - v_o)/20\text{k} = [v_I - (-v_I)]/20\text{k} = v_I/10\text{k},$$

$$I = i_1 + i_2 = v_I/10\text{k} + v_I/10\text{k} = v_I/5\text{k},$$

$$R_i = v_I/(v_I/5\text{k}) = 5\text{ k}\Omega.$$

Electronics: Problem 5.8.

SITUATION:

A two-stage BJT amplifier has already been designed and is shown in the following circuit:

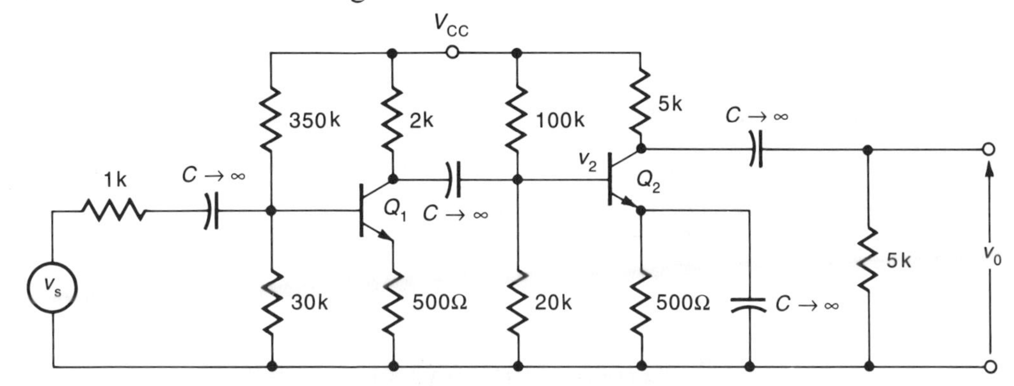

The h-parameters associated with this design are given as follows:

$h_{ie}\,(h_{11}) = 1\text{k}\Omega$, $h_{fe}\,(h_{21}) = 50$, $h_{re}\,(h_{12}) = 2\text{x}10^{-4}$

$h_{oe}\,(h_{22}) = 20\text{x}10^{-6}\,\text{mho}$

REQUIREMENTS:

Using an h-parameter model, you are to determine the overall mid-frequency gain, v_o/v_s. Any assumptions made should be clearly stated.

SOLUTION

Using the h parameter model we will first find the voltage gain of stage 2, v_0/v_2 and the input impedance looking into Q_2. Then we can find the gain of stage 1 as loaded by stage 2.

For stage 2, the equivalent circuit in h parameter form:

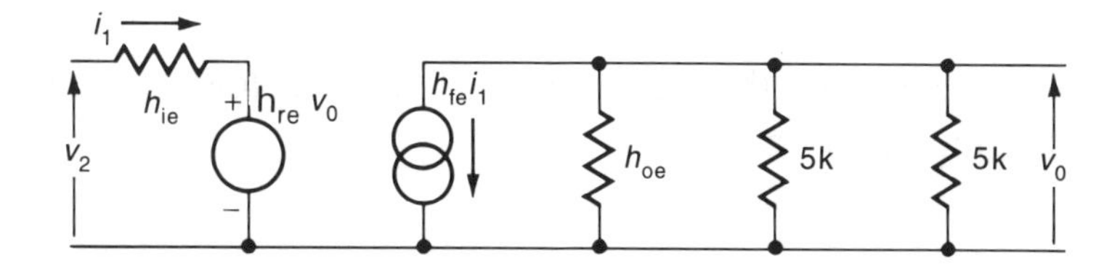

Let $Y_L = \dfrac{1}{5k\|5k}$ $\qquad Y_L = 400\times10^{-6}$ mhos

I $\qquad v_0 = \dfrac{-h_{fe}i_1}{h_{oe}+Y_L} = -119\times10^3 i_1$

II $\qquad v_2 = h_{ie}i_1 + h_{re}v_0$

Then $\quad v_2 = i_1 h_{ie} - \dfrac{i_1 h_{fe} h_{re}}{h_{22}+Y_L} = i_1\left[h_{ie} - \dfrac{h_{fe}h_{re}}{h_{22}+Y_L}\right]$

$$z_{in_2} = h_{ie} - \frac{h_{fe}h_{re}}{h_{22}+Y_L} = 1000 - 119\times10^3\left(2\times10^{-4}\right) = 976\Omega$$

From I above, we have $\quad i_1 = -\dfrac{v_0\left(h_{oe}+Y_L\right)}{h_{fe}} = -v_0\left(8.4\times10^{-6}\right)$

Substituting into II gives:

$$v_2 = h_{ie}\left[-\frac{v_0\left(h_{oe}+Y_L\right)}{h_{fe}}\right] + h_{re}v_0$$

$$= -(1k)\left(8.4\times10^{-6}\right)v_0 - 2\times10^{-4}\,v_0$$

$$= -8.6\times10^{-3}\,v_0$$

$$\frac{v_0}{v_2} = -116$$

Now we take an equivalent circuit for the input stage including the loading by the input to the second stage and the biasing networks.

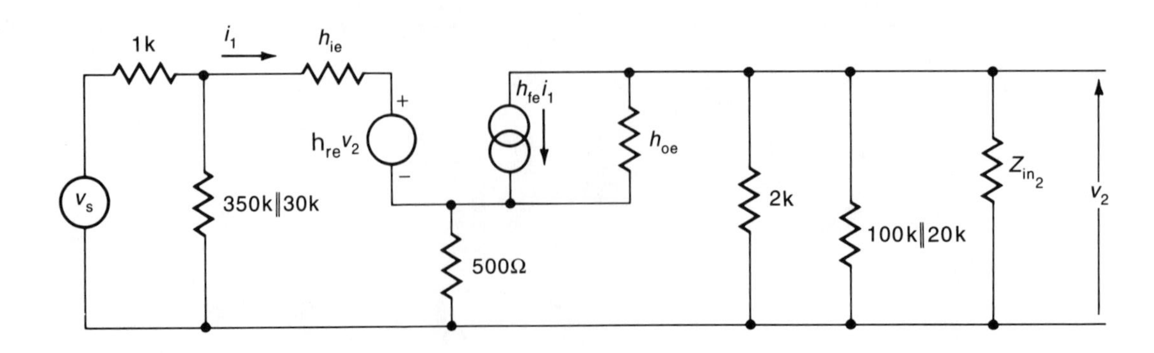

We note the fact that the 2k collector load is in parallel with Z_{in2} (976Ω) and these are in parallel with the bias network for transistor #2.
Note that 100k / 20k = 16.7k while 2k / 976Ω =656Ω.
Therefore assume bias network for stage #2 can be neglected.

The unbypassed emitter resistor in stage #1 presents a very high input impedance to the source:

$(Z_{in} = h_{ie} + (h_{fe} + 1)R_E)$.

However, the h_{oe} of the transistor is high compared to the 500Ω so we may assume that we can neglect h_{oe} as being high resistance [1/ h_{oe} =50k] compared to either the load (656Ω) or the 500Ω R_E. Finally, we will incorporate the bias network for transistor #1 into the source by taking a Thevenin equivalent CKT.

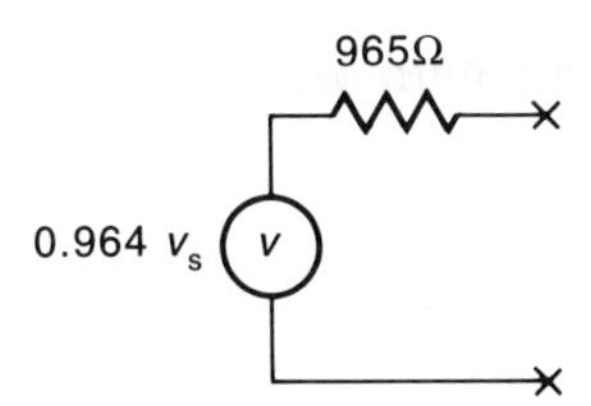

By the previous discussion we will refer the R_E (500Ω) resistor to the input by

$R_E' = (h_{fe} + 1)\, R_E = 25.5k\Omega$

Now our equivalent circuit looks like this:

$$i_1 = \frac{0.964 v_S - h_{re} v_2}{965 + h_{ie} + 25.5k} \qquad v_2 = -h_{fe} i_1 (656)$$

$$i_1 = \frac{-v_2}{(656)(50)} = -30.5 \times 10^{-6} v_2$$

$$-30.5 \times 10^{-6} v_2 = \frac{0.964 v_S - 2 \times 10^{-4} v_2}{965 + 1k + 25.5k}$$

$$-838 \times 10^{-3} v_2 = 0.964 v_S - 2 \times 10^{-4} v_2$$

$$-v_2 \left(838 \times 10^{-3}\right) = 0.964 v_S$$

$$\frac{v_2}{v_S} = \frac{-0.964}{838 \times 10^{-3}} = -1.15$$

Then the total gain $\frac{v_0}{v_S} = \frac{v_0}{v_2} \times \frac{v_2}{\gamma_S} = (-1.15)(-116) = 134$

The gain of this amplifier could be estimated reasonably as follows:

Stage #1 has strong current - series feedback (the un - bypassed emitter resistor). Thus its gain is very nearly $\frac{-R_{L_{eff}}}{R_{E_{eff}}}$ where $R_{L_{eff}}$ is the 2k collector load (in parallel with h_{ie} of Q_2) and the $R_{E_{eff}}$ is the R_E plus h_{ib},

h_{ib} is approximately $\frac{h_{ie}}{h_{fe}} = 20\Omega$

Thus stage #1 has an approximate gain of:

$$\text{Gain} = \frac{-667}{500+20} = -1.28$$

For stage #2 the gain is approximately $\frac{-h_{fe}R_{L_{eff}}}{h_{11}}$ where

$R_{L_{eff}}$ is $5k \| 5k$ or $2.5k\Omega$.

$$\text{Then Gain \#2} \approx \frac{-(2.5k)(50)}{1k} = -125$$

Finally, the voltage divider of the input stage and the source resistance.

$Z_{in}\#1 \approx h_{11} + (h_{fe}+1)R_E \approx 1k + (51)(5k) = 27k.$

The biasing network to stage #1 was found to be about 27k so that the combined parallel input impedance is about 13.5k.

$$\text{Then} \quad v_i \approx \frac{v_S(13.5k)}{13.5k+1k} = 0.93v_S$$

Then the overall gain is estimated as follows:

$$G = \frac{v_0}{v_S} \approx 0.93(-125)(-1.28) = 149$$

This is about 11% higher than found by the more accurate procedure - well within the tolerance of the known parameters.

ELECTRONICS PROBLEM 5.9

SITUATION

The circuit shown below is an N-channel Junction Field-Effect Transistor with self-bias and a pinch-off voltage of -3 volts. At that value of pinch-off voltage, the current is 6 mA. The breakdown voltage for this transistor is 30 volts.

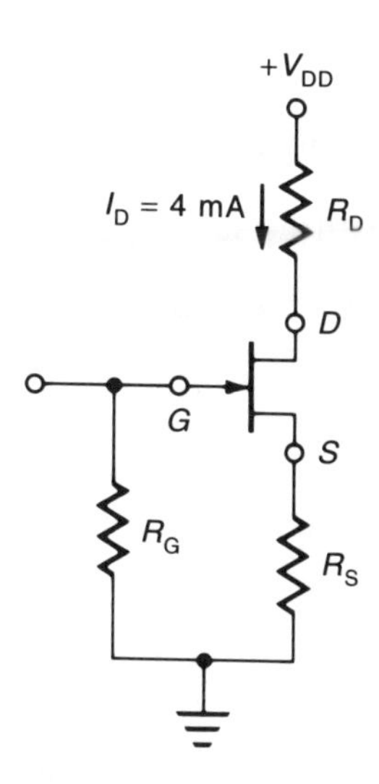

REQUIRED

Design the above circuit so that the device will be biased at approximately 10V drain-to-source and have a channel current of approximately 4 mA.

SOLUTION

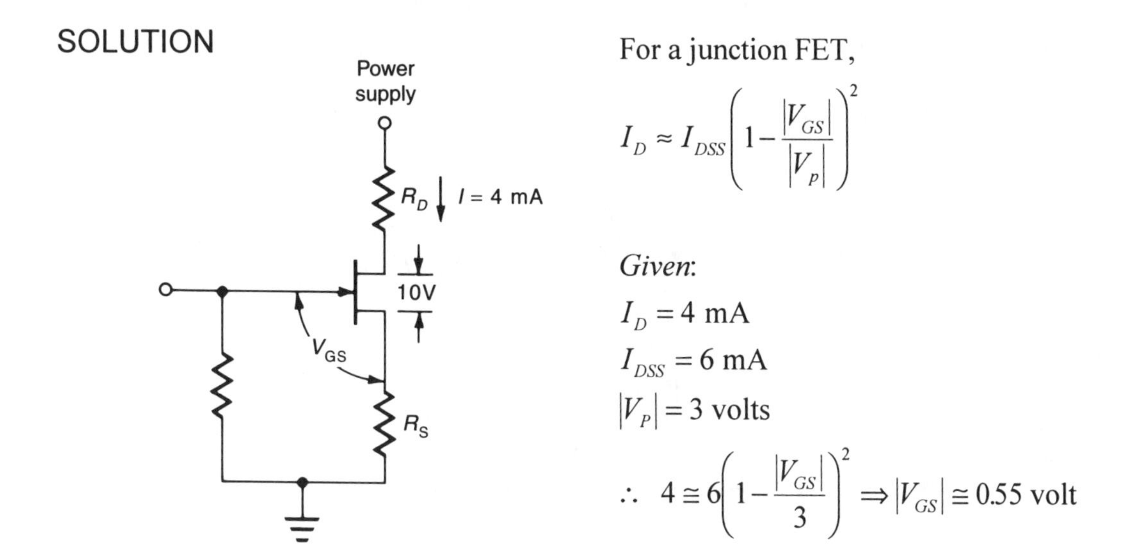

For a junction FET,

$$I_D \approx I_{DSS}\left(1-\frac{|V_{GS}|}{|V_P|}\right)^2$$

Given:

$I_D = 4$ mA

$I_{DSS} = 6$ mA

$|V_P| = 3$ volts

$$\therefore \quad 4 \cong 6\left(1-\frac{|V_{GS}|}{3}\right)^2 \Rightarrow |V_{GS}| \cong 0.55 \text{ volt}$$

Gate leakage current I_{GSS} is typically of the order of nanoamperes. Choose $R_G = 1\ \text{M}\Omega$, so that the gate remains within millivolts of ground potential. Choose R_S to obtain the proper V_{GS}.

$$|V_{GS}| = 0.55 \text{ volts}, \quad I_D = 4 \text{ mA} \Rightarrow R_S = \frac{0.55}{4 \times 10^{-3}} = 140\Omega$$

Since the breakdown voltage is 30 volts, choose $V_{DD} = 24$ volts.

$$V_{DS} = V_{DD} - I_D(R_D + R_S)$$

$$10 = 24 - 4 \times 10^{-3}(R_D + R_S)$$

$$R_D + R_S = \frac{14}{4 \times 10^{-3}} = 3500\Omega$$

$$\therefore R_S = 150\Omega \text{ and } R_D = 3.3k\Omega$$

If 10% resistors are used, the available values are:

$$R_S = 150\Omega \quad \text{and } R_D = 3.3k\Omega$$

ELECTRONICS PROBLEM 5.10

SITUATION

A two stage BJT has already been designed, but a complete analysis is needed for gain and losses.

REQUIRED

Calculate the power gain of the two-stage amplifier shown below. Show the gain of the two individual stages, the interstage losses and the prestage losses. The ground-based '*h*' parameters for the transistors used in both stages are:

h_{ib} = 50 ohms; $h_{rb} = 5\text{x}10^{-4}$; h_{fb} = -0.97 and h_{ob} = 10-6 mho

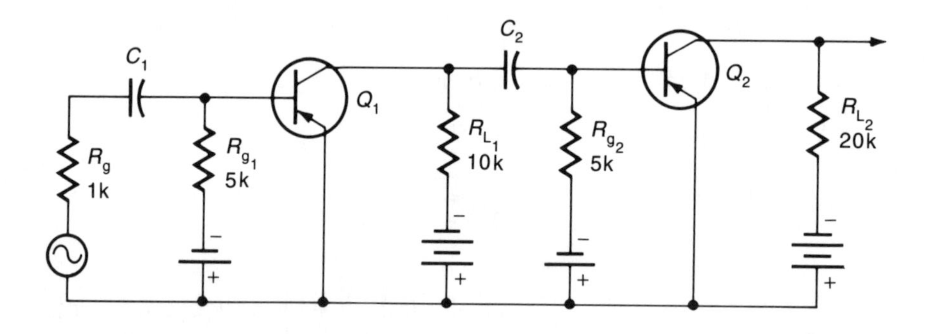

Assume reactance to be negligible.

SOLUTION

Solution based upon the following assumptions:

1. Mid-band frequency -- reactance of capacitors is negligible.
2. Power gain defined as power delivered to load divided by power delivered from source (this is power delivered to R_{L2} divided by power delivered of the voltage node at the junction of C_1 and R_{g1}).
3. Power losses are defined as:
   a) Pre-first stage -- power lost in bias resistor R_{g1}.
   b) Inter-stage -- power lost in R_{L1} and R_{g2}.
   c) Power losses -- at signal frequency only, dc bias losses not considered.

Parameters are given in common base configuration; since transistors are operated in common emitter orientation, the parameters must be converted to common emitter form. Also, gain and impedance equations must either be derived or found in a transistor handbook.

Handbook conversion tables (see Table 5.1) give the following relationships:

$$h_{ie} = \frac{h_{ib}}{1+h_{fb}} = 1670\Omega \qquad h_{re} = \frac{h_{ib}h_{ob}}{1+h_{fd}} = 11.7\times10^{-4}$$

$$h_{fe} = \frac{-h_{fd}}{1+h_{fb}} = 32 \qquad h_{oe} = \frac{h_{ob}}{1+h_{fd}} = 3.3\times10^{-5}mho$$

and (from Fig 4.14):

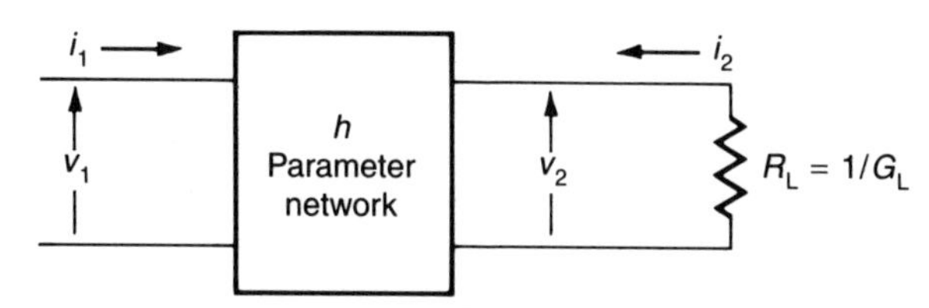

I. $$Z_{in} \cong \frac{v_1}{i_1} = \frac{h_{ie}h_{oe} - h_{fe}h_{re} + h_{ie}G_L}{h_{oe}+G_L} = h_{ie} - \frac{h_{fe}h_{re}}{h_{oe}+G_L}$$

II. $$A_V \cong \frac{v_2}{v_1} = \frac{h_{fe}R_L}{h_{ie} + R_L\left(h_{ie}h_{oe} - h_{fe}h_{re}\right)} = \frac{-h_{fe}}{h_{ie}G_L + \left(h_{ie}h_{oe} - h_{fe}h_{re}\right)}$$

The gain of each stage may be calculated by breaking the circuit as follows:

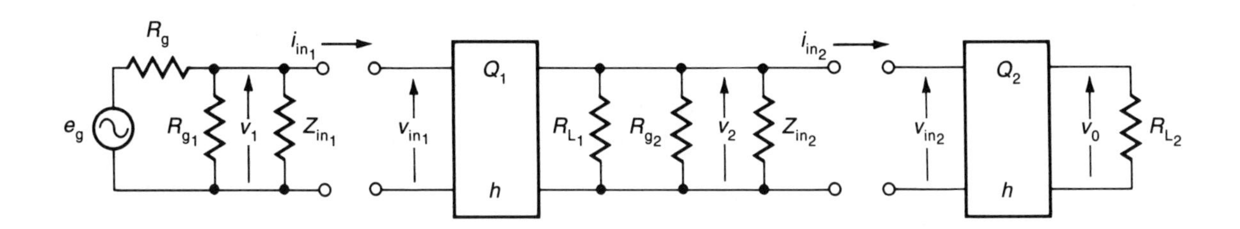

Definitions:

v_1 = voltage at base of Q_1 $\qquad$ v_0 = voltage across load R_{L_2}

$v_1 = v_{in_1}$ $\qquad$ $v_1 = e_g \dfrac{\left(R_{g1} \text{ in parallel with } Z_{in_1}\right)}{R_{g_1} + \left(R_{g1} \text{ in parallel with } Z_{in_1}\right)}$

$v_2 =$ voltage at collector of Q_1

$v_2 = v_{in_2}$

Calculation of Stage Q_2 (Eqn II)

$$\frac{v_0}{v_{in_2}} = A_V = \frac{-h_{fe}}{h_{ie}G_L + \left(h_{ie}h_{oe} - h_{fe}h_{re}\right)}$$

$$= \frac{-32}{(1670)(0.5\times10^{-4}) + (1670)(3.3\times10^{-5}) - (11.7\times10^{-4})(32)} = -320$$

To calculate gain of first stage, the impedance Z_{in_2} must first be obtained. This appears in parallel with R_{L_1} and R_{g_2}, all of which appear as a parallel load for stage Q_1. Then Z_{in_2} (from Eqn I) is given as:

$$Z_{in_2} = h_{ie} - \frac{h_{re}h_{fe}}{h_{oe} + G_L} = 1670 - 450 = 1220\Omega$$

Then the effective load for Q_1 is:

$$G'_{L_1} = \frac{1}{R'_{L_1}} = \frac{1}{R_{g_2}} + \frac{1}{R_{L_1}} + \frac{1}{Z_{in_2}} = \frac{1}{5K} + \frac{1}{10K} + \frac{1}{1.22K} = \frac{1}{895} = 1.12\times10^{-3}\,mho$$

$R'_{L_1} = 895$ ohm

Then (from Eqn II):

$$\frac{v_2}{v_{in_1}} = A_V = \frac{-h_{fe}}{h_{ie}G'_L + \left(h_{ie}h_{oe} - h_{oe}h_{fe}\right)}$$

$$= \frac{-32}{(1670)(1.12\times10^{-3}) + (1670)(3.3\times10^{-5}) - (11.7\times10^{-4})(32)} = -16.9$$

The total voltage gain from v_1 to v_0 is:

$$\frac{v_0}{v_1} = \left(\frac{v_{in_2}}{v_1}\right)\left(\frac{v_0}{v_{in_2}}\right) = (-320)(-16.9) = 5400$$

To find the power delivered by the source, Z_{in_1} must be calculated (Eqn I):

$$Z_{in_1} = \frac{V_{in}}{i_{in_1}} = h_{ie} - \frac{h_{re}h_{fe}}{h_{oe} + G_{L_1}}$$

$$= 1670 - \frac{(11.7\times10^{-4})(32)}{(3.3\times10^{-5}) + (1.12\times10^{-3})} = 1670 - 30 = 1640\Omega$$

Then the power delivered by the generator is:

$$P_{in} = \frac{v_1^2}{\text{Parrel combination of } R_{g_1} \text{and } Z_{in_1}} = \frac{v_1^2}{\frac{(5\times10^3)(1640)}{(5\times10^3)+1640}} = \frac{v_1^2}{1.25\times10^3}$$

and the power delivered to the load is:

$$P_0 = \frac{v_0^2}{R_{L_2}} = \frac{v_0^2}{20\times10^3}$$

Then the power gain is:

$$G = \frac{P_0}{P_{in}} = \frac{\frac{v_0^2}{20\times10^3}}{\frac{v_1^2}{1.25\times10^3}} = 0.0625\left(\frac{v_0}{v_1}\right)^2 = 0.0625(5400)^2$$

$$= 1.82\times10^6$$

$$G_{db} = 10\log G = 62.6db$$

Now consider the inter - stage power loss (lost in parallel combination of R_{L_1} and R_{g_2}):

$$\frac{(R_{L_1})(R_{g_2})}{R_{L_1}+R_{g_2}} = \frac{(10\times10^3)(10\times10^3)}{(10\times10^3)+(5\times10^3)} = 3.33\times10^3\Omega$$

$$P_{I.L.} = \frac{v_2^2}{3.33\times10^3} = \frac{(16.9v_1)^2}{3.33\times10^3}$$

The total loss is then:

$$P_{Losses} = \frac{v_1^2}{5\times10^3} + \frac{(16.9v_1)^2}{3.33\times10^3}$$

but $v_1 = 0.556\ e_g$

$$P_{Losses} = \frac{(0.556\ e_g)^2}{5\times10^3} + \frac{(16.9(0.556\ e_g))^2}{3.33\times10^3}$$

$$= 27.7\times10^{-3} e_g^2 \text{ watts}$$

This page intentionally left blank.

Control Systems Prob 6.1

SITUATION

The following system has already been designed, however the performance parameters need to be specified.

REQUIREMENT

Determine whether the following system is stable and predict the closed loop pole location for the system for K=4. Also, find the system error.

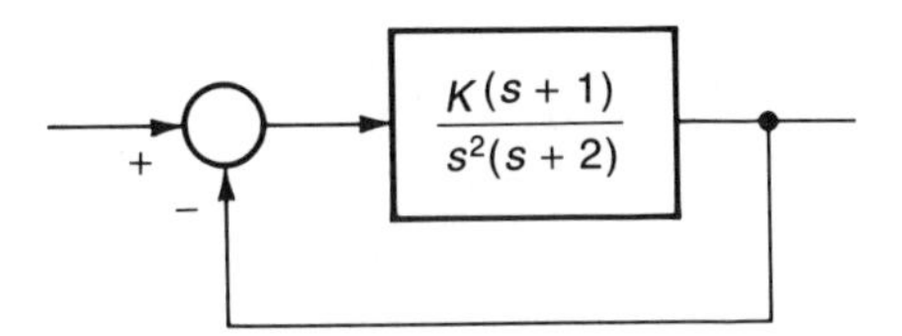

SOLUTION

For system stability use Routh-Hurwitz method:

$$G_{Syst} = \frac{G}{1+G} = \frac{K(s+1)}{s^2(s+2)+K(s+1)} = \frac{K(s+1)}{s^3+2s^2+Ks+K}$$

Characteristic Polynomial: s^3+2s^2+Ks+K

Routhian Array:

$$\begin{array}{c|cc} s^3 & 1 & K \\ s^2 & 2 & K \\ s^1 & x & \\ s^0 & y & \end{array} \quad \text{where } x = \frac{(2)(K)-(K)}{2} = \frac{K}{2}$$

$$y = \frac{\left(\frac{K}{2}\right)(K)-0}{\frac{K}{2}} = K$$

The first column is positive for all positive values of K and the system is stable. The root locus is sketched by setting G= -1 (that is, using the basic rules of root locus, $|G| = 1$, $\phi_G = \pm n180°$ with n being any odd iteger)
one obtains the following sketch:

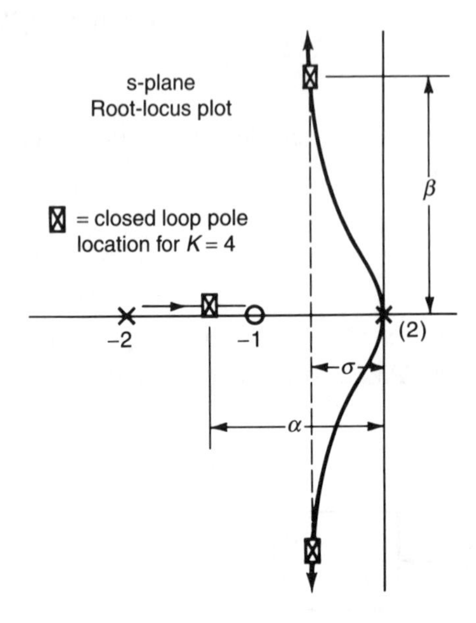

The system is stable for all values of positives K's.

$$G_{Syst} = \frac{K}{s^3 + 2s^2 + Ks + K} = \frac{4}{(s+\alpha)[(s+\sigma \pm j\beta)]}$$

$$\text{with } \mathrm{K} = 4 \; ; \quad \alpha \cong 1.25 \quad , \; \sigma \cong 0.45 \; , \; \beta \cong 1.8$$

The system error is zero for both a step and ramp input, but is finite for an acceleration input (a/s^3):

$$e\Big|_{t\to\infty} = \lim_{s\to 0} s\left[\frac{R}{1+G}\right] = \left[\frac{\left(\frac{a}{s^3}\right)}{1+\frac{4(s+1)}{s^2(s+2)}}\right]$$

$$= \lim_{s\to 0} s\frac{a}{s^2 + \frac{4(s+1)}{s+2}} = \frac{a}{\left(\frac{4}{2}\right)} = 0.5a$$

Therefore the error is 0.5 of the acceleration input.

CONTROL SYSTEMS PROB. 6.2

SITUATION

The open loop transfer function for a control system is approximated by:

$$G(s) = \frac{C(s)}{E(s)} = \frac{K(s-3)}{(s+0.5)(s+7)}$$

It is desired to make the output signal (C) correspond as nearly as possible to some input signal, (R), in steady state, at the same time keeping the system stable.

REQUIRED

(a) Sketch a block diagram for a feedback control system to accomplish the given objective. Carefully label the summation polarity of all signals coming into the feedback junction summing point. (Note that the given transfer function has peculiar properties.)

(b) Select a value of K which assures system stability and at the same time brings the ratio C/R in steady state as close to +1.0 as possible. (Note that the properties of G(s) are such that it is advisable to make a very careful check on the requirements for closed-loop system stability.)

SOLUTION

(a) The block diagram could be given as follows (with either positive or negative feedback):

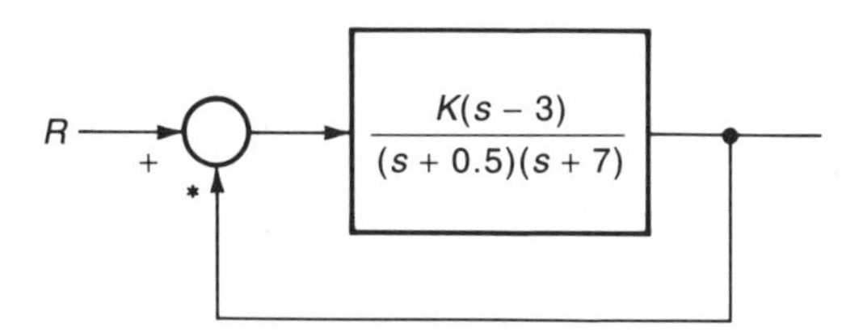

* For negative (-) feedback, the system function is:

$$G_{Syst} = \frac{G}{1+G} = \frac{K(s-3)}{(s+0.5)(s+7)+K(s-3)}$$

$$= \frac{K(s-3)}{s^2+(7.5+K)s+(3.5-3K)}$$

The denominator (which determines the "character" of the response) must not have any negative factors (indicating closed loop poles in the right half plane --or an unstable system). A simple test for stability is the Routh criterion; or in this case, (for a simple second-order system) all of the coefficients of the denominator polynomial must be positive, therefore:

$$G_{Syst} = \frac{G}{1+G} = \frac{K(s-3)}{(s+0.5)(s+7)+K(s-3)}$$
$$= \frac{K(s-3)}{s^2+(7.5+K)s+(3.5 - 3K)}$$

Here, for stability, again the denominator polynomial must be positive, thus:

$$3K < 3.5 \text{ , } K < 1.167$$

Of course the root locus method of analysis may also be used:

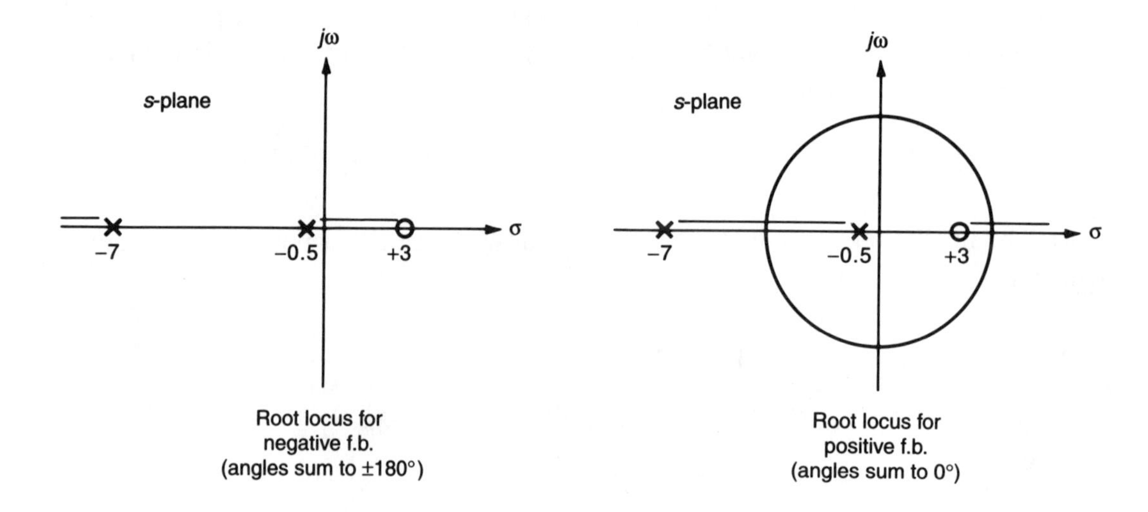

(b) For C/R to be as close to unity as possible, then it is only necessary to minimize E=R-C (since it is a unity feedback system):

For negative feedback:

$$E = \frac{R}{1+G}$$

Now assume a step input $\left(R = \frac{r_0}{s} \right)$ and that one is interested in the error after a long period of time--such that the final value theorem may be applied:

$$e\Big|_{t\to\infty} = \lim_{s\to 0} sE = \lim_{s\to 0} s\left[\frac{\frac{r_0}{s}}{1+\frac{K(s-3)}{(s+0.5)(s+7)}} \right]$$

$$= \frac{r_0}{1+\frac{K(-3)}{(0.5)(7)}} = \frac{r_0}{1-.857K}$$

To minimize this equation, the term ,1-0.857K, should be as large as possible. For system stability the maximum value that K can have is 1.167. If K is at this maximum value then 0.857K is equal to 1.0 and the term is zero. This causes e to approach infinity. If K is zero then the transfer function, g, goes to zero. Hence it is obvious that K must be between 1.167 and zero. If K were near zero, then there would be no forward path for the signal and the system would be useless. Referring to the left-hand s-plan root locus plot on the previous page, if K equals 1.167 then the close loop pole is at the origin and the system would act as an integrator (meaning that the error would go to infinity after a long period of time). Some value of K between the extremes would cause the error to be greater than the input step. One may conclude that the real path for the root locus is between the -0.5 pole and the zero origin.

This problem suggests that one must use some kind of compensation network, or, better, to review the original open loop transfer function for possible hardware changes such that the negative zero perhaps could be made positive. An ideal location of the zero would be to the left of the open loop pole (at -7) and then a wide range of positive K's could be selected.

Author's note: On an actual multiple choice examination , one would not expect a problem that does not have a good solution. If a problem needed a compensator to make it work properly, the question undoubtedly would be worded such that a partial solution had been obtained and your work would be from that point onward.

CONTROL SYSTEMS PROB. 6.3

SITUATION

Consider the following simplified diagram for a control system:

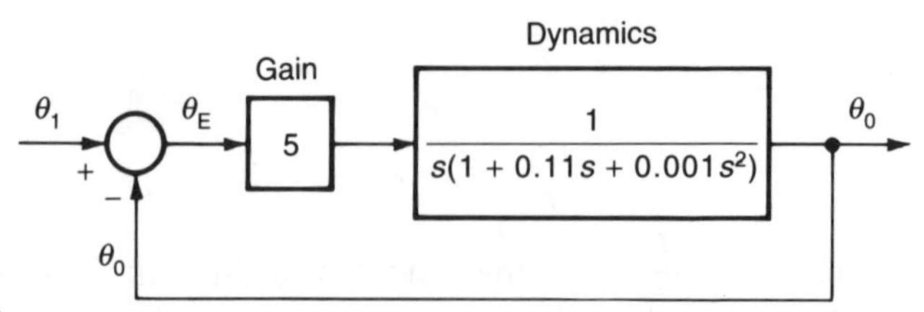

REQUIREMENT

(1) Determine whether the system is stable by any standard criterion.
(2) What range of real values of gain will enable the system to be stable?

SOLUTION

There are several methods of determining system stability, two of which are:

1. Routh's criterion
2. Root locus

All of these methods are based upon solving the equivalent closed loop "characteristic equation" for positive roots, i.e., all solutions that would indicate whether the resulting exponential terms were expanding with time.

Routh Criterion: Solve for the closed loop transfer function:

$$KG_{CL} = \frac{KG}{1+KG} = \frac{5}{0.001s^3 + 0.11s^2 + s + 5}$$

and the characteristic equation (denominator)

can be formed into a triangular array as follows:

Characteristic Equation:

$$A_n s^n + A_{n-1}s^{n-1} + A_{n-2}s^{n-2} + \cdots A_0$$

Then:

s^n	A_n	A_{n-2}	A_{n-4}	...
s^{n-1}	A_{n-1}	A_{n-3}	A_{n-5}	...
s^{n-2}	x_a	x_b	x_c	...
.	.			
.	.			
s^0	Z_a			

where $x_a = \dfrac{A_{n-1}A_{n-2} - A_n A_{n-3}}{A_{n-1}}$ etc.

All coefficients of the first coefficient column must be of the same sign for stability.

Thus:

s^3	0.001 1
s^2	0.11 5
s^1	0.955
s^0	5

where:

$$X_1 = \frac{(.11)(1) - (0.001)(5)}{0.11} = 0.955$$

$$Y_1 = \frac{(0.955)(5) - 0}{.955} = 5$$

Therefore system is stable for the given value of gain.

(2) To determine the range of values of gain for the system to be stable, merely let the "5" in the Routh-Hurwitz array be a variable K and solve:

s^3	0.001 1
s^2	0.11 K
s^1	x_1

$$x_1 = \frac{(.11)(1) - (0.001)K}{0.11}$$

and since the requirement for x_1 is that it be positive for stability:

$$0.11 > 0.001K$$

$$K < 110 \quad \text{for stability}$$

Solution based upon root - locus method:

$$KG = \frac{K}{s(1 + 0.11s + 0.001s^2)} = \frac{\frac{K}{1000}}{s(1000 + 110s + s^2)}$$

$$= \left(\frac{K}{1000}\right)\left[\frac{1}{s(s+10)(s+100)}\right]$$

The root-locus plot will then be:

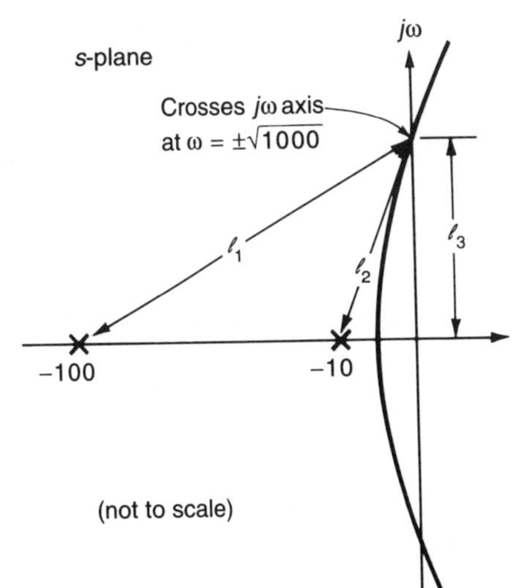

$$K_0 = (\ell_1)(\ell_2)(\ell_3) = 11.0 \times 10^4$$

$$K = \frac{K_0}{1000} = 110$$

Thus for any value of K_0 less than 110, the system would be stable.

CONTROL SYSTEMS PROB. 6.4

SITUATION

A viscous-damped servomechanism with proportional control, stiffness K, and damping coefficient B, is found to exceed the allowing tracking error by 10 times.

The latter is corrected by altering the stiffness, and by adding derivative control to provide the original degree of underdamping.

REQUIRED

Find the new proportional control stiffness and the derivative action time required.

SOLUTION

A simple proportional control servomechanism might be represented as:

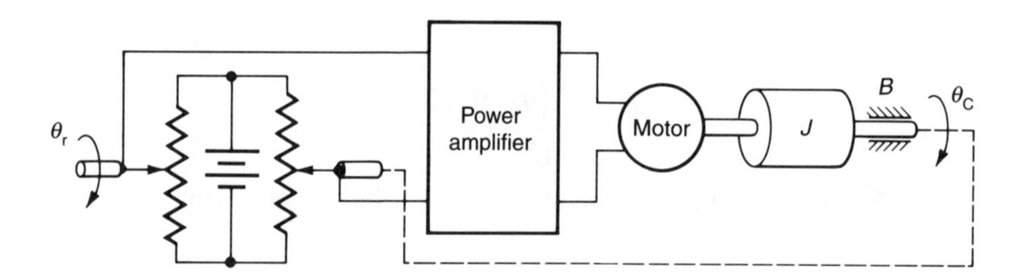

The equation describing this system could be written as:

$$K(\Theta_r - \Theta_c) = J\ddot{\Theta}_c + b\dot{\Theta}_c$$

(Equation assumes an oversimplified relationship where the motor torque is directly proportional to the input voltage.)

And the Laplace transformed system then could be shown to be:

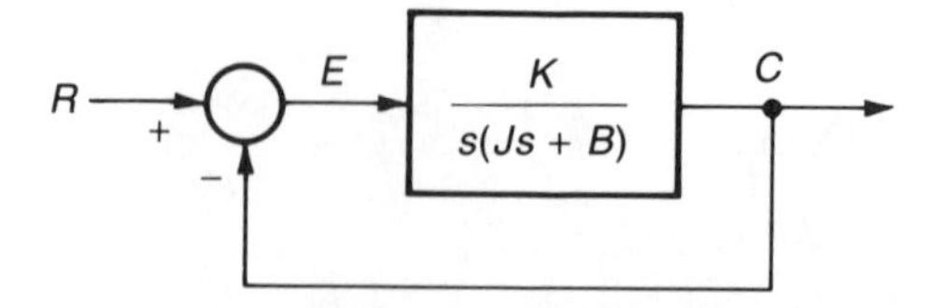

For ease of computing, let J = 1, then the system transfer function becomes:

$$G_{Syst} = \frac{K}{s^2 + Bs + K} = \frac{\omega_n^2}{s^2 + s\zeta\omega_n s + \omega_n^2}$$

where $\omega_n = \sqrt{K}$

$$\zeta = \frac{B}{2\sqrt{K}}$$

The system tracking error (assuming a unit ramp input) then may be found using the final value theorem:

$$e\Big|_{t\to\infty} = \lim_{s\to 0} sE = \lim_{s\to 0} s\left[\frac{R}{1+G}\right] = \lim_{s\to 0} s\left[\frac{\frac{1}{s^2}}{1+\frac{K}{s(s+B)}}\right] \to \frac{B}{K}$$

And, the root locus will then be (assuming an underdamped system):

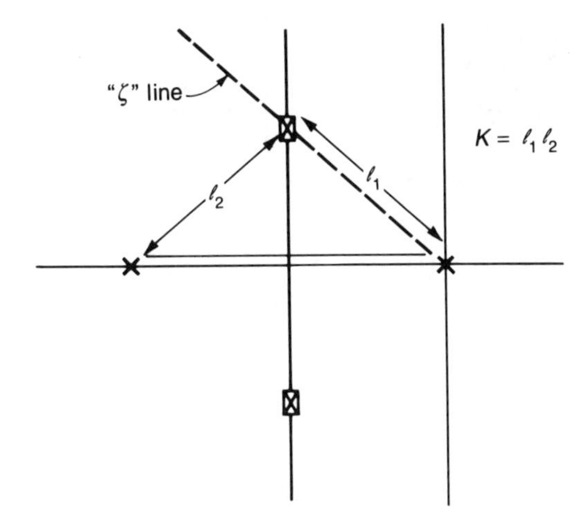

To reduce the system error to 0.1 of the original error, add an ideal derivative compensator;*

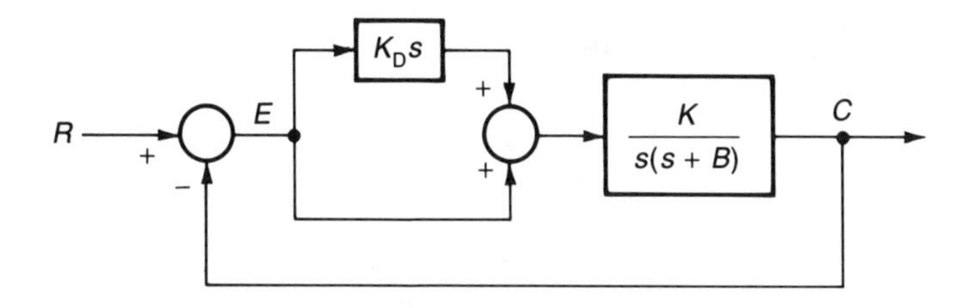

The new forward transfer function then becomes:

$$G_{fwd} = \frac{(1 + K_D s)K}{s(s+B)}$$

* This portion of the problem requires a somewhat lengthy solution and a graphical presentation of the method would probably satisfy the exam grader.

And the error again may be found using the final value theorem:

$$e\Big|_{t\to\infty} = \lim_{s\to 0} s\left[\frac{\frac{1}{s^2}}{1+\frac{(1+K_D s)K}{s(s+B)}}\right] \to \frac{B}{K}$$

To reduce the new error by a factor of 10, try increasing the stiffness, K, to give:

$$e_{new} = 0.1e_{old} = \frac{B}{10K}$$

The new root locus will then have a zero located at $-1/K_D$ (yet to be determined). Since the amount of damping is to remain the same (i.e., the "ζ" line to remain the same), the locus will be of the form:

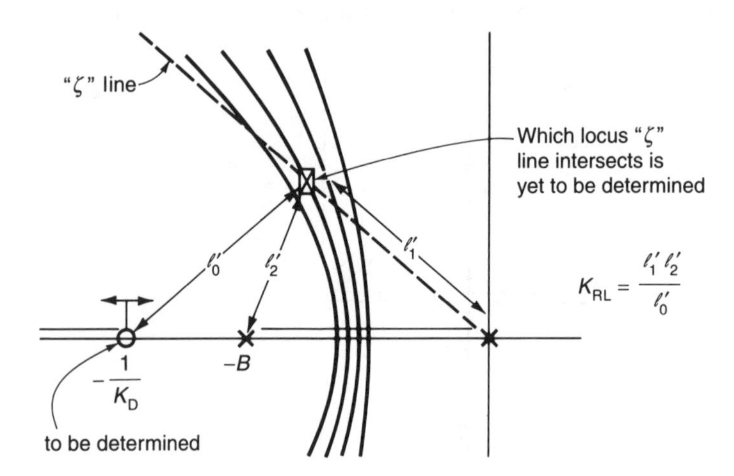

The requirement of error, being K_{new}=10 $K_{original}$, and locating the new closed loop pole of the "ζ" line will give the relationship of:

$$K_{\text{root locus}} = \frac{\ell'_1 \ell'_2}{\ell'_0} = K_D(10K) = K_D 10\ell_1\ell_2$$

Using this relationship, one may, by trial and error techniques, find the value of K_D.

However a more straight forward technique not using root locus methods will allow us to approximate K_D directly (this will be approximate as the zero in the closed loop response will alter the damping).

Consider:

$$G_{Syst} = \frac{(10K)(1+K_D s)}{s^2 + Bs + K_D(10K)s + (10K)} = \frac{\omega_n^2(1+K_D s)}{s^2 + 2\zeta\omega_n s + \omega_n^2}$$

where $\omega_n = \sqrt{10\text{K}}$

$$2\zeta\omega_n = \text{B} + 10\text{K K}_\text{D}$$

$$\zeta = \frac{\text{B} + 10\text{K K}_\text{D}}{2\sqrt{10\text{K}}}$$

but the new "ζ" and the original "ζ" are required to be the same,

Giving: $\text{B} = \dfrac{\text{B} + 10\text{KK}_\text{D}}{\sqrt{10}}$

As an example, let $\text{K} = \text{B} = 2$ and "ζ" $= 0.707$

Then $2 = \dfrac{2 + (10)(2)K_D}{\sqrt{10}}$ Giving $\text{K}_\text{D} = 0.216$

Or, the zero (for the root locus plot) is located approximately at $-1/K_D$, which gives - 4.6 for the numbers previously used.
Therefore the new stiffness should be 10 times the original and the "time constant" of the derivative compensator should equal K_D

CONTROL SYSTEMS PROB. 6.5

SITUATION

A servo system for the positional control of a rotatable mass is stabilized by means of viscous damping. The amount of damping is less than that required for critical damping.

REQUIRED

Calculate the amount (percent) of the first overshoot, if the input member is suddenly moved to a new position, the undamped natural frequency is 5 Hz, and the viscous friction coefficient is a fifth of that required for critical damping.

SOLUTION

Assume an ideal controller with torque directly proportional to error, E, (T = KE), and that the system is of 2nd order:

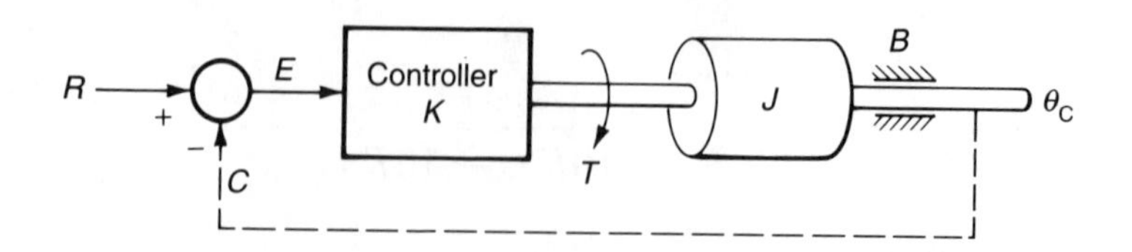

$$T = KE = K(R - C)$$

$$T = (Js^2 + Bs)\Theta$$

$$G_{Syst} = \frac{K}{Js^2 + Bs + K} = \frac{\frac{K}{J}}{s^2 + \frac{B}{J}s + \frac{K}{J}}$$

$$= \frac{\omega_n^2}{s^2 + 2\zeta\omega_n s + \omega_n^2}$$

$\omega_n =$ natural frequency $= 2\pi f_n = 2\pi 5 =$ constant

$2\zeta\omega_n = \frac{B}{J}$, $\therefore \zeta \propto$ B since both K and J are constants.

For critical damped system, $\zeta = 1$.

For $B = 1/5\ B_{critical}$, $\zeta = 0.2$

From "standardized" 2nd Order Curves:

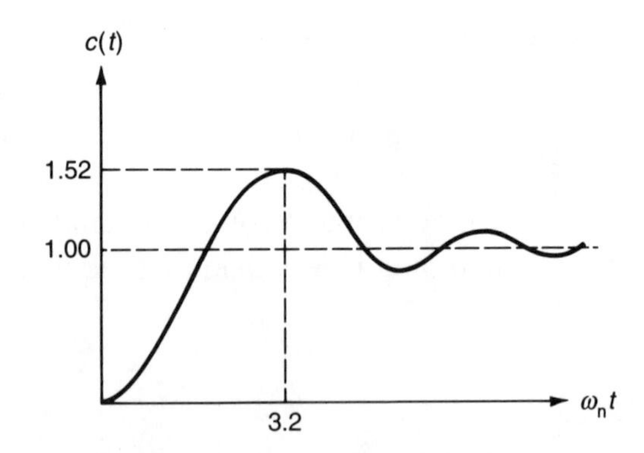

Therefore % overshoot = 52% and time-to-first peak, t_p,

$$t_p \approx \frac{3.2}{\omega_n} \approx 0.1 \text{ seconds}$$

CONTROL SYSTEMS PROB. 6.6

SITUATION

An ac system (a suppressed carrier system operating at 400 Hz) is already designed; and, when connected in closed loop, it may be approximated by an ideal second order system. The phase-shifting capacitor for the control motor has been selected such that current through the main winding and the current from the control winding has a +/- 90° phase difference (yielding an optimal motor torque at speeds near zero rpm). The error detector, G_{ED}, is a pair of potentiometers supplied from the same carrier voltage source.

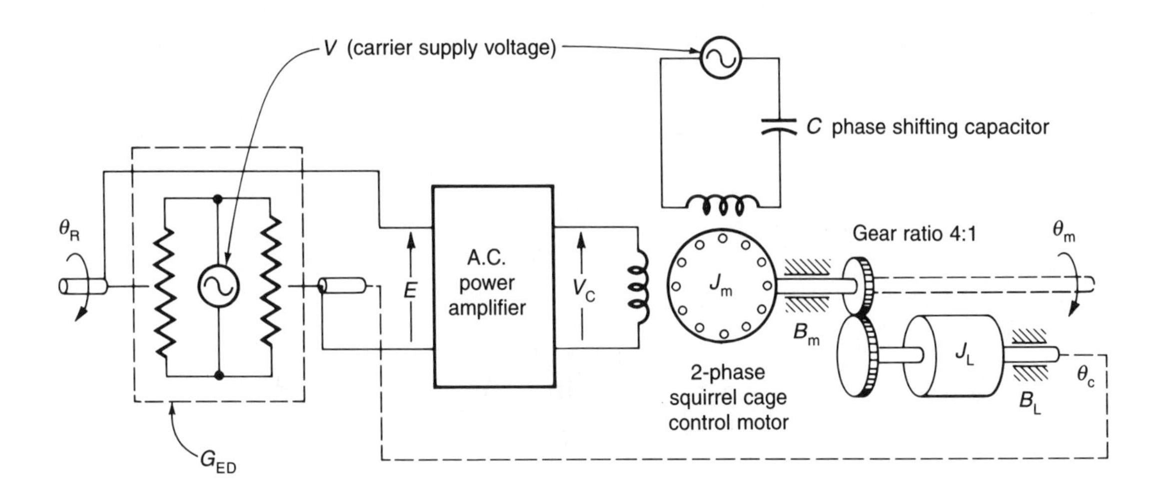

REQUIREMENT

It will be necessary to find the overall transfer function of the system for possible future simulation runs. Assume the error detector transfer function, G_{ED}, is a constant whose gain is unity. The control motor (or, servo motor) name plate gives the following:

Voltage (rated for both windings) = 100 volts
Developed Stall (or Starting) Torque = 50 oz-in. *
No-load speed = 3,820 rpm (= 400 rad/sec)
Bearing Friction: Negligible
Polar Moment of Inertia, J, of rotor = 0.2 oz-in.sec.^{2} *

The polar moment of inertia of the load is 3.2 oz-in.sec^2 * with the bearing friction again being negligible.

*Assume these are the correct units when working with radian/sec as speed.

1. Determine the total effective polar moment of inertia, J_{eff}, of the motor-load combination (as referred to the motor shaft when connected through the 4:1 (a=4) ideal gear train.

2. Determine the transfer function of the motor-load combination, G_{M-L}. The transfer function (here, assume the power amplifier has a gain of unity, then the error voltage, E, equals amplifier control voltage, V_{cont}) is given as:

$$G_{M-L} = \Theta_M/V_{cont} = \Theta_m/E = K/[s(\tau s+1)]$$

3. Show the system block diagram (ready for simulation) with the numerical values for the transfer function(s).

4a. For a unit step input of one radian applied to Θ_R and for a different power amplifier gain (other than unity), the error voltage as viewed on a strip-chart recorder is shown as:

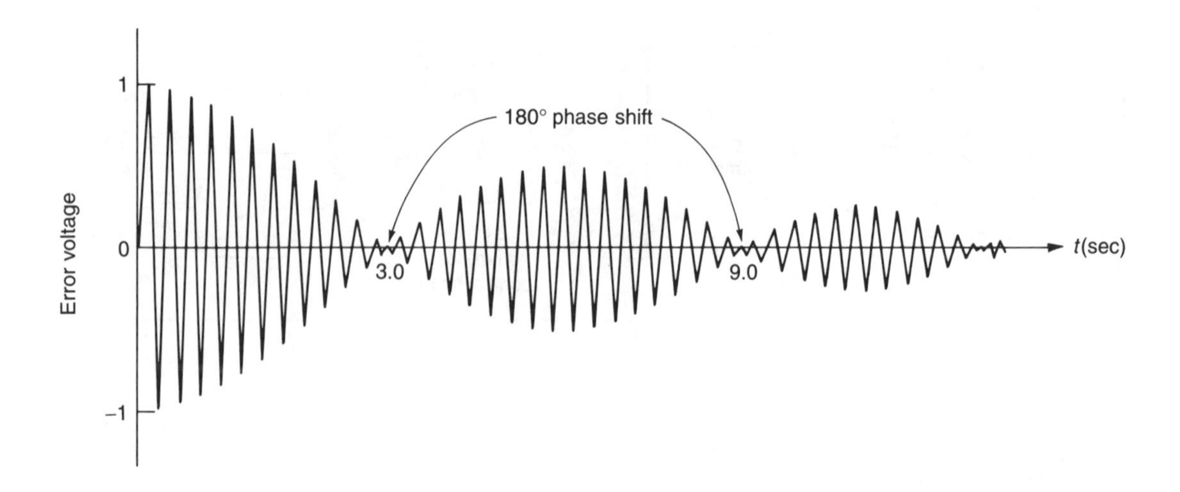

Determine approximate percent overshoot and time-to-first peak for the closed loop system.

4b. From your calculations for part 4a do you expect the gain was greater or less than unity? Why.

SOLUTION

1. The effective polar moment of inertia (as referred to the motor side of the gear train) is determined as,

$$J_{Eff} = J_M + (1/n)^2 J_L$$

$$= 0.2 + (1/4)^2 x3.2 = 0.4 \text{ units (oz-in.sec}^2.)$$

2. Although the control motor torque vs speed curve(s) are nonlinear, the

distinguishing feature of this kind of squirrel cage induction motor is that it has a relatively higher resistance to inductance ratio. This brings the peak starting torque (or "pull out" torque) to a point near zero speed (or even slightly to the left of zero). This makes characteristic torque vs speed curve closer to a straight line as shown:

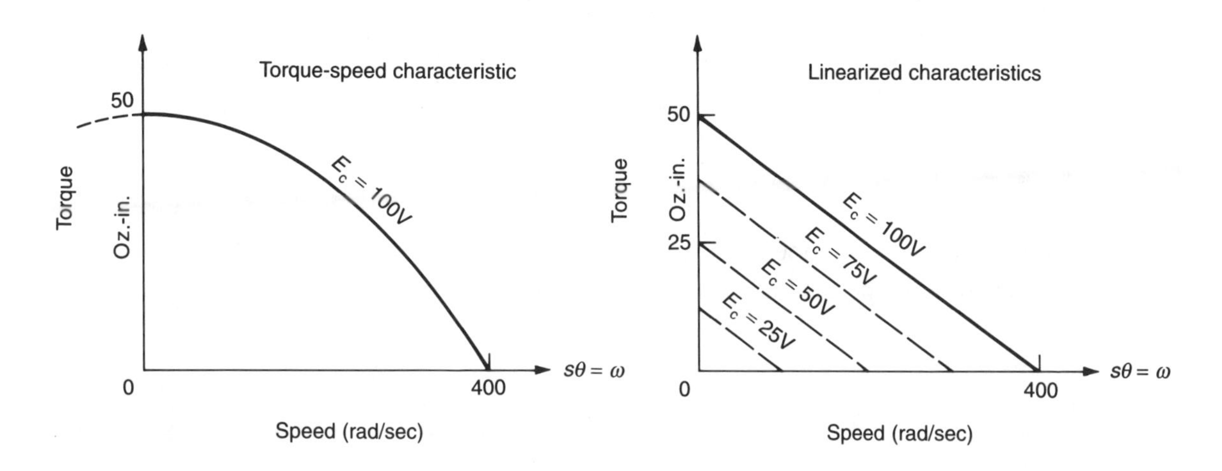

For values of control voltage other than full rated, the approximation lines are given as percentages of rated voltage and torque as shown.

Recall the equation of a straight line on an x-y axis: y=mx+b. Here, y is the torque developed, x the speed, m* is the slope, and b is the y intercept which is the percentage starting torque for the same percentage of the control voltage.

The approximate equations of interest are easily found as,

$$T_{Developed} = ms\Theta + T_o = ms\Theta + K_V V_C,$$

where m = slope = T_o/speed (no-load) = - 50/400 = -0.125, and
$K_V = T_o$/voltage(rated) = 50/100 = 0.5.

The developed torque is used to accelerate the polar moment of inertia and overcome bearing friction (zero in this case), thus,

$$T_{Developed} = J_{Eff}s^2\Theta + \text{friction} = ms\Theta + K_V V_C,$$

$$J_{Eff}s^2\Theta + \text{friction} - ms\Theta = K_V V_C$$

*Some control systems engineers take m as 1/2 the actual torque vs speed slope at zero speed as a better approximation to the nonlinear curve.

$0.4s^2\Theta + 0 - (-0.125)s\Theta = 0.5V_C$, $0.125[(0.4/0.125)s^2 + s]\Theta = 0.5V_C$

$$\Theta/V_C = K/[s(s\tau + 1)] = 4/[s(3.2s+1)]$$

3. System block diagram:

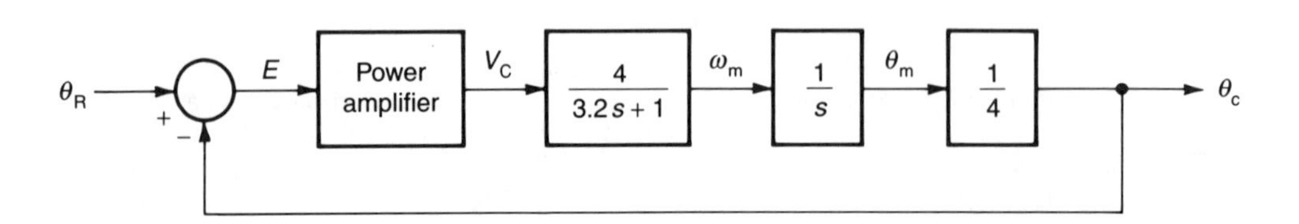

4a. Sketching the envelope of the modulated sine wave as an error voltage, one may determine the percent overshoot and time-to-first-peak directly as:

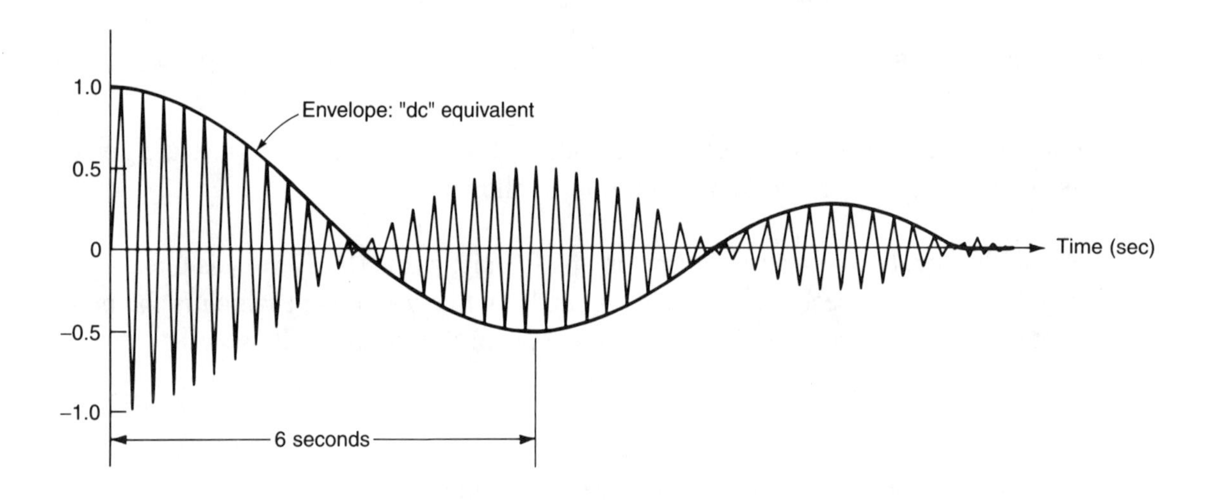

The percent overshoot is over 50% and t_p is approximately six seconds.

4b. From the original block diagram (with unity gain) one may easily obtain the closed loop transfer function as,

$$\Theta_L/\Theta_R = 1/(3.2s^2 + 1s + 1) = 0.312/(s^2 + 0.312s + 0.312),$$

here, $\omega_n = \sqrt{0.312} = 0.559$, $\zeta = 0.279$ (well under damped). From standardized second ordered curves, a zeta of 0.279 gives approximately 40% of overshoot.

The percentage overshoot of approximately 50% for part 4a compared with the 40% (using the original unity gain setting) is 10% higher, therefore the new gain of part 4a must have been increased slightly.

CONTROL SYSTEMS PROB. 6.7

This is a 10 question multiple choice problem. It is considerably longer and more detailed than might be expected on the actual exam.)

SITUATION

The positional movement of a robot arm has a theoretical open loop transfer function of:

$$G(s) = C(s)/E(s) = K(s+4)/(s(s+1)(s+2))$$

REQUIRED

QUESTION 1.

If an optical sensor is used to monitor the arm position such that a closed loop system is achieved, what is the closest gain, K, that would just cause system instability (assume the electronic optical sensor unit is equivalent to unity feedback)?

(a) Zero
(b) Infinity
(c) 2.0
(d) 4.0
(e) 6.0

Solution: Use Routh-Hurwitz method:

$$G_{Syst} = \frac{G}{1+G} = \frac{K(s+4)}{s^3+3s^2+(2+K)s+4K}$$

s^3	1	(2+K)
s^2	3	4K
s^1	x_1	
s^0		

$$x_1 = \frac{3(2+K)-4K}{3} > 0$$

$$\therefore 6-K>0$$

$$K<6$$

Answer is (e).

QUESTION 2 In open loop the arm transfer function has a frequency response such that the phase shift will be -180° at a frequency nearest to what value (in rad/sec)?

(a) Zero

(b) Infinity
(c) Unable to determine since K has not been specified.
(d) 3.0
(e) 5.0

SOLUTION
Use the Bode phase approximation method; sketch on semi-log paper, or use the direct calculation method with two or three "guessed" trial frequencies:

$$G = \frac{K(s+4)}{s(s+1)(s+2)} = \frac{2K(0.25s+1)}{s(s+1)(0.5s+1)}$$

$$s \rightarrow j\omega$$

$$\angle\phi = \tan^{-1} 0.25\omega - 90° - \tan^{-1}\omega - \tan^{-1} 0.5\omega \Rightarrow -180°$$

Try $\omega = 2$:

$$\angle\phi = 14°\text{-}90°\text{-}63.4°\text{-}26.6° = \text{-}166° \neq 180°$$

Try $\omega = 3$:

$$\angle\phi = 36.9°\text{-}90°\text{-}71.6°\text{-}56.3° = -181° \approx -180°$$

Answer is (d).

QUESTION 3 In closed loop (with the unity feedback sensor connected) the percentage overshoot of the output is to be near 16% for a step input; what would be the approximate damping ratio, zeta, (if the system may be approximated by a second ordered one).

(a) Over damped (nonexistent).
(b) Very lightly under damped.
(c) 0.5
(d) 0.7
(e) 1.0

Solution: Since the system is approximated by a second-ordered one, the normalized, standardized curves give a family of plots that relate zeta to percent overshoot. Or, the damping ratio may be calculated directly:

$$\%O.S. = 100e^{-\zeta\pi/\sqrt{1-\zeta^2}} \quad , \quad 0.16 = e^{-\zeta\pi/\sqrt{1-\zeta^2}}$$

$$\ln 0.16 = -1.833 = \frac{\zeta\pi}{\sqrt{1-\zeta^2}} \quad , \quad (1.833)^2 = \frac{\zeta^2\pi^2}{1-\zeta^2} \quad , \quad \zeta \cong 0.5$$

Answer is (c).

QUESTION 4 In closed loop (with the unity feedback sensor connected) and with an increased gain setting, it is found that the system will just break into oscillation. What is the closest natural frequency (in rad/sec) of this oscillation?

(a) Eventually reaches infinity
(b) Will never break into oscillation
(c) 1.0
(d) 3.0
(e) 6.0

Solution: From the root locus plot, find the frequency where the curve just passes through the imaginary axis; or, the value may be calculated directly from the closed loop characteristic polynomial (CP) equation:

$$C.P. = s^3 + 3s^2 + (2+K)s + 4K = 0 \ , \quad \text{for } s = 0 + j\omega \text{(on axis)}$$
$$= -j\omega\omega^2 - 3\omega^2 + j\omega(2+K) + 4K = 0$$
$$= \left(-3\omega^2 + 4K\right) + j\omega\left(-\omega^2 + 2 + K\right) = 0$$
$$= \quad 0 \quad + \quad 0 \quad = 0$$
$$\therefore \omega^2 = 8 \ , \quad \omega = \sqrt{8} \approx 3$$

Answer is (d).

QUESTION 5 For the system in closed loop (through the optical sensor, for unity feedback) the gain has been set such that the damping ratio (zeta) of 0.5, what percent overshoot may be expected for a step input? (Assume the system is approximated by a second-ordered one.)

(a) 16%
(b) 25%
(c) 37%
(d) 50%
(e) Unable to determine since K is not given.

SOLUTION

This is a "standard" graphical relationship that that is given in almost any text on control systems that is plotted for an ideal second-order curve. Or it may be calculated from:

$$\%O.S. = 100e^{-\zeta\pi/\sqrt{1-\zeta^2}} \quad = 100e^{-.5\pi/\sqrt{1-\zeta^2}}$$
$$= 16.3 \approx 16\ \%$$

Answer is (a).

QUESTION 6

For the system in closed loop (unity feedback) to have a damped natural frequency of 1.0 rad/sec, what is the closest gain setting for K?

(a) 6.0
(b) 4.0
(c) 2.0
(d) 0.5
(e) Unable to determine since zeta is not given.

Solution: Carefully sketch the root locus in the region near 1.0 rad/sec (at the intersection of the crude root locus and a straight line passing through the imaginary axis at 1.0). Once the correct locus is determined, K is the product of the line lengths from the open loop poles divided the line length of the open loop zero:

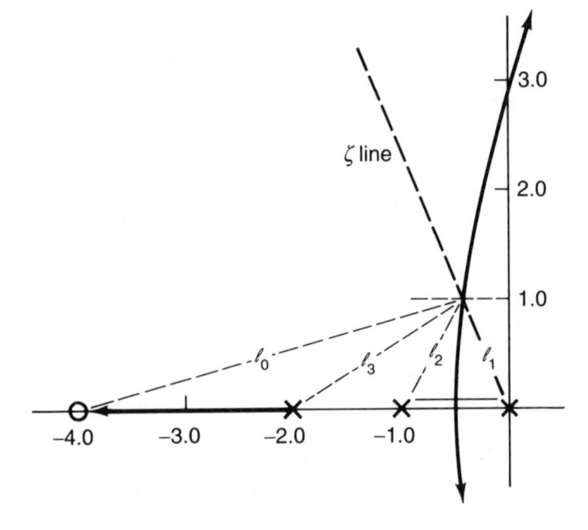

$$K = \frac{\ell_1 \ell_2 \ell_3}{\ell_0}$$

$$= \frac{1.1 \times 1.2 \times 1.8}{3.8} = 0.6$$

Answer is (d).

QUESTION 7

The arm (in closed loop) is to follow a moving object with a slew rate (ramp) of 2.0 rad/sec. What is the closest steady state error of the arm and the object for a system gain K=4.0? The arm (in closed loop) is to follow a moving object with a slew rate (ramp) of 2.0 rad/sec. What is the closest steady state error of the arm and the object for a system gain K=4.0?

(a) Zero
(b) Infinity (after infinite time)
(c) 0.25
(d) 4.0
(e) 10.0

Solution: Since this is a type I system (i.e., one integration in the loop equation, the error will be finite and is found from the final value theorem:)

$$E = \frac{1}{1+G} R$$

$$e(t)\Big|_{t\to\infty} = \lim_{s\to 0} s\left[\frac{1}{1+\dfrac{K(s+4}{s(s+1)(s+2)}}\left(\frac{2}{s^2}\right)\right] \to \frac{2}{\dfrac{4(4)}{(1)(2)}} = 0.25$$

Answer is (c).

QUESTION 8

The arm (in closed loop) is to follow a moving object with an acceleration of 1.0 rad/sec/sec. What is the closest steady state error of the arm and the object for a system gain K=1.0?

(a) Zero
(b) Infinity
(c) 1.0
(d) 4.0
(e) Unable to calculate since transfer function contains a zero in the numerator.

Solution: Since the input is an acceleration ($R(s)=1/s^3$), the loop itself has only one integration, the error will have to approach infinity.

Answer is (b).

QUESTION 9

For the closed loop system and for a particular gain the yields a damped natural frequency of 1.0 rad/sec, the system is too slow, but has an acceptable damping ratio. A faster system with a higher gain and damped natural frequency of 1.5 rad/sec meets time specifications but has too much overshoot. To remedy the situation, a phase lead compensator (a zero and pole combination) is to be used in a feedforward configuration. The compensator zero is to have a value of 1.0 (i.e., -1 in the s-plane), what is the approximate value of the compensator pole such that the damping ratio, zeta, is near that of original system (i.e., when the frequency was 1.0, yet the frequency is 1.5)?

(a) Unable to calculate as K is unknown.
(b) 1.0
(c) 2.5
(d) 4.0
(e) 25

SOLUTION

This is somewhat lengthy to solve (i.e., time wise, it may pay to skip this question). One method is to use the original root locus to locate the original uncompensated zeta for a damped frequency of 1.0 (i.e., a horizontal line passing through the imaginary axis at 1.0 and the root locus); using this zeta line, locate the desired new locus location on the zeta line and a horizontal line passing through the imaginary axis at 1.5. This new locus location will be incorrect by a certain angle (i.e., all correct locus locations should add to +/-n180°); find the difference between this certain angle and that of -180° (this difference will be approximately 35°). Then, by trial and error, locate the compensator pole on the real axis such that the new locus angles will sum to -180° (a location of -2.6 for this pole will satisfy the angle relationship).

Answer is (c)

QUESTION 10 The open loop function has a state variable representation of:

$$\dot{x} = \underline{A}x + \underline{B}e, \quad c = \underline{C}x = y.$$

Note: Do not confuse the lower case "c" (i.e., the output of the transfer function) with the upper case "$\underline{C}$" (i.e., the connecting matrix).

And, if it is known that $\mathbf{A} = \begin{bmatrix} 0 & 1 & 0 \\ 0 & 0 & 1 \\ 0 & -2 & -3 \end{bmatrix}$, then the matrix **C** that

is representative of the original transfer function is closest to which of the following row matrices?

(a) [4K K 0]
(b) [K 4K 0]
(c) [1 2 3]
(d) [0 -2 -3]
(e) None of above since order of **A** matrix is incorrect.

SOLUTION

General the **A** matrix may take several forms, however here it is given, the **C** matrix relating the output to the state variables may be found directly as follows:

$$G(s) = \frac{K(s+4)}{(s^3+3s^2+2s)} = \frac{(Ks^{-2}+4Ks^{-3})}{(1+3s^{-1}+2s^{-2}+0s^{-3})}$$

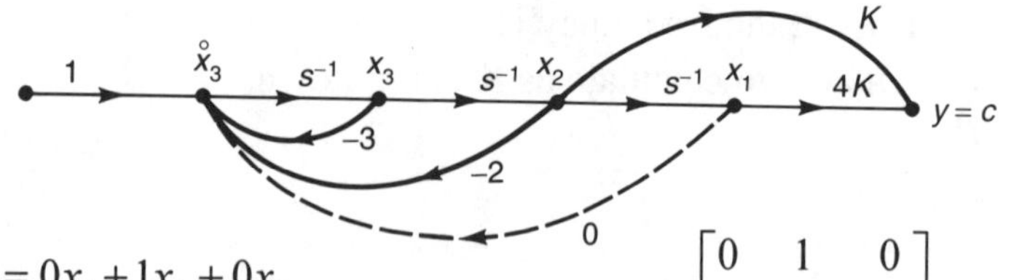

$$\dot{x}_1 = 0x_1 + 1x_2 + 0x_3$$
$$\dot{x}_2 = 0x_1 + 0x_2 + 1x_3$$
$$\dot{x}_3 = 0x_1 - 2x_2 - 3x_3 + 1e$$
$$y = c = 4K + Kx_2 + 0x_3$$

$$\underline{\dot{x}} = \begin{bmatrix} 0 & 1 & 0 \\ 0 & 0 & 1 \\ 0 & -2 & -3 \end{bmatrix} \underline{x} + \begin{bmatrix} 0 \\ 0 \\ 1 \end{bmatrix} e$$

$$y = \begin{bmatrix} 4K & K & 0 \end{bmatrix} \underline{x} = c$$

Answer is (a)

CONTROL SYSTEMS PROB. 6.8

SITUATION

A small subsystem has a transfer function of:

$$G_{ss} = K/(s^2+2s+2).$$

The subsystem is to be part of a somewhat larger system in that a loop is closed around the subsystem through an error detector of unity gain with a unity (H=1) feedback path.

REQUIREMENT

a) For the subsystem itself and for K= 2.0 is this a stable system? For what positive values of K would this subsystem be stable?

b) For the complete system (with the unity feedback connected), determine the value of K such that the system would have a zeta, ζ, of 0.4. Also determine the percent overshoot (to a unit step input) the system would have for this same value of K.

c) If, for the complete system, a compensator consisting of a pure integration (1/s) is inserted in the forward path (either just ahead or just following the subsystem), what value of K would cause the system to be marginally stable? What is the natural frequency of this marginally stable system?

SOLUTION

a) The subsystem is an ideal second order system with poles located at -1 + j1 and -1-j1. Obviously, since the poles are located in the left hand plane of the s-plane, it is a stable system. Also, since the poles are fixed, any positive real value of K would result in a stable system.

b) For the complete system the loci of the roots would emanate from the subsystem poles by going straight up and straight down. A zeta line radiating for the origin of the s-plane at an angle of $\cos^{-1} 0.4 = 66.4°$ (from the real 180° axis), would intersect the root locus at -1 +j2.33. Since K is the product of the line lengths from the open loop poles to the closed loop pole, it is found to be:

$$K = \ell_1 \times \ell_2 = 1.33 \times 3.33 = 4.43$$

Again, since the new system is also an ideal second ordered one, the percent overshoot may be calculated (or found from any standard set of second order curves for a particular zeta) as follows:

$$\text{Percent overshoot} = 100\ e^{-\zeta\pi\sqrt{1-\zeta^2}} = 25.6\%$$

c) The forward path equation is merely $K/[s(s^2+2s+2)]$, and, since H=1, the characteristic equation of the new system is:

$$s^3+2s^2+2s+K.$$

And a Routh-Hurwitz array yields,

s^3	1 2
s^2	2 K
s^1	X
s^0	Y

where X = (2x2-1xK)/2; for a marginally stable system X = 0, therefore K = 4.

Another way of finding the value of K for a marginally stable system is to allow s--->jω (i.e., the root locus crosses the jω axis for this condition), then substitute jω for s in the characteristic system equation (this method also determines the natural frequency):

$$(j\omega)^3 + 2(j\omega)^2 + 2j\omega + K = 0 = (-2\omega^2+K) + j\omega(2-\omega^2)$$

since both $(-2\omega^2+K) = 0$ and $j\omega(2-\omega^2) = 0$, therefore:

$$j\omega = 0, \text{ or } \omega = \sqrt{2} \text{ rad/second}$$

and,

$$K = 2\omega^2 = 4.$$

CONTROL SYSTEMS PROB. 6.9

SITUATION

The following system is to be stabilized by the addition of tachometer feedback, K_t.

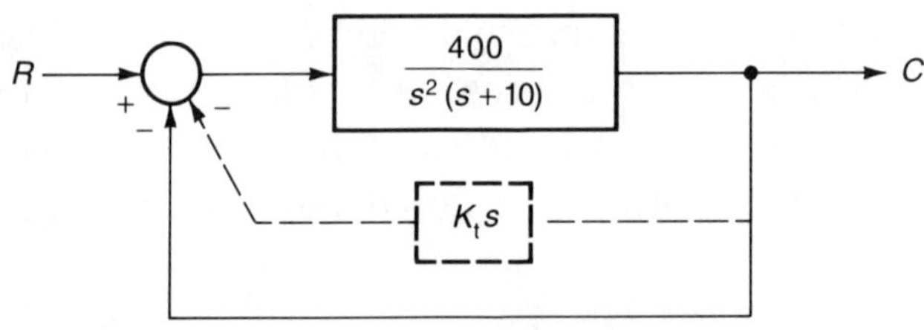

REQUIRED

(a) Find the minimum value of K_t such that the system will just be stable.

(b) If the value of K_t (found in Part (a) were increased by a factor of 1.25 [i.e., 1.25 x K_t min], determine the approximate step response characteristics. (Hint: one method of an approximate solution is to use only a reasonably accurate root-locus sketch and then approximate with "standardized" 2nd order curves.)

SOLUTION

(a) For stability use Routh-Hurwitz criterion for the characteristic polynomial:

$$G_{Syst} = \frac{C}{R} = \frac{400}{s^2(s+10)+(1+K_t s)400}$$

Since equivalent block diagram may be given as:

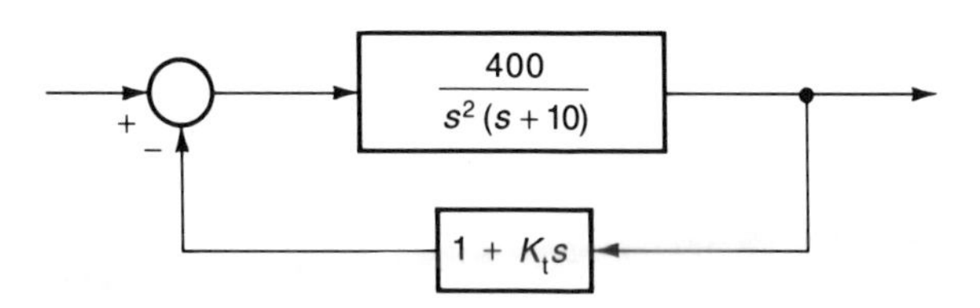

Characteristic Polynomial $=s^3+10s^2+400K_t s+400$

Routh-Hurwitz Array:

s^3	1	$400K_t$
s^2	10	400
s^1	x_1	

$$x_1 = \frac{(10)(400K_t)-(1)(400)}{10}$$

$\therefore K_t > 0.1$ for $x_1 > 0$

$(K_t \text{ minimum} = 0.1)$

(b)

Plot root-locus for HG:

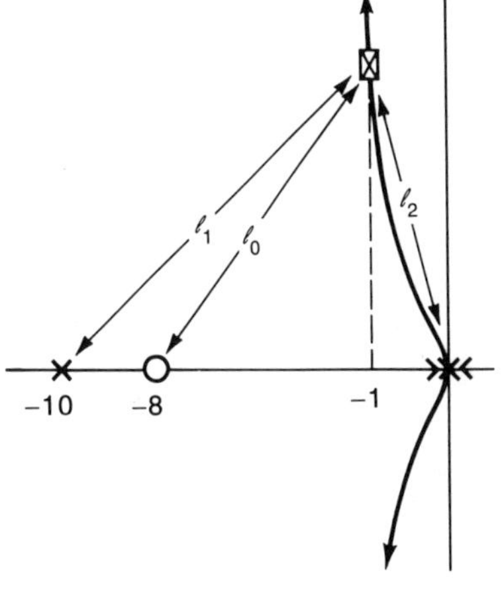

let $\text{H} = 1 + K_t s = 1 + 0.125s = 0.125(8+s)$

$$\therefore \ \text{HG} = \frac{(0.125)(400)(s+8)}{s^2(s+10)} = \frac{50(s+8)}{s^2(s+10)}$$

Here K(root-locus) = 50.

Since solution is only approximate, let $\ell_1 \approx \ell_0$

and $\ell^2{}_2 = \omega_n^2$

then zeta line $\approx 80°$ and $\omega_n = \sqrt{50}.$

Zeta ≈ 0.17

From any standard 2nd order transient response curves, one can easily determine percent overshoot and time to first peak, t_p.

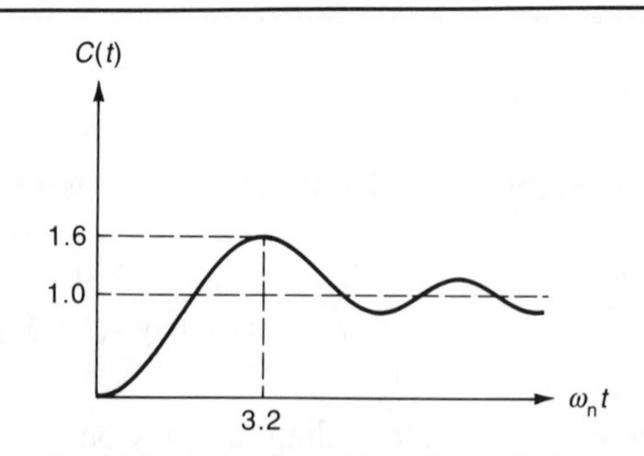

%O.S. =60%

t_p=3.2/ω_n=0.45 seconds.

COMPUTING PROB. 7.1.

SITUATION

A control system design, in block diagram form, has been submitted for performance analysis by analog computer simulation techniques or by use of a packaged digital simulation program. The performance of the system should include the displacement and the velocity (the step and ramp response) of the output shaft for several different values the forward gain, K, and also for several different values of the velocity feedback gain, K_t, for different settings of K. Also it will be desired to monitor the system error, e, as the simulation is in progress. The control system block diagram is as follows:

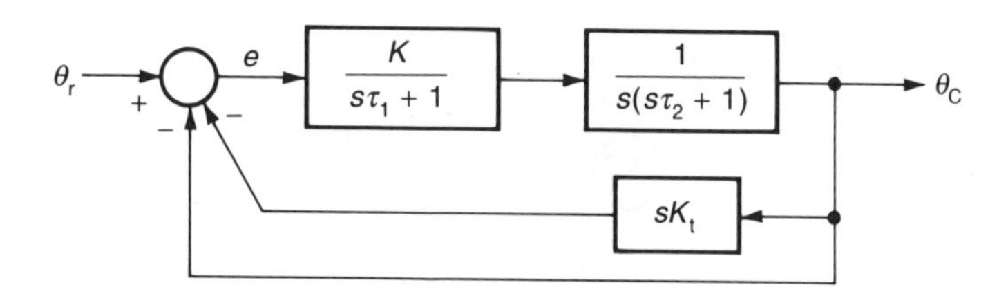

REQUIREMENT

1. Convert the system block diagram for an equivalent analog computer schematic.

2. Show any schematic that might be needed for both a step and ramp input.

3. Should the simulation be done with a digital packaged program, determine a suitable step size that might be considered for integration routines.

SOLUTION

Since the system has a differentiating block, sK_t, (which is very difficult to simulated by both analog and digital methods) the block diagram is rearranged so that this block is ahead of the integrator as follows:

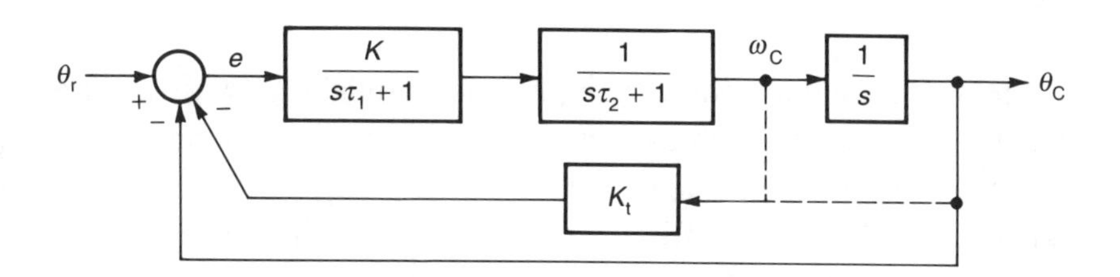

1. The analog computer equivalent schematic is arranged to have a pick off point for the system error at e, and the various gain controls may be easily manipulated for several different settings. Also, since all op-amps operate in the inverted mode and all loops are for negative feedback, all loops are to have an odd number of op-amps in a loop (this, of course, is not a requirement in a digital simulation program). Also, since no numerical values are specified, no provisions are made for magnitude or time scaling -- which, again, is not a factor in digital simulation.

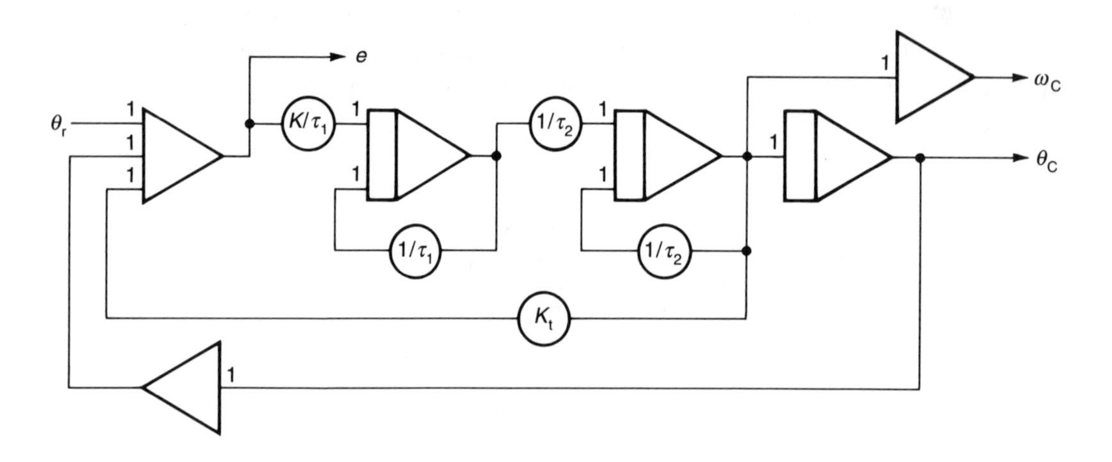

2. No special circuitry is needed to obtain a step response, however for a ramp input, an integrator block is placed ahead of the summing block.

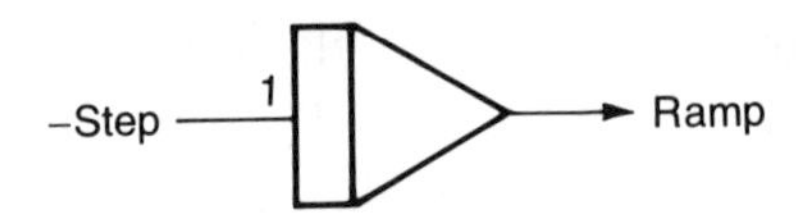

3. Most packaged digital simulation programs use an automatic variable step size routine for the actual integration, however some of the earlier ones required the programmer to enter this value. As a rule of thumb this value of delta time would be less than one-tenth of the shortest time constant in the system.

COMPUTERS PROB. 7.2.

SITUATION

A certain requirement for a magnetic circuit (an iron toroidal ring with an air gap cut into it) has a unique shape for the cross section of the iron; it is triangular in shape. Obviously the iron faces on each side of the air gap is triangular and the area of the air gap is triangular (neglecting any fringing effect). One of the design requirements is to find the amount of magnetic flux in the air gap (same as in the ring) from certain required flux densities, **B**, and various triangular dimensions. (Recall, the flux

equation is: $\phi = \mathbf{B} \times A$.)
The lengths of the sides of a triangle are given by the values of the variables X, Y, and Z, then the area of the triangle can be computed from:

$$\text{AREA} = \sqrt{W(W-X)(W-Y)(W-Z)} \qquad \text{where } W = \frac{X+Y+Z}{2}$$

REQUIREMENT

You are to write a BASIC or FORTRAN computer program to do the following:
a) Input values of X, Y, and Z(in meters) and **B** (in Tesla) from the keyboard
b) Compute the area of the triangle and the flux.
c) Output a heading identifying X, Y, Z and FLUX, followed by their values in an exponential format.

SOLUTION (in BASIC)
(a & b)

```
10 INPUT "INPUT X,Y,Z IN METERS, SEPARATED BY COMMAS",X,Y,Z
15 INPUT "INPUT FLUX DENSITY,B(IN TESLA)", B
20 W=(X+Y+Z)/2.
30 AREA=(W*(W-X)*(W-Y)*(W-Z))^.5
36 FLUX=B*AREA
40 PRINT "SIDE X","SIDE Y","SIDE Z","AREA","FLUX"
50 PRINT USING "#.###^^^^    ";X,Y,Z,AREA,FLUX
60 END
```

c)
Note: A typical output for a test input of X=0.002, Y=0.00282, Z=0.002, and B=1.0 (this portion , of course, would not be part of the exam) gives,

```
RUN

INPUT X,Y,Z IN METERS, SEPARATED BY COMMAS  .002, .00282, .002
INPUT FLUX DENSITY, B (IN TESLA) 1.0

SIDE X     SIDE Y     SIDE W     AREA      FLUX
0.200E-02  0.282E-02 0.200E-02   0.200E-05  0.200E-05
```

COMPUTERS PROB. 7.3

SITUATION

A tentative design for a control system to position a radar antenna has been designed. The system has been designed without any feedback compensation (such that minimal cost and weight factors may be achieved); however, if need be, this feedback compensation could be added in by a value of K_T. Before the system is built, it is decided to test the design by computer simulation to check the performance of the system without any feedback compensation (i.e., K_T=0). The tentative design block diagram is given as:

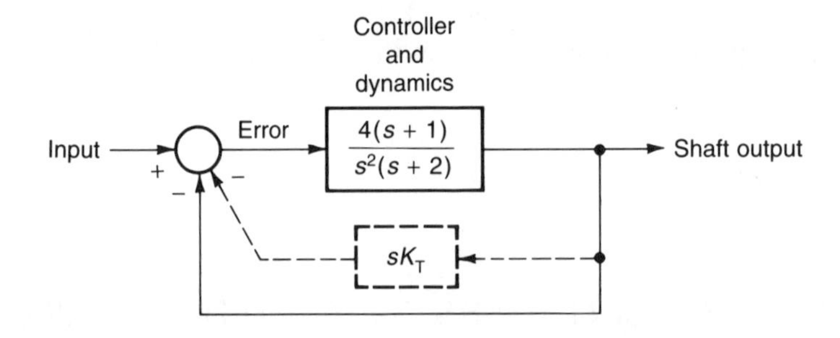

REQUIREMENT

Of the many high level simulation languages available, it is decided to use CSMP with the default method of integration. Also, since the programmer is not familiar with optimization techniques of analysis for the feedback compensation, the solution for the output vs. time for a step input will be programmed for K_T equal to zero; then if the results are not satisfactory, K_T will be set in and incremented until the desired results are achieved.

SOLUTION

Because of the possible use of the derivative function used for the feedback compensation, the block diagram is first rearrange to avoid this operation.

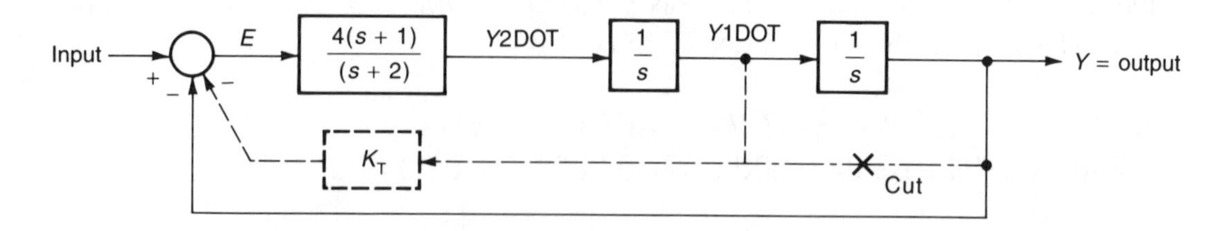

Or, for programming conceptualization, again the diagram is modified as:

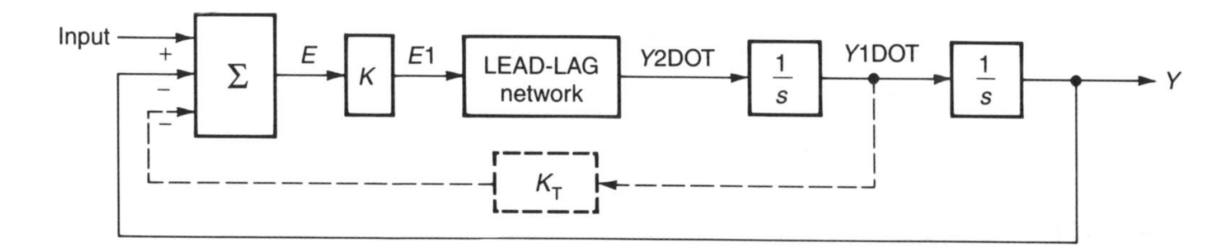

It should be noted from a CSMP manual that a lead-lag network is given by:

$Y(s)/X(s) = (P_1s+1)/(P_2s+1)$, or $P_2(dy/dt)+y = P_1(dx/dt)+x$.

Then $4(s+1)/(s+2) = 2(s+1)/(0.5s+1)$

Thus, one possible simulation program could be:

```
LABEL  SIMULATION OF AN ANTENNA POSITION CONTROL SYSTEM
  INITIAL
    CONSTANT K=2.0, KT=0.0, P1=1.0, P2=0.5
  DYNAMIC
    INPUT=STEP(0.0), E=INPUT-Y-KT*Y1DOT, E1=K*E
    Y2DOT=LEADLAG(P1,P2,E1)
    Y1DOT=INTGRL(0.0,Y2DOT)
       Y=INTGRL(0.0,Y1DOT)
   OUTPUT=Y
  TERMINAL
   TIMER FINTIME=5.0, PRTDEL=0.05, PRTPLT OUTPUT
  END
  STOP
  END JOB
```

The computer will then print the output for these conditions and, if the results are satisfactory, the problem is solved; if, however the results are poor, different values of KT may be tried. One way of obtaining the step results for these values of added feedback compensation is simply to change KT=0 to several other increasing values and repeat the operation. A better way would be to initially delete KT from the CONSTANT line, then follow with another statement line such as:

PARAMETER KT=(0.0, 0.2, 0.4, O.6, 0.8, 1.0)

Still another way would be to do this with a FORTRAN statement along with a NOSORT notation. The FORTRAN statement could be based upon a conditional requirement of some parameter of the output response and then looped until that requirement is achieved; this method depends on the sophistication of the programmer with regard to both FORTRAN and also his knowledge of control system optimal requirements.

COMPUTER PROB. 7.4

SITUATION

An ac circuit problem is to be solved for two specific conditons, thus a computer solution is appropriate.

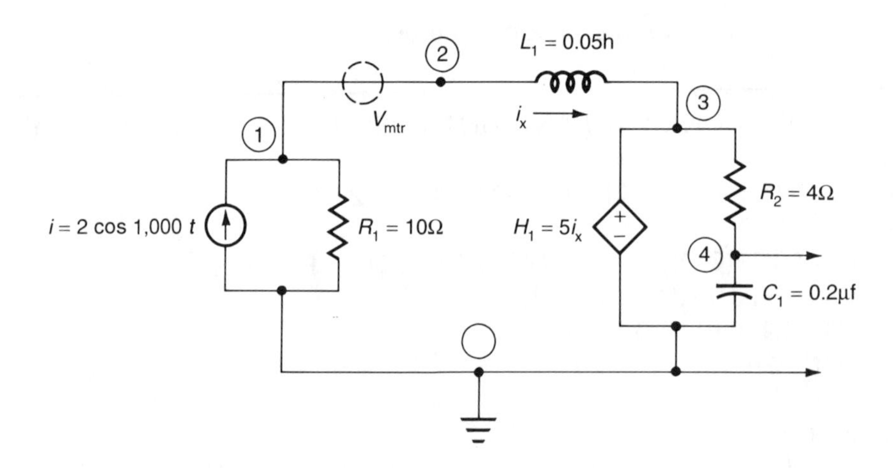

REQUIREMENT

Using any popular packaged circuit program, state the name of the program you choose and write out your program(s) including any remark statements that would be suitable. Here, assume another person will actually run the program to obtain the solution and thus your remark statements will be needed for clarification. Your program(s) will be to find the steady state voltage and phase angle (with respect to current source) across the capacitor for two conditions:

1) For a fixed current source of $i(t)$= 2cos 1000t.

2) Repeat (1) except the frequency will be a vary from 500 to 2,000 rad/sec in steps of 250 rad/sec.

SOLUTION

One very popular digital package circuit program that has been used by a large number universities over the last number of years is PSPICE (see pages 7.11-7.15 in the first volume) and will be used here for demonstration.

Part 1: Fixed parameters.

```
CIRCUIT PROBLEM, ac_prob.cir
I1      0 1    AC     2a     0deg
Vmtr    1 2           0
* Vmtr is a dummy voltage source of 0 volts that reads current.
H1      3 0    Vmtr   5
R1      1 0    10ohm
R2      3 4    4ohm
* As a check for XL and XC recover from w=1000 r/s.
* Thus XL=wL=50 ohm---> L= 0.05 henry.
* And XC=1/wc=5 ohms--->C= 200 microfarad.
L1      2 3    0.05h
C1      4 0    0.2E-6
* f= w/(2 pi) = 159.155
.AC    LIN    1     159.155      159.155
.PRINT AC      VM(4)  VP(4)
.END
```

After the program is ran successfully, the solution will appear in a file called "ac_prob.out" and the printout gives:

```
FREQ       VM(4)       VP(4)
1.592E2    1.916E00    -7.335E1
```

Of course the solution is 28.28/-45° volts.

Part 2: For a variable frequency of 500 to 2,000 in steps of 250 r/s (4 steps), merely substitute the ac frequency line as:

```
* f= w/(2 pi) from 500 to 2,000 r/s= 79.58 to 318.3 Hz
.AC    LIN    4      79.58       318.3
.PRINT AC      VM(4)  VP(4)
.END
```

The solution printout will be as follows:

FREQ	VM(4)	VP(4)
7.958E+01	3.430E+00	-5.905E+01
1.592E+02	1.916E+00	-7.332E+01
2.387E+02	1.307E+00	-7.872E+01
3.183E+02	9.890E-01	-8.151E+01

End

DIGITAL SYSTEMS PROB. 8.1

SITUATION

A single sensor has been used to detect excessive levels of a contaminant in a chemical process. Its output is "0" for normal conditions and "1" when the impurity level becomes too high. An evaluation program shows that the sensor is subject to occasional false alarms; also, it sometimes fails to operate when the contaminant level is high.

To improve the situation, it is decided to use three sensors and to disregard the indication of any single sensor whose output differs from the other two sensors. It is desired to have a single "correct signal" output which will be "0" for normal conditions and "1" for an excessive contaminant level, and "alarm" output which will be "0" when all sensor outputs are identical ("0" or "1") and which will latch into a "1" output at any time when the three sensor outputs become not identical. Also, an "alarm reset" input is to be provided such that a "1" input to it will reset the alarm output to zero after an alarm condition ceases to exist.

REQUIRED

Design a logic system to meet the above requirements. The system is to be implemented with NOR logic. There are no "fan in" or "fan out" limitations, that is, any NOR gate can have as many inputs as may be needed and can drive any necessary number of gates from its output.

Your solution should include:

(a) A truth table for the "correct signal" and "alarm" output.

(b) The logic equations.

(c) A logic circuit diagram showing how the sensors and NOR gates are to be connected.

SOLUTION

Let sensor outputs be A, B, and C; Then a truth table may be formed with f_0 being the contaminant signal and f_A being the alarm.

Term	Sensors			Outputs	
m	A	B	C	f_0	f_A
0	0	0	0	0	0
1	0	0	1	0	1
2	0	1	0	0	1
3	0	1	1	1	1
4	1	0	0	0	1
5	1	0	1	1	1
6	1	1	0	1	1
7	1	1	1	1	0

$$\therefore f_0 = \sum m(3,5,6,7)$$

$$\therefore f_A = \sum m(1,2,3,4,5,6)$$

Then simplify by use of Karnaugh Map.
for f_0:

$$\therefore f_0 = AB + BC + AC$$

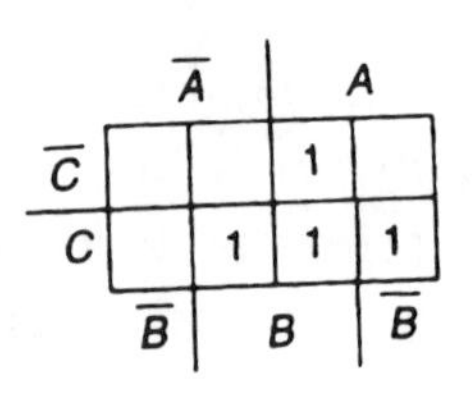

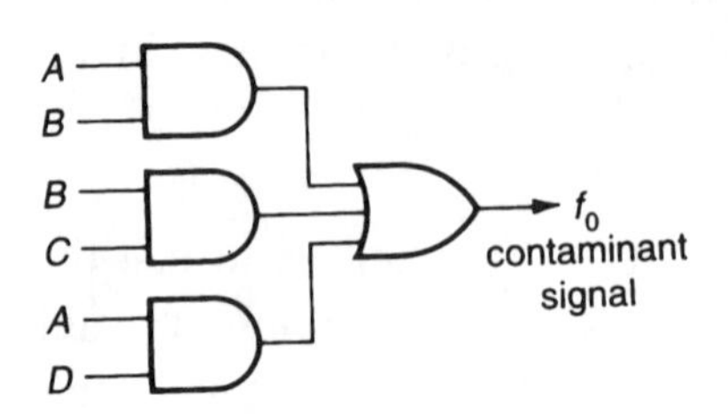

for f_A:

$$\therefore f_A = \overline{A}B + \overline{B}C + A\overline{C}$$

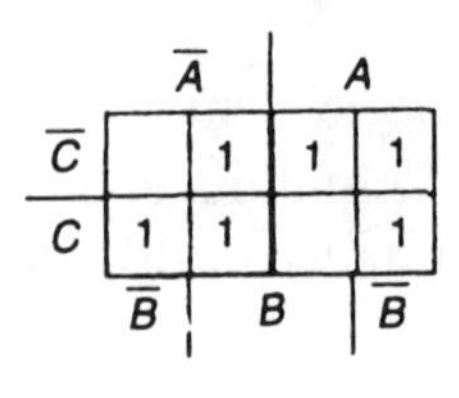

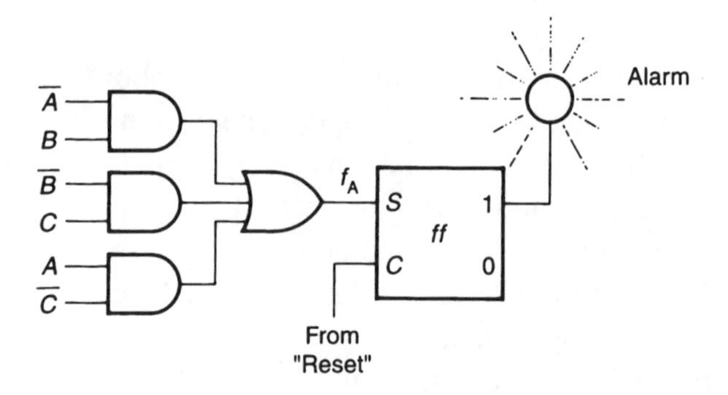

One could have simplified the Karnaugh Map in terms of "0's" as:

$$\overline{f_0} = \overline{B}\,\overline{C} + \overline{A}\,\overline{B} + \overline{A}\,\overline{C}$$

$$\therefore f_0 = \overline{\overline{B}\,\overline{C} + \overline{A}\,\overline{B} + \overline{A}\,\overline{C}}$$

$= (B + C) \cdot (A + C) \cdot (A + B)$ to go directly to NOR logic:

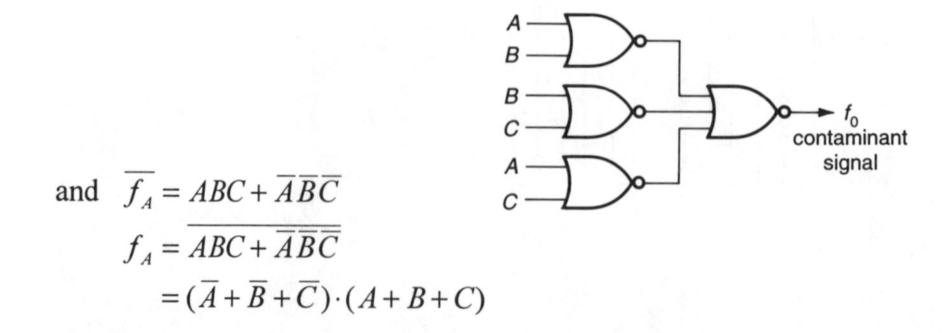

and $\overline{f_A} = ABC + \overline{A}\,\overline{B}\,\overline{C}$

$$f_A = \overline{ABC + \overline{A}\,\overline{B}\,\overline{C}}$$

$$= (\overline{A} + \overline{B} + \overline{C}) \cdot (A + B + C)$$

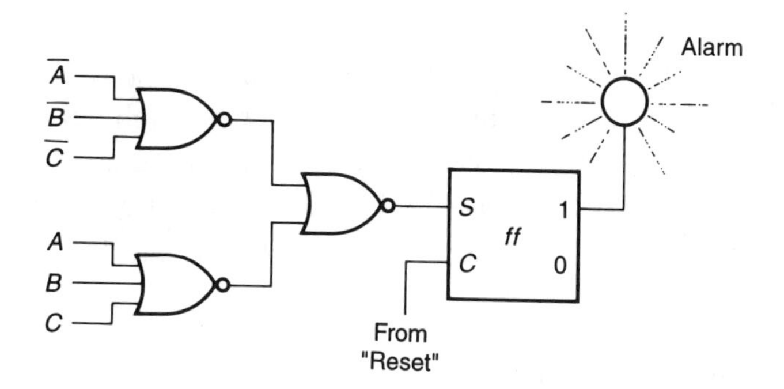
$\overline{A}$
$\overline{B}$
$\overline{C}$
A
B
C
S
ff
1
C
0
From
"Reset"
Alarm

DIGITAL SYSTEMS PROB. 8.2

SITUATION

A particular memory requires a RAM of 1Kx4 bits. Unfortunately only 1Kx1 bit RAMs (say, IC #2125A's)* are available for this particular application. The data from this RAM structure is to be made available along with another identical RAM structure to be added through a group (4) of full adders. The output from these full adders is to be stored for later use in 5 edge triggered D type flip flops.

REQUIREMENT

a) Show the interconnections necessary for the equivalent 1Kx4 bit RAMs (from the 1Kx1 bit RAMs). Call this equivalent circuit RAM1 and another identical unit RAM2

b) Show the interconnections for the 4 full adders, the RAM's, and the 5 D type flip flops. Assume all data is in the natural binary format.

c) Assume now that data stored in the RAM's is known to be in the binary coded decimal format and that data stored in the D flip-flops is to be made available to 2 seven segment drivers and LED displays. Show the interconnections necessary for correcting the full adders to correctly read the BCD sums and the connections to drivers and the LED displays.

*2125A Chip Logic

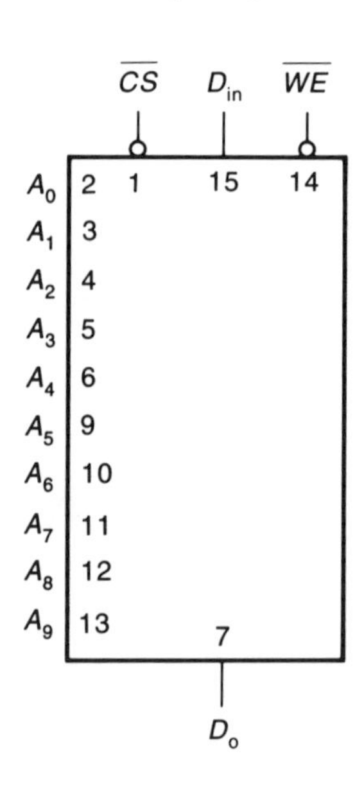

SOLUTION

a)

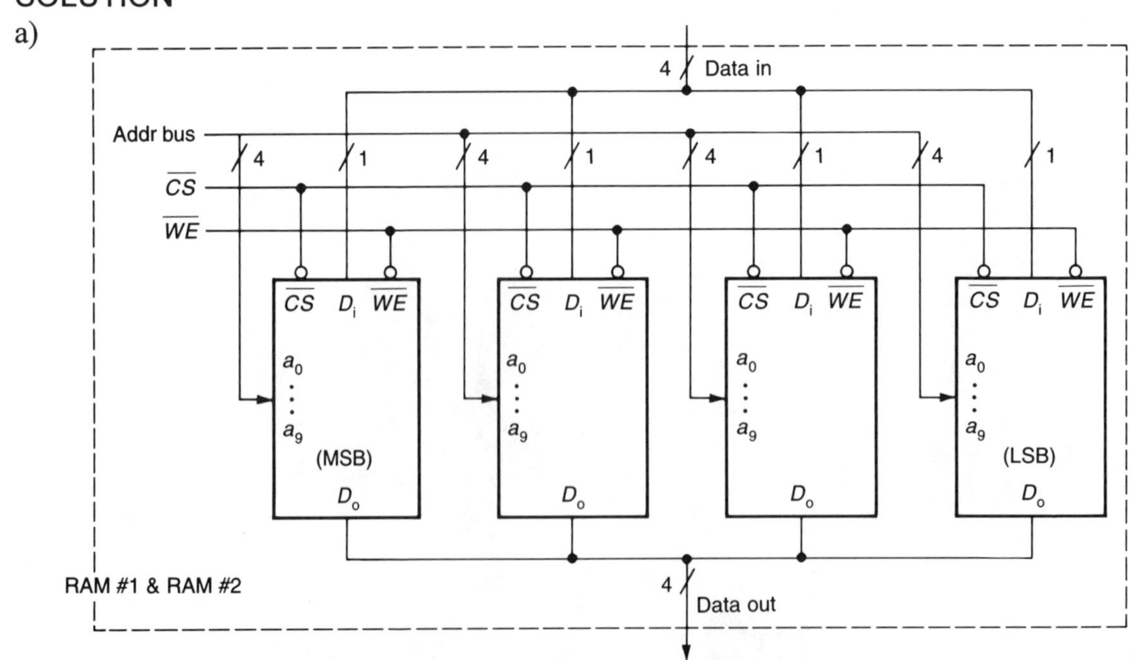

(b).

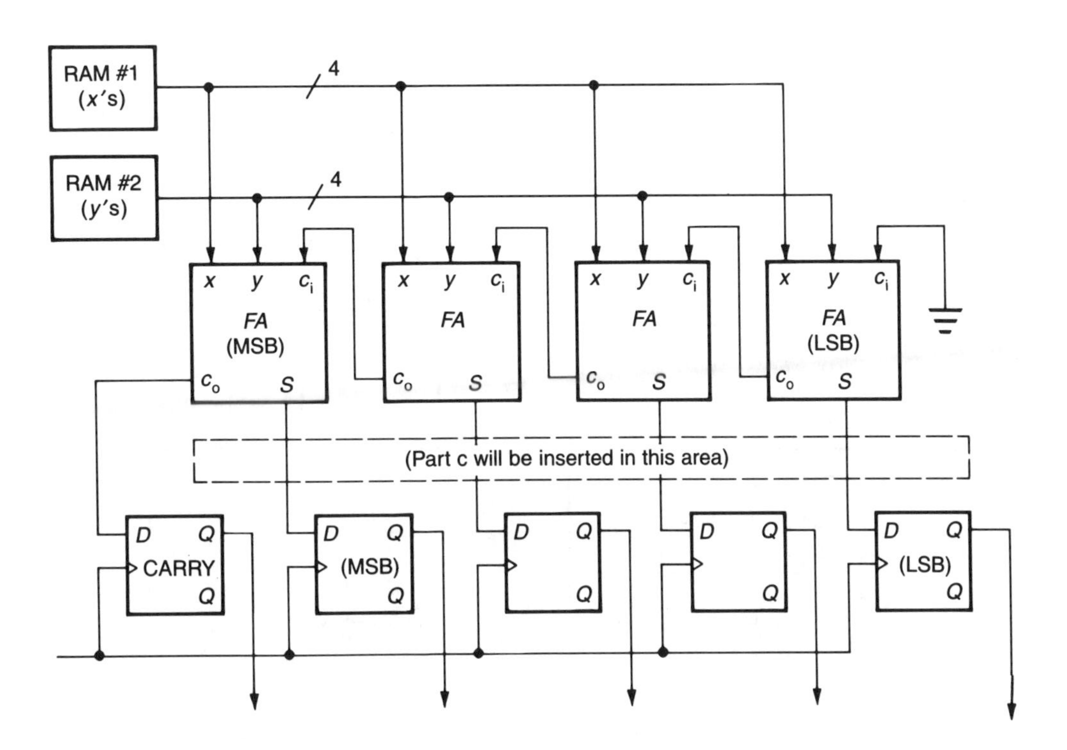

c) To correct the outputs of the natural binary signals to that of binary coded decimal, it must be recalled that natural binary includes numbers between 0 and 15; while BCD numbers are limited to 0 through 9. As long as the sum does not exceed 9 the outputs are correct. However, if the sum exceeds 9, then 6 must be added to the natural binary sum to yield two BCD numbers. As an example, suppose 7 and 5 are added together, the natural binary sum is $(1100)_2$ while if we add 6 more, the new sum is $(0001\ 0010)_{BCD}$.

```
  7 ---->   0111
  5 ---->   0101
            1100       Sum is greater than 9!
            0110       Therefore add 6.
12---->0001 0010       Answer in BCD.
```

Thus it is necessary to detect if outputs of S's of the full adder exceed 9, if so add 6 with the additional circuitry. This detection is easily accomplished if the 8 line is high AND either the 4 OR 2 line is high (and, of course, OR if the c_o output of the MSB FA is high).

The two 7 segment driver/decoders are connected only to show a 0 or a 1 for the MSB and a 0-->9 for the LSB.

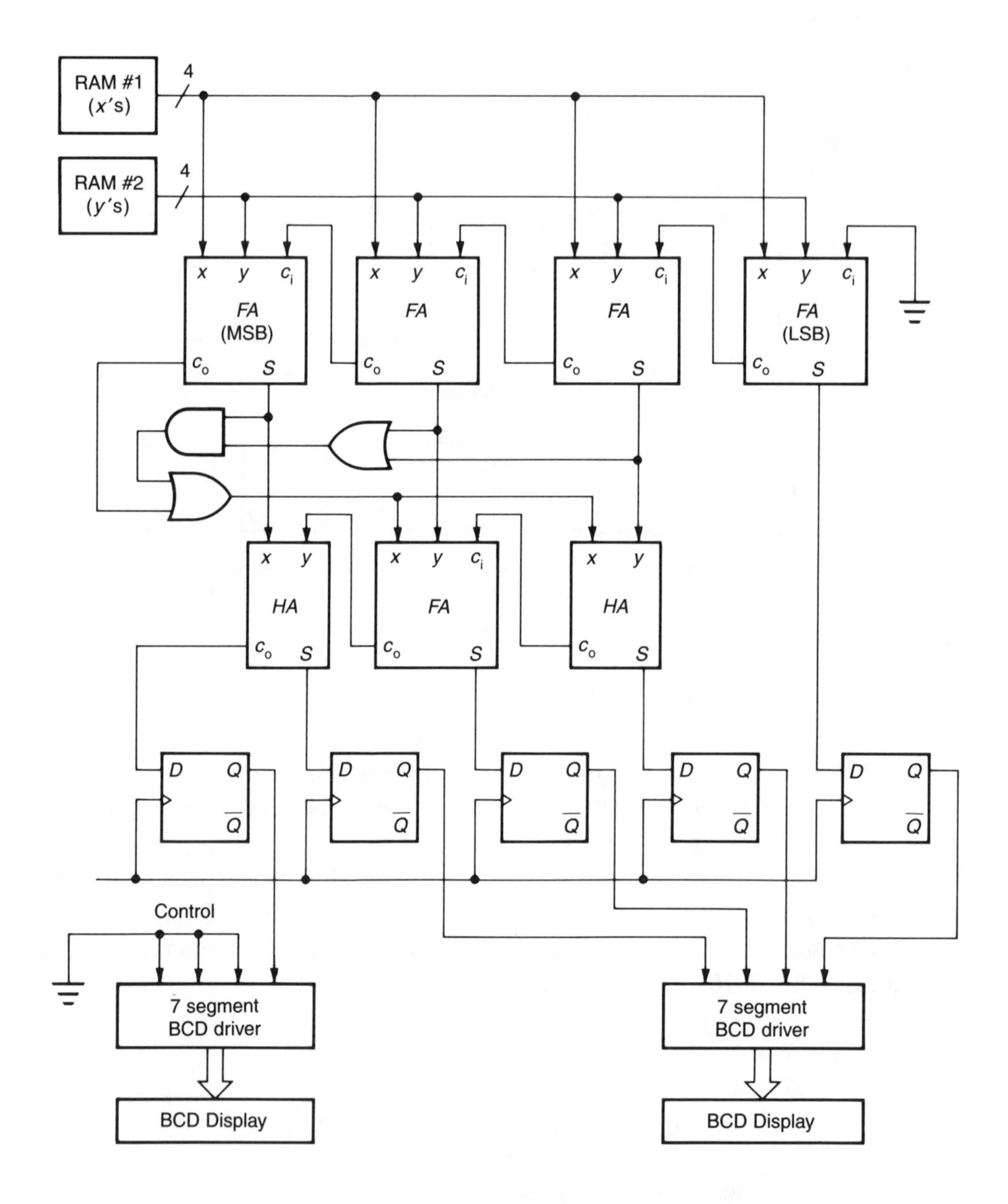

RAM #1
(x's)
RAM #2
(y's)
4
4
x y ci
FA
(MSB)
co S
x y ci
FA
co S
x y ci
FA
co S
x y ci
FA
(LSB)
co S
x y
HA
co S
x y ci
FA
co S
x y
HA
co S
D Q
Q
Control
7 segment
BCD driver
BCD Display
7 segment
BCD driver
BCD Display

DIGITAL SYSTEMS PROB. 8.3

> This is a 10 question problem and may be longer than found on the actual PE exam.

SITUATION

An Off-ON type sensor (OFF=0 volts, ON= 5 volts) is used to measure the vibration level in a rocket motor (ON if vibration level exceeds a preset threshold); when ON, a warning device is activated. However, because overly high vibrations can sometimes change the preset value of the sensor, it is proposed to replace the one sensor with three (the line of reasoning being that if only one sensor is on, it is probably out of correct calibration). A logic circuit is designed such that the warning device should be activated only if two or more sensors are ON. Assume the sensors are A, B, and C.

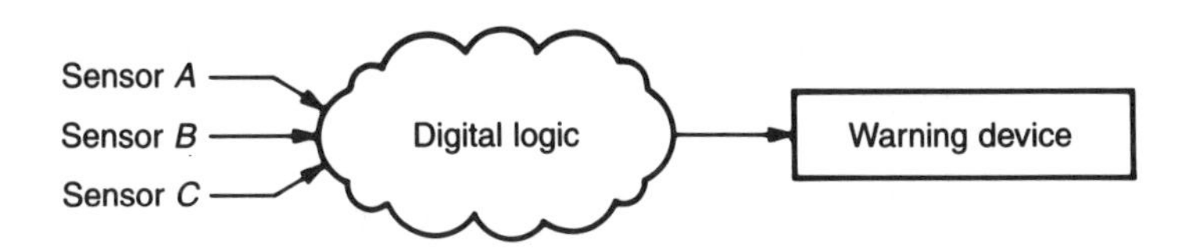

REQUIRED

QUESTION 1 The nominal function (i.e., canonical form) for the warning device to be activated is:

a)

$$f = \overline{A}B\overline{C} + A\overline{B}C + \overline{A}BC + AB\overline{C} + \overline{A}\overline{B}\overline{C}$$

b)

$$f = \overline{A}BC + A\overline{B}C + AB\overline{C} + ABC$$

c)

$$f = \overline{AB}C + A\overline{B}C + A\overline{B}\overline{C}$$

d) Correct function not given since only two sensors need to be activated.

e) All functions listed are correct since only two sensors need to be activated.

SOLUTION

Use the truth table to form the function on two or more 1's for the sensors.

m	A	B	C	f
0	0	0	0	0
1	0	0	1	0
2	0	1	0	0
3	0	1	1	1
4	1	0	0	0
5	1	0	1	1
6	1	1	0	1
7	1	1	1	1

$$f_{Alarm} = \overline{A}BC + A\overline{B}C + AB\overline{C} + ABC$$

Answer is (b).

QUESTION 2

The minimal sum of products expression for the warning device to be activated is:

a)

$f = AB + AC + BC$

b)

$f = \overline{A}BC + A\overline{C}$

c)

$f = AC + A\overline{B} + B\overline{C}$

d)

$f = A\overline{C} + \overline{B}C$

e) Correct function not given since only two sensors need to be activated.

SOLUTION

Form a Karnaugh map for $f = \Sigma m(3,5,6,7)$.

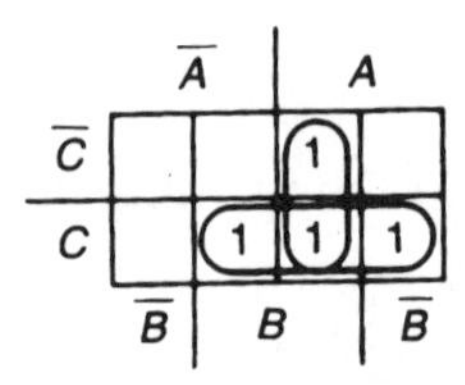

$f = AB + AC + BC$

Answer is (a).

QUESTION 3.

Another logic circuit is designed such that a second warning device will come ON if all three vibration sensors do not agree. The minimal product of sums expression for this second warning device is given by:

a) $f = ABC + \overline{A}B\overline{C} + A\overline{B}C$

b) $f = (A+\overline{B})(\overline{B}+C)(A+C)$

c) $f = (A+\overline{C})(\overline{B}+C)(\overline{A}+B)$

d) $f = (A+\overline{B}+C)(\overline{A}+B+\overline{C})$

e) Correct function not given because second warning light will always be ON.

SOLUTION

From the same type of truth table as Q-1, form the function when A, B, and C agree (i.e., terms 0 and 7).

	$\overline{A}$		A	
$\overline{C}$	1	0	0	0
C	0	0	1	0
	$\overline{B}$	B		$\overline{B}$

$$\overline{f}(\text{ activate 2nd alarm }) = \overline{A}C + B\overline{C} + A\overline{B}$$

$$\overline{\overline{f}}(\text{ activate 2nd alarm }) = \overline{\overline{A}C + B\overline{C} + A\overline{B}}$$

$$f(\text{ activate 2nd alarm }) = (A+\overline{C})(\overline{B}+C)(\overline{A}+B)$$

Answer is (c).

QUESTION 4

It has been determined that vibration spikes that exceed the preset values of the sensors that only last for 50 ms or less may be neglected and should not turn on the first warning device. A logic circuit is designed to be inserted in the line coming from the previously designed logic (i.e., the output activating circuit) and the warning device. This new circuit should not allow the activating signal to pass to the warning device for the first 50 ms of a high vibration and then reset itself. One such design is as follows (it is known that a J-K latch "flips" if the "J" terminal exceeds 2.9 volts and its input impedance is negligibly high before triggering):

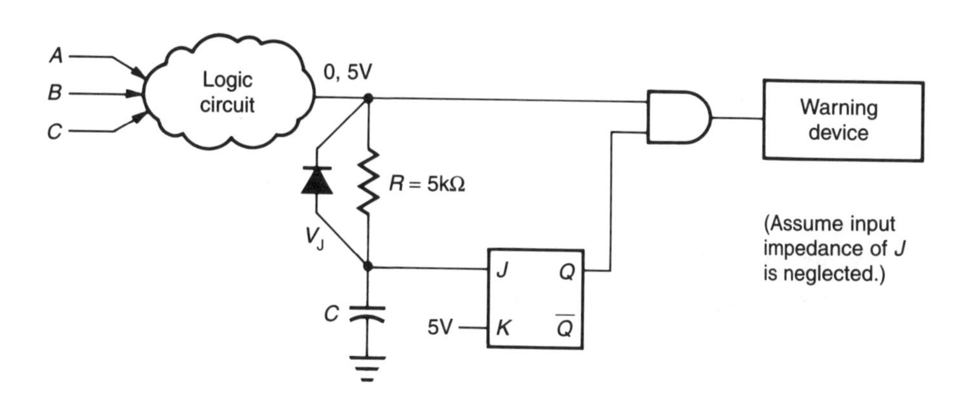

For the RC circuit, a value of 5k ohms is selected for R; what should be the nearest value of C?

a) C = 5 ufd
b) C = 10 ufd
c) C = 50 ufd
d) C = 100 ufd
e) Not possible for circuit to operate (for any value of C) within specifications since "K" is held at 5 volts.

SOLUTION

$V_J=2.9=5(1-e^{-(1/RC)t})=5(1-e^{-x})$, x=0.867

0.867=(1/RC)t, for R=5K, C=11.5 ufd

Answer is (b).

QUESTION 5

The circuit for part 4) proves to be too inaccurate and it is decided to use the following counting circuit instead, what is the nearest value of clock frequency that should be used?

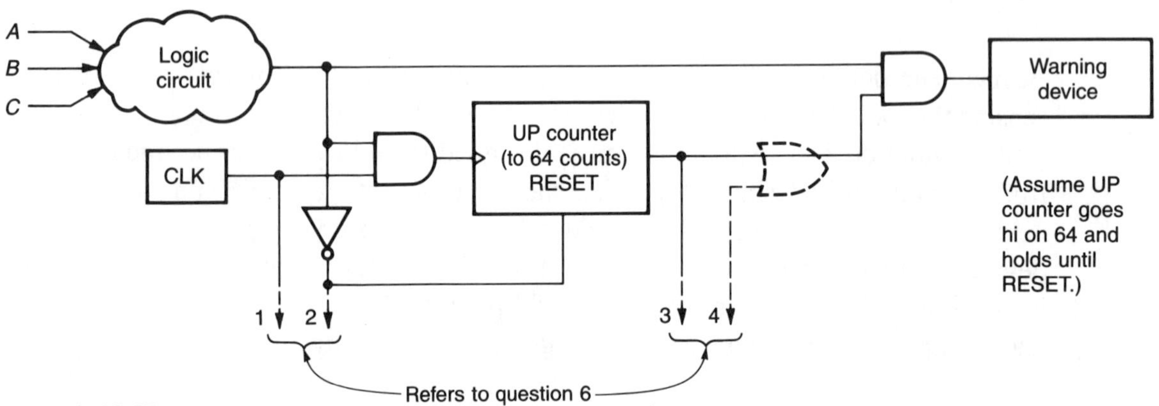

a) 12 Hz
b) 1.2 kHz
c) 64 kHz
d) 50 MHz
e) 3.2 kHz

SOLUTION

64 counts must occur in 50 ms

$64/50\text{ms}=1.28\text{x}10^3$, clock=1.28 kHz

Answer is (b).

QUESTION 6

Although the circuit of question 5 satisfies the new specifications, it is again later decided that if a high vibration has a duration of less than 50 ms (such that the warning device is not activated), then a window of another 50 ms (for low vibrations less than the preset value) should pass before another delay of 50 ms may be tolerated (i.e., warning device should be activated should another high vibration occur before the 50 ms has elapsed). The following circuit with another 64 count counter is added (at the dashed lines) to that of question 5:

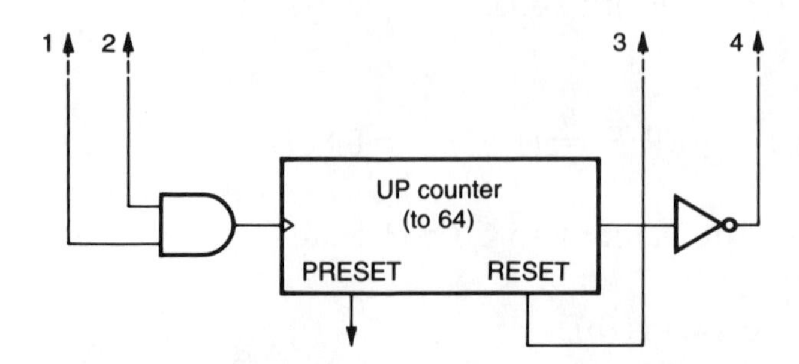

a) For correct operation PRESET should be tied hi.
b) " " " " " " " lo.
c) " " " " " " left floating.
d) " " " " " " to output of right hand AND gate.
e) Circuit is incorrect and can't satisfy these new specifications.

SOLUTION

Sketching a timing diagram will show that the added counter output should always start hi otherwise warning device will stay hi.

Answer is (a).

QUESTION 7

Refer to original problem with only the one sensor but whose output is an analog voltage (rather than OFF/ON) such that the output voltage is linear from 0 to 5 volts (with 0 corresponding to no vibration and 5 volts being the highest vibration expected. A standard 8-bit successive approximation type A/D converter is used to interface a small digital computer. For a vibration level corresponding to mid-range (i.e., 2.5 volts), the nearest value of the digital out (after the end-of-conversion signal is sent) is nearest to which of digital outputs expressed in hexadecimal values:
a) 55
b) F7
c) 7F
d) 127
e) 2.5

SOLUTION

Note that $(255)_{10}$ is for full scale for an 8-bit converter. Mid range corresponds to $(127)_{10}$ which, in turn, corresponds to $(7F)_{16}$.

Answer is (c)

QUESTION 8

Again it is decided to activate an alarm device if the output of the A/D converter exceeds a given value. The design of the circuit that is connected to the output lines of the A/D converter is such that the MSB and the next MSB lines are taken to an AND gate and the output this gates turns on the alarm.
The analog voltage output of the sensor is approximately which of the following voltages to just turn on the alarm?
a) 0.5 volts.
b) 0.75 " .
c) 1.8 " .
d) 3.8 " .
e) 4.8 " .

SOLUTION

The MSB corresponds to 128 and the next MSB is 64, and 128+64=192, therefore (192/255)5.0=3.75 volts.

Answer is (d).

QUESTION 9

Assume that in addition to the alarm circuit of question 8, an extra EMERGENCY ALARM is to be activated if the analog voltage level from the sensor exceeds 3.92 volts. The circuit chosen is that of question 8 along with another AND gate (whose output will activate this new alarm); one of its inputs will come from the output of the AND gate (of question 8), and other will come from:

a) the output of the sensor.
b) the output of the LSB line from the converter.
c) the output of the MSB line from the converter.
d) the output of the next LSB line from the converter.
e) the output of the 3rd next LSB line from the converter.

SOLUTION

For the configuration of question 8 the "weight" of the two lines are equivalent to $(192)_{10}$ or 3.75 volts, therefore 3.90-3.75=0.15 volts, or (0.15/5.0)255=7.65 (nearest integer value=8). Therefore the line that represents 2^3 is correct.

Answer is (e).

QUESTION 10

If the desired A/D converter (for question 8) is an 8-bit type, but only a 16-bit type is available, then one may substitute the 16-bit converter by:

a) Doubling the clock rate.
b) Using the MSB eight bits of the data bus and floating the lower eight bits (assuming the same reference voltage).
c) Using the lower eight bits and grounding the upper eight bits (assuming the same reference voltage).
d) Using the lower eight bits and grounding the upper eight bits (but doubling the reference voltage).
e) Using the middle eight bits and grounding the others (assuming the same reference voltage).

Answer is (b).

DIGITAL SYSTEMS PROB. 8.4

SITUATION:

A Programmable Peripheral Interface Adapter is to be added to a digital system such that this PIA will furnish three added 8-bit ports to the system. Port A is to be an input, port B is to be an output, and port C is to be an input port. Assume the PIA chosen is an 8255 and the register for port B of the PIA is preloaded with its data. The mode definition format (assuming an ACTIVE Mode Set Flag) is given as:

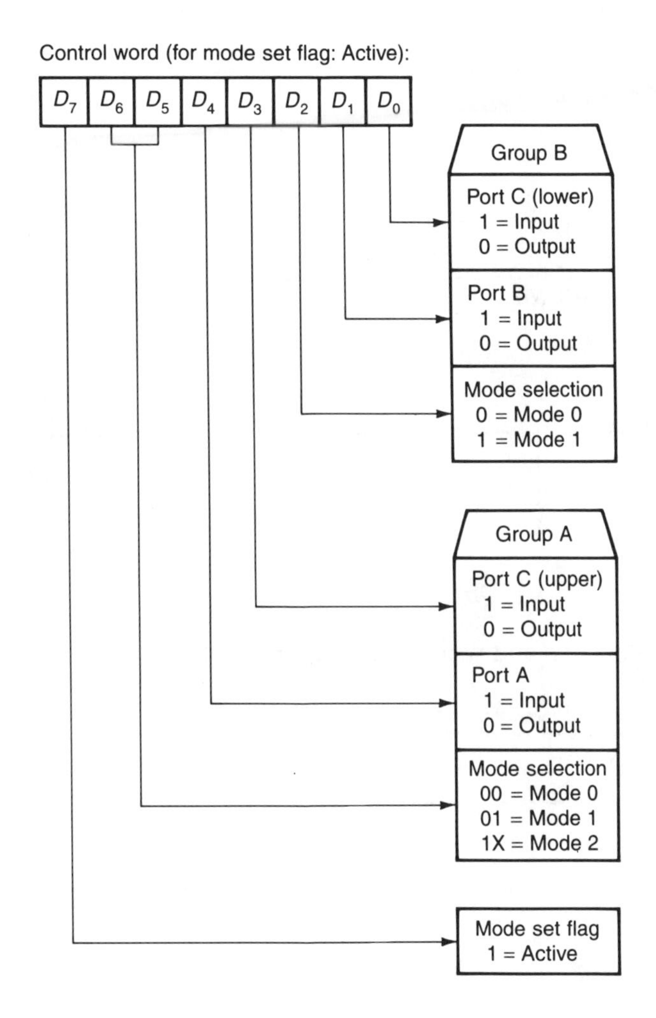

REQUIREMENT:
Determine the control word necessary for correct operation of the PIA.

SOLUTION

A restatement of the problem for using an 8255 PIA (assuming a Mode Set Flag = Active) is:

Input Ports (8-bits each):

Port A will operate in mode 0.

Port C will make use of both upper (4-bits) and lower (4-bits) as both inputs in mode 0.
Output Port (to operate in mode 1):

Port B (8-bits) will operate in mode 0.

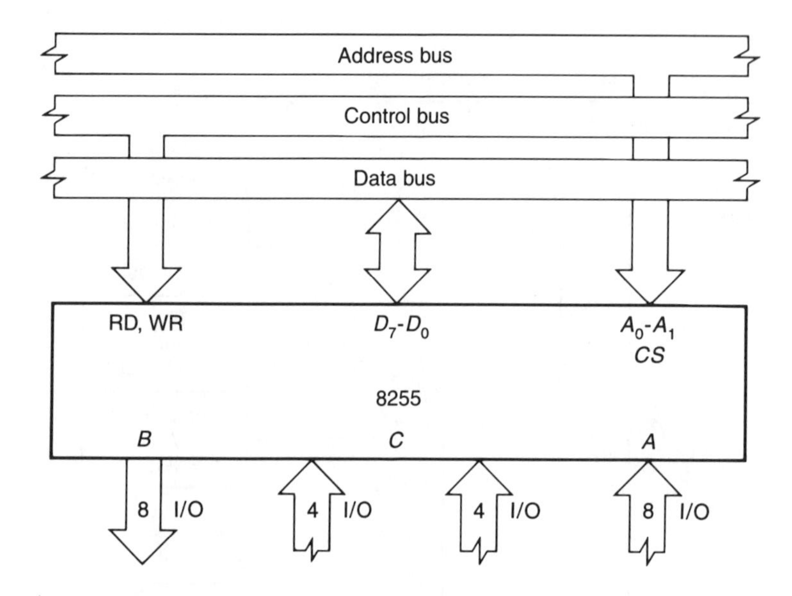

From these statements and with reference to the mode definition format, the correct control word is:

$D_7 \ldots D_0 = 1\,0\,0\,1\,1\,1\,0\,1.$

TRANSMISSION LINES PROB. 9.1

SITUATION

A transducer whose internal impedance is 600 ohms resistive generates 0.1 volt rms at a frequency of 1000 Hz. It is desired to transmit this signal over a telephone cable whose length is 2.5 miles (recall, 1 neper = 0.115 dB). The characteristic impedance of the line, Z_0, is 600 ohms, and its loss is 0.40 neper per mile. The phase shift is 0.1255 radians per mile. All parameters are measured at 1000 Hz. The input to the line is to be amplified by amplifier A so as to feed the line at a level of 0 dbm. The output of the line is to be amplified by amplifier B so as to provide 2.0 watts to a recorder. Amplifiers A and B each have an input impedance of 600 ohms.

REQUIRED

(a) What is the gain of amplifier A in decibels?
(b) What is the gain of amplifier B in decibels?
(c) What is the delay in microseconds between the input and output of the line?
(d) What would be the least expensive way to reduce the 1000 Hz loss in the telephone line?
(e) What effect, if any, would this measure in (d) above have on the line's delay?

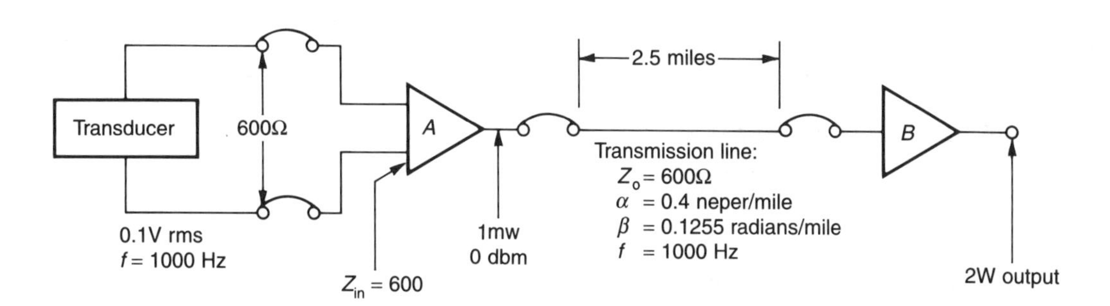

SOLUTION

(a) P_{in} level $= \dfrac{E^2}{R} = \dfrac{\left[\frac{1}{2}(0.1)\right]^2}{600} \times 10^3 = \dfrac{0.0167}{4}$ mv

P_{out} level is given: 0 dbm = 1 mw

$$\text{db Gain}_A = 10 \log_{10} \frac{P_{out}}{P_{in}} = 10 \log_{10} \frac{1}{\left(\frac{0.0167}{4}\right)} = 10 \log_{10} 240$$

$$= 10 \times 2.38 = 23.8 \text{ db}$$

(b) Loss in cable: 0.4 nepers / mile x 2.5 miles = 1 neper

1 neper x 8.686 db / nepers = 8.686 db

P_{in} level = 0 - 8.686 = -8.686 dbm

$$P_{out} \text{ level} = 10 \log \frac{2}{1 \times 10^{-3}} = 10 \log 2 \times 10^{3} = 10 \times 3.3 = 33 \text{ dbm}$$

db Gain_B = 33.0 - (-8.686) = 41.686 db. ANSWER

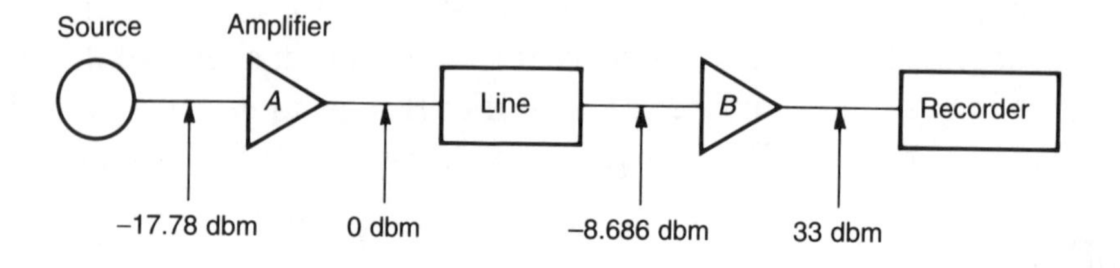

$$\text{(c) Delay} = \frac{0.1255 \text{ rad / mi x } 2.5 \text{ mi x } 10^{6} \text{ microsec / sec}}{2\pi \text{ x } 1000 \text{ Hz}} = 50 \text{ microseconds. ANS.}$$

(d) The loss at 1000 Hz could be reduced the least expensively by loading (series) the telephone line; i.e., inserting inductance coils at regularly spaced intervals.

(e) The delay of the transmission line will increase when load coils are added, since we increased the phase shift.

TRANSMISSION PROB 9.2

SITUATION

A short transmission line (less than a full wave length) is being considered for use in a project; it will be used to connect a voltage source to a resistive load. The problem is that the characteristic impedance of the line doesn't match either the sending or receiving end and therefore will require a full analysis of parameters of the line before making an engineering decision to use this particular configuration.

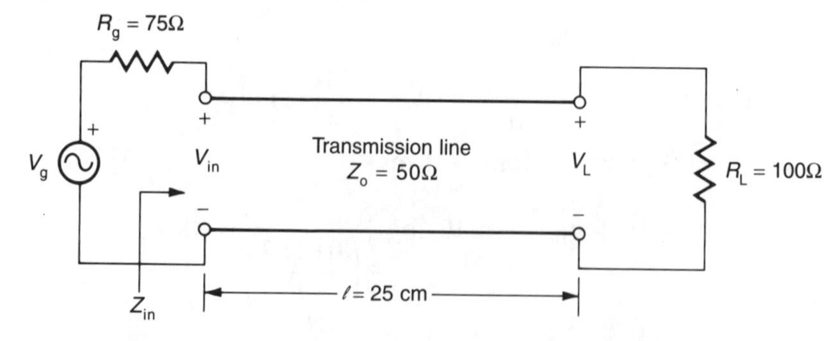

A voltage generator, V_g, produces 100 volts (rms) with an internal resistance, R_g, of 75 Ω. It is connected to a load, R_L= 100 Ω, through a section of a lossless transmission line length, ℓ , of 25 cm and whose characteristic impedance, Z_o is 50

Ω. The frequency of the generator is 300MHz and the phase velocity, u, on the line is 300x10⁸ m/s.

REQUIREMENT

The parameters needed for consideration are as follows; select the correct answers:

1. The capacitance per unit length of the line is:

   (a) 66.7 μF (b) 33.4μF (c) 50.3 μF (d) 50.3 pF (e) 66.7 pF

2. The inductance per unit length of the line is:

   (a) 250 μH (b) 166.7 μH (c) 25 μH (d) 166.7 nH (e) 25 nH

3. The attenuation constant of the line is:

   (a) 1.0 (b) Infinity (c) 0 (d) 50 (e) 100

4. The phase constant of the line is:

   (a) 6 rad/m (b) 2π rad/m (c) 12π rad/m (d) 4π rad/m (e) 0 rad/m

5. The input impedance, Z_{in}, looking toward the load is:

   (a) 25 Ω (b) 250 Ω (c) 25 kΩ (d) 2.5 Ω (e) 0 Ω

6. The input voltage at Z_{in} is:

   (a) 250 V (b) 25 V (c) 50 V (d) 100 V (e) 0 V

7. The reflection coefficient, Γ , is:

   (a) 0 (b) 3 (c) 1/3 (d) 1/6 (e) 1

8. The standing wave ratio, SWR , is:

   (a) 1 (b) 1/2 (c) 4 (d) 2 (e) 0

9. The load voltage, V_L, is:

   (a) 0 V (b) -37.5$\underline{/-90^\circ}$ V (c) 5$\underline{/-60^\circ}$ V (d) 5$\underline{/-90^\circ}$ V (e) 50$\underline{/-90^\circ}$ V

10. The average power delivered to the load is:

    (a) 0.5 w (b) 1.25 w (c) 12.5 w (d) 0 (e) 25 w

SOLUTIONS

1) $Z_o = \sqrt{L/C}$; $m = 1/\sqrt{LC}$; $Z_o m = 1/C$.

$C = 1/(Z_o\mu) = 1/(50x3x10^8) = 0.06667x10^{-9} = 66.67$ pF

Answer (e)

2) $Z_o^2 C = L$; $L = 2500x0.06667x10^{-9} = 166.67$ nH

Answer (d)

3) $\alpha = 0$ (since the line is lossless). Answer (c).

4) $\beta = \omega\sqrt{LC} = \omega/\mu$; $\omega = 2\pi f$; $\beta = (2\pi x300x10^6)/(3x10^8) = 2\pi$ rad/m

Answer (b)

5) $\lambda = C/f = (3x10^8)/(300x10^6) = 1$ m; $\ell = 25$cm $= \lambda/4$.

The line is a quarter wave length long, therefore: $Z_{in} = Z_o^2/R_L = 25\ \Omega$.

Answer (a)

6) $V_{in} = V_g(Z_{in})/(R_g+Z_{in}) = 25$ V;

Answer (b)

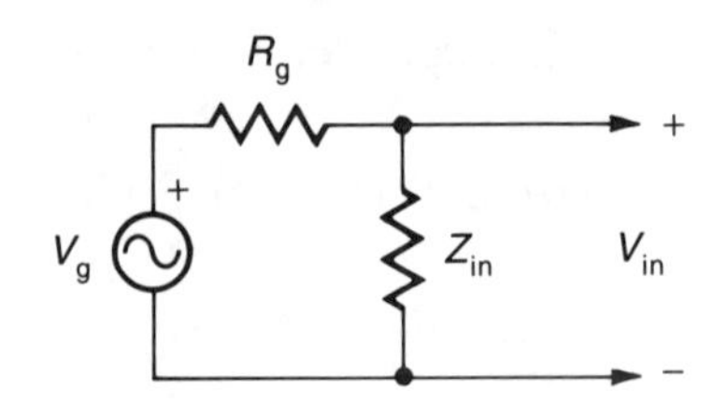

7) $\Gamma = (R_L - Z_o)/(R_L + Z_o) = (100-50)/(100+50) = 1/3$.

Answer (c)

8) SWR $= (1+|\Gamma|)/(1-|\Gamma|) = [1+(1/3)]/[1-(1/3)] = 2$.

Answer (d)

9) $V(z) = V_o(e^{-j\beta z} + e^{j\beta z})$; $V_L = (1+\Gamma)V_o$, where z=0 is the reference at the load end.

$V_o = V_{in}/(e^{j\beta l} + e^{-j\beta l}) = 25/(j1 - j\Gamma) = -j37.5$ V

$V_L = (-j37.5)(1+1/3) = -j50 = 50\underline{/-90^o}$ V.

Answer (e)

10) $P_L = 0.5V_L^2/R_L = 12.5$ watts.

Answer (c)

TRANSMISSION LINES PROB. 9.3

SITUATION

A section of nonloaded telephone cable, located in a rural area, has the following characteristics per loop mile at a frequency of 1000 Hz:

Series resistance, 85.8 ohms
Inductance, 1.00 millihenry
Capacitance, 0.062 microfarad
Shunt conductance, 1.50 micromho

The cable consists of 400 pairs of No. 19 AWG, is shielded, and is 3.06" o.d.

REQUIRED

Compute the following parameters for this cable at 1000 Hz:

(a) characteristic impedance
(b) attenuation in decibels per mile
(c) phase shift per mile
(d) velocity of propagation

Note: For full credit, you must show all of your work. Answers taken directly from reference books are not acceptable.

SOLUTION

(a)

The characteristic impedance of a line, or Z_c is a complex expression composed on:

z = series impedance per unit length, per phase = $R + j \times 2\pi f L$
y= shunt admittance per unit length, per phase to neutral = $G + j2\pi f_c$

Converting to henry, farad and mho (multiply by 10^{-3} or 10^{-6})

$$z = 85.8\,j\,2\pi \text{x} 1000 \text{ x } 10^{-3} = 85.8 + j6.28 = 85.9\angle 4° \text{ ohm/mile}$$

$$y = (1.5 + j2\pi \text{ x } 1000 \text{ x } 0.062) \text{ x } 10^{-6} = (1.5 + j389)\times 10^{-6} =$$

$$= 389\times 10^{-6}\angle 89.9° \text{ mho/mile}$$

$$Z_c = \sqrt{\frac{z}{y}} = \sqrt{\frac{85.9\angle 4°}{389\times 10^{-6}\angle 89.9°}} = \sqrt{0.221\times 10^{6}}\angle\frac{4°-89.9°}{2}$$

$$= 4.7\times 10^{2}\angle -43° = 470\angle -43° \text{ ohms ANS.}$$

(b)

Let us denote a complex quantity γ as the propagation constant and ℓ the line length in miles:

$$\gamma = \sqrt{z\ell y\ell} = \sqrt{zy\ell^2} = \ell\sqrt{zy}$$

The real part of the propagation constant γ is called "attenuation constant" β and is measured in nepers(recall, 1 neper $= 0.115$ dB) per unit length. The quadrature part of γ is called the "phase constant" β and is measured in radians per unit length; Thus,

$$\gamma\ell = \ell\sqrt{yz} = 1.0 \times \sqrt{85.9 \times 389 \times 10^{-6}} \angle \frac{4° + 89.9°}{2}$$
$$= 0.182\angle 47° = 0.124 + j0.133(radians) = \alpha + j\beta$$

Converting nepers into decibels, we obtain:
α=0.124 nepers/mile = 0.124 x 8.686 = 1.08 decibels/mile   ANS.

(c) Converting radians into degrees: 180^0 = 3.1416 radians, then 0.133 radians $=7.6^0$, thus phase shift or $\beta = 8.6^0$/mile   ANS.

(d) A wavelength is the distance along a line between two points of a wave which differ in phase by 360^0 or 2π radians. If λ is the phase shift in radians per mile, the wavelength in miles is:

$$\lambda = 2\pi/\beta = 6.28/0.133 = 47.2 \text{ miles}$$

Velocity of propagation is the product of the wavelength in miles and the frequency in hertz, or

$$\text{Velocity} = \lambda f = 47.2 \times 1000 = 47{,}200 \text{ miles/second.}$$

Reference: *Elements of Power System Analysis*, by William D. Stevenson, Jr. McGraw-Hill. p 100-106. Also see the new 1994 revised edition referenced at the end of Chapter 4.

TRANSMISSION LINES PROB. 9.4

SITUATION

Shown below is a diagram of a coaxial transmission line in which the circuit elements have the values indicated, and E_g is the open circuit generator voltage, Z_0 is the characteristic impedance of all transmission lines, and L_2 is greater than one wavelength.

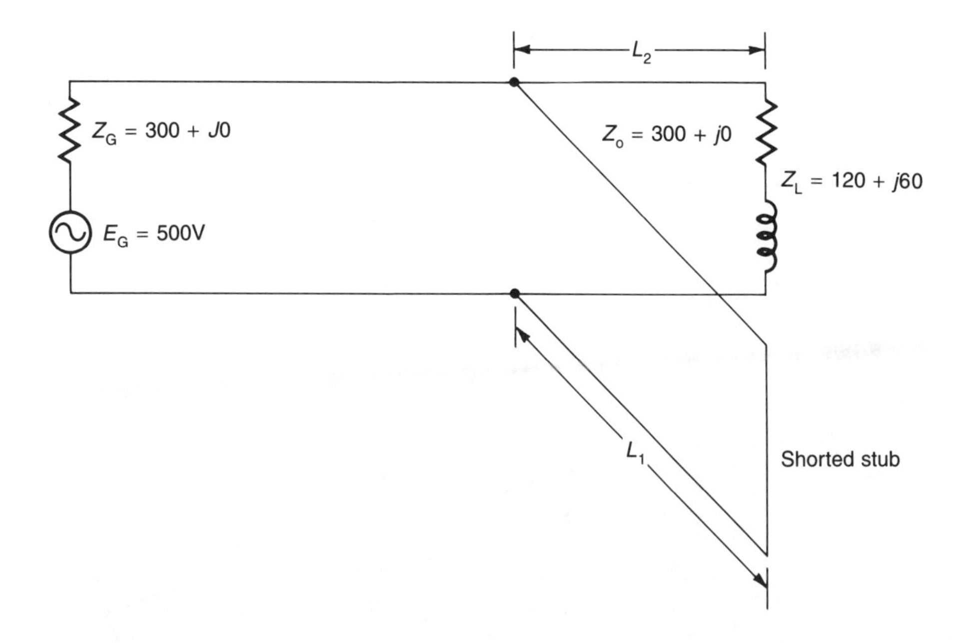

REQUIRED

(a) Determine the length L_2, in terms of a distance which is greater than one wavelength, and is the proper distance so that a shorted stub attached at this point cause the line to be matched to Z_0.
(b) Determine the length of the shorted stub L_1 needed to match the load to the line at this point.
(c) Determine the VSWR on the unmatched portion of the line L_2.
(d) Determine the power that will be supplied to the load under the matched conditions.
(e) Determine the greatest voltage that will appear across the transmission line in the unmatched section L_2.

Note: A Smith* Chart is included to aid in the solution, or a straight analytical solution may be employed.

*"SMITH" is a registered trade mark of the Analog Instruments Co, and is reproduced by courtesy of Analog Instruments Co., Box 808, New Providence, NJ 07974.

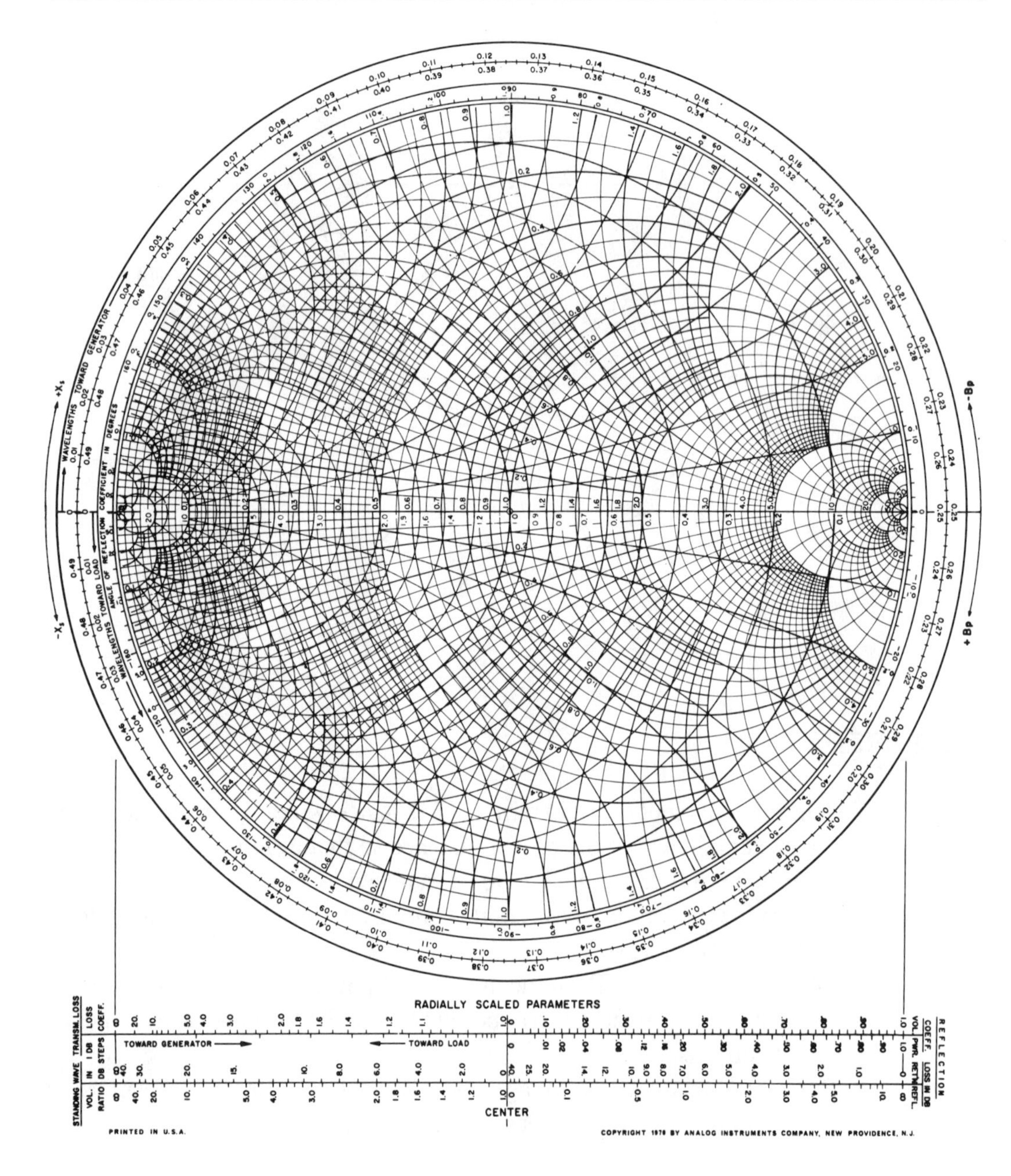

SOLUTION

As a first step, the Y_γ/Y_0 ratio has to be determined:

$$Y_\gamma = \frac{1}{Z_\gamma} = \frac{1}{120+j60} = \frac{1}{120+j60} \times \frac{120-j60}{120-j60} = 0.0067 - j0.0033$$

$$Y_0 = \frac{1}{Z_0} = \frac{1}{300} = 0.0033$$

$$\frac{Y_\gamma}{Y_0} = \frac{0.0067 - j0.0033}{0.0033} = 2 - j1$$

This ratio is point A on the Smith Chart.

(a) To determine the length L_2 the stub is placed at the point where $G=1/Z_0$ or at

G=1 which is point B on the Smith Chart.

However, L_2 must be larger than one wavelength, so that location is 360^0 further away from load than indicated.

Initial location:	-26.5^0
Final location:	-61.8^0
Difference	35.3^0

$$35.3^0/2 = 17.65^0$$

The proper distance of the stub: $17.65^0 + 360^0 = 377.65^0$ ANS.

(b) The length of the shorted stub L_1 is such that reactance of type opposite to line is required.

On Smith chart start at U_1 ($Y\gamma_{stub}/Y_0 = \infty$), move CCW to U_2 where the imaginary component of G + jB is opposite that of stub location but equal in magnitude.

Initial location	0^0
Final location	270^0
Difference	270^0

The length of the stub: $270^0/2 = 135^0$ ANS.

(c) To determine the VSWR on the unmatched portion of the line L_2, from point C on Smith Chart, we obtain:

$N = 2.6$

This value can also be obtained as follows:

$$\rho = \frac{Z_L - Z_0}{Z_L + Z_0} = \frac{120 + 6j0 - 300 - j0}{120 + j60 + 300 + j0} = \frac{-180 + j60}{420 + j60} \times \frac{420 - j60}{420 - j60} = 0.4 + j0.2$$

$$|\rho| = \sqrt{0.4^2 + 0.2^2} = 0.446$$

$$VSWR = N = \frac{1 + |\rho|}{1 - |\rho|} = \frac{1 + 0.446}{1 - 0.446} = \frac{1.446}{0.554} = 2.6 \quad \text{q.e.d.}$$

(d)

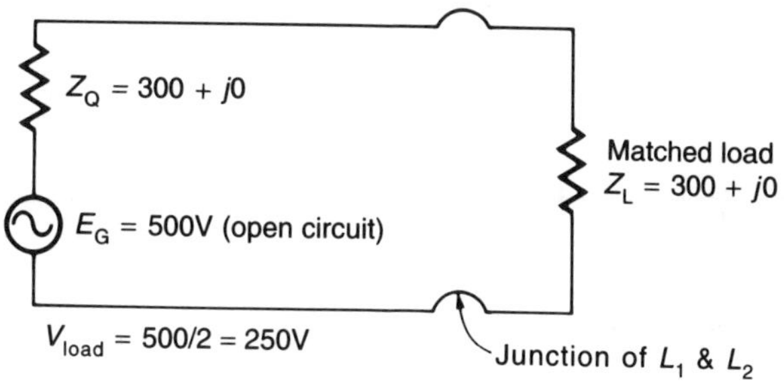

$Z_Q = 300 + j0$

Matched load $Z_L = 300 + j0$

$E_G = 500V$ (open circuit)

$V_{load} = 500/2 = 250V$

Junction of L_1 & L_2

The power then will be supplied to the load under the matched condition:

$$P_{load} = \frac{E^2}{R} = \frac{(250)^2}{300} = 208 \text{ watts} \qquad \text{ANS.}$$

(e)

$$V_{\max} = V_{load}\left(1+|\rho|\right) = 250\times 1.446 = 361 \text{ volts} \qquad \text{ANS.}$$

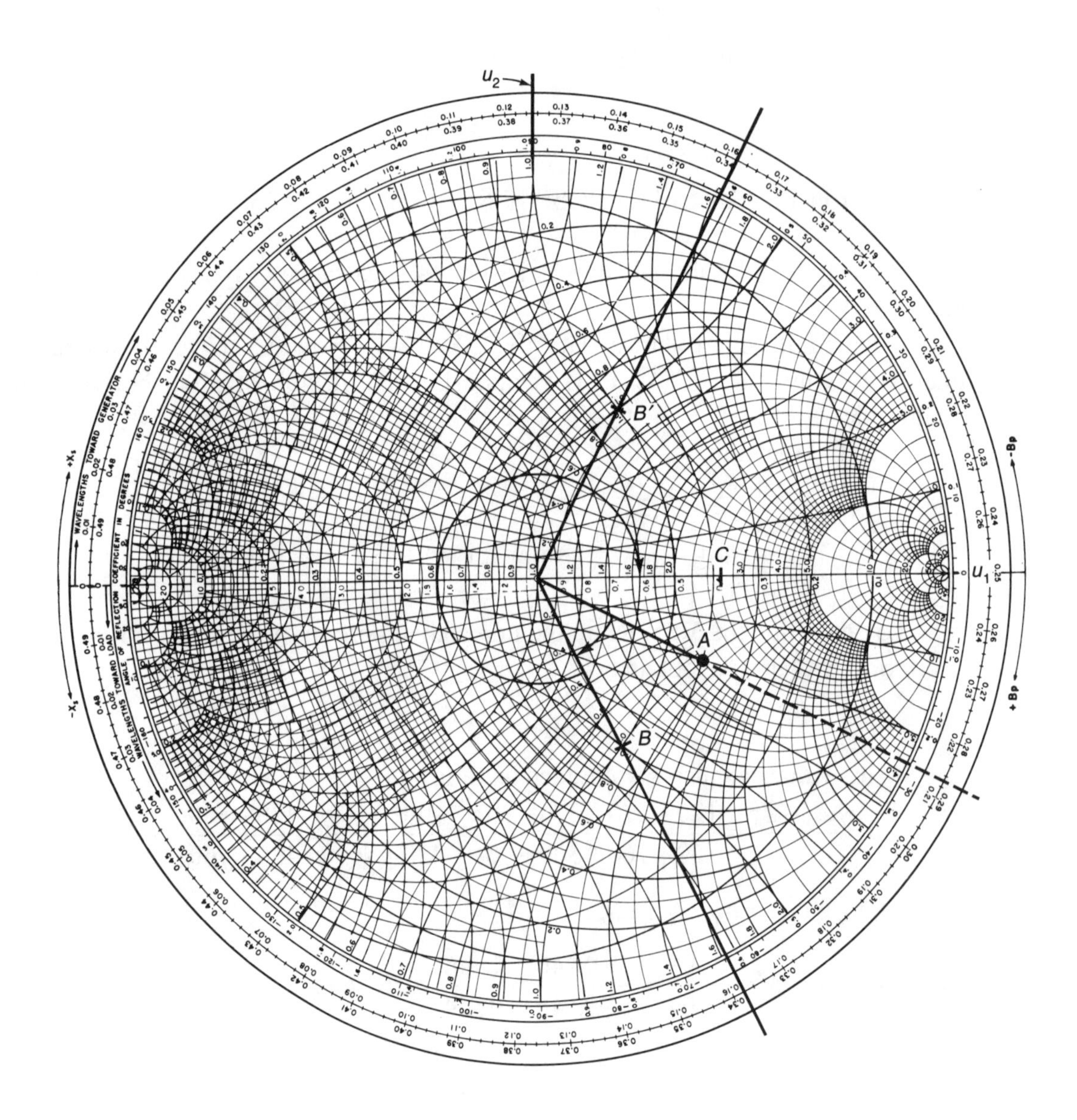

MICROWAVE ANTENNA PROB. 9.5

SITUATION:

Two microwave stations operating at 5 GHz are 50 kilometers apart. Each has an antenna whose gain is 50 dB greater than isotropic.

REQUIREMENT:

If 6 watts is applied to the input of the transmitting antenna, what is the signal level at the output terminal of the receiving antenna under free-space conditions? What is the path loss?

SOLUTION:

$$G = 50 \text{ dB } = 10^5 \text{ , } f = 5\times10^9 \text{ Hz, } P_T = 6 \text{ watts}$$

$$\lambda = \frac{v}{f} = \frac{3\times10^8}{5\times10^9} = 6\times10^{-2} \text{ m}$$

$$A = \frac{\lambda^2 G}{4\pi} = \frac{(6\times10^{-2})^2 10^5}{4\pi} = 28.65 \text{ m}^2 \text{ effective antenna area}$$

$$\text{watts / m}^2 \text{ @ 50 kM } = \frac{P_T G}{\text{surface area of sphere}} = \frac{6\times10^5}{4\pi(50\times10^3)^2}$$

$$= 1.91\times10^{-5} \text{ watts / m}^2$$

$$P_{RCVR} = A\left[\text{w / m}^2 \text{@}50kM\right] = 28.65\times1.91\times10^{-5} = 5.47\times10^{-4} \text{ watt ANS.}$$

$$\text{Path Loss } = -10 \log \text{G} = -10 \log \frac{4\pi}{\lambda^2}(4\pi r^2) = -10 \log\frac{4\pi r^2}{\lambda}$$

$$= -20 \log \frac{4\pi\times50\times10^3}{6\times10^{-2}} = -20 \log 1.05 \times 10^7$$

$$= -140 \log 1.05 = 140.4 \text{ dB ANS.}$$

ANTENNAS PROB. 9.6

SITUATION

A phased array consists of two half-wave antennas located a half-wave apart, as shown in the figure below:

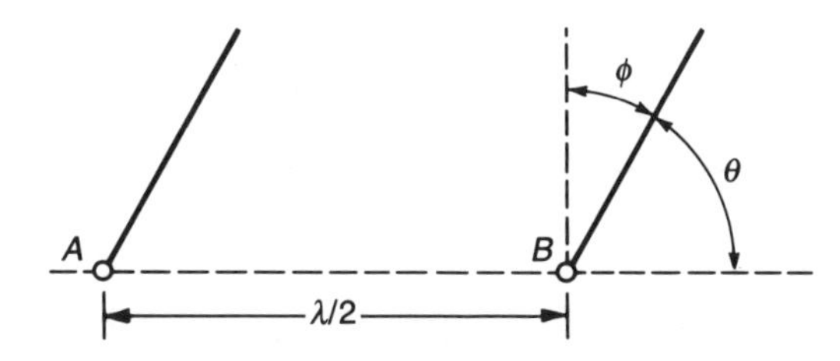

Antenna currents, I_A and I_B are equal, and I_A leads I_B by 90^0. Maximum field strength of each individually excited antenna is 250 mv/m at a distance of 40 kM. Radiation resistance of each antenna is 73.1Ω (the same as a half-dipole).

REQUIREMENT

Determine the angle, Θ, at which the resultant field strength is maximum, and calculate the field strength of the array for Θ = 90^0 at a distance of 25 kM.

SOLUTION

The rms value of the resultant electric field is given by the formula:

$$\mathcal{E}_r = \mathcal{E}_{rms}\cos\left(\pi n\sin\phi + \frac{\delta}{2}\right)$$

where

δ is the phase angle between I_A and I_B , and is positive when I_A leads I_B

n is the distance between A and B in wavelengths, and is usually fractional

Angle ϕ is measured from the normal to the line of the antennas

Thus,

$$\mathcal{E}_r = \mathcal{E}_{rms}\cos\left(\frac{\pi}{2}\sin\phi + \frac{\pi}{4}\right)$$

For maximum $\mathcal{E}_r$,

$$\cos\left(\frac{\pi}{2}\sin\phi + \frac{\pi}{4}\right) = 1$$

or

$$\left(\frac{\pi}{2}\sin\phi + \frac{\pi}{4}\right) = 0$$

$$\sin\phi = 0.5\,,\quad \phi = 30°$$

$$\therefore\ \Theta = 60° \quad \text{ANS.}$$

At Θ = 90°,

$$\cos\left(\frac{\pi}{2}\sin 0° + \frac{\pi}{4}\right) = \cos\frac{\pi}{4} = 0.707$$

The field strength of the array at 25 kM is:

$$\mathcal{E}_i = 0.707\times 250\frac{mv}{m}\left[\frac{40}{25}\right]^2 = 452.5\ \text{mv/m} \qquad \text{ANS.}$$

COMMUNICATIONS PROB. 9.7

SITUATION

In the modulating circuit sketched below, the modulating signal (see spectrum sketch) is limited to angular frequencies ω where

$$\omega_{m_1} < \omega < \omega_{m_2} <<<< \omega_c$$

Where ω_m = modulating frequency; and ω_c = carrier frequency.

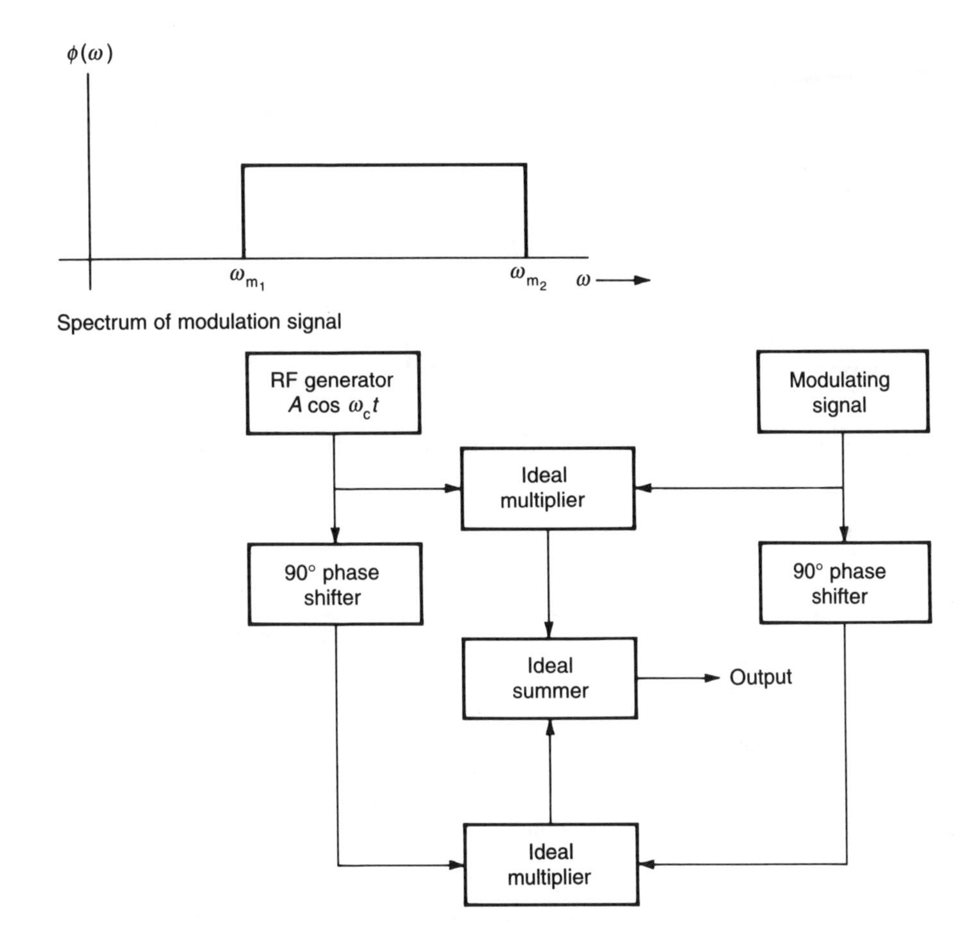

REQUIRED

(a) Sketch the spectrum of the output.
(b) What is this type of modulated signal called?
(c) Sketch another circuit which would produce the same output spectrum.

SOLUTION

(a) The in-phase input is as follows:

$$(A\cos\omega_C t)(\phi\cos\omega_m t) =$$
$$\frac{A\phi}{2}\left[\cos(\omega_C - \omega_m)t - \cos(\omega_C + \omega_m)t\right]$$

The 90° phase shift is as follows:

$$(A\sin\omega_C t)(\phi\sin\omega_m t) =$$
$$\frac{A\phi}{2}\left[\cos(\omega_C - \omega_m)t + \cos(\omega_C + \omega_m)t\right]$$

The sum of above two expressions is:

$$A\phi\cos(\omega_C - \omega_m)t$$

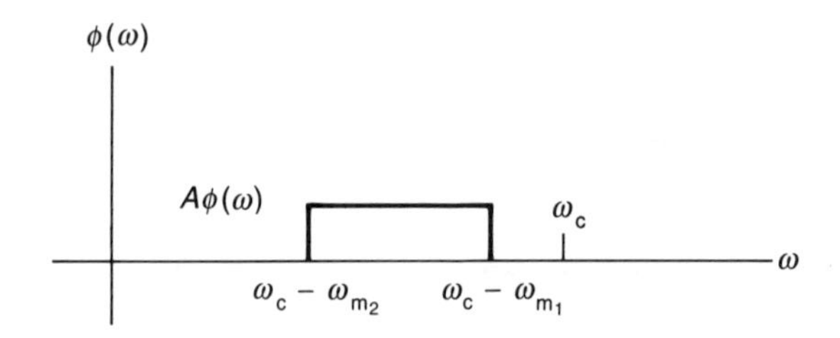

(b) This is a single side band AM modulated signal.

(c) A balanced modulator and sideband filter would produce the same output spectrum:

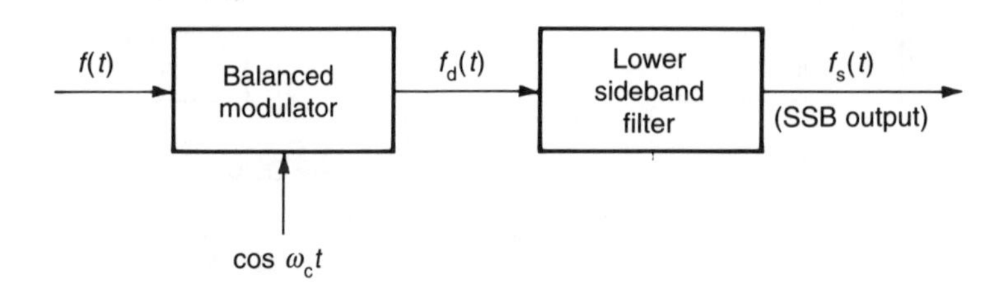

Reference: " Information Transmission, Modulation and Noise" by Schwartz, McGraw-Hill p. 106,107.

MODULATION PROB. 9.8

SITUATION

A certain transmitter has an effective radiated power of 9 kW with the carrier unmodulated and 10.125 kW when the carrier is modulated by a sinusoidal signal.

REQUIRED

(a) Determine the percent modulation at 10.125 kW output.
(b) Determine the total effective radiated power if in addition to the sinusoidal signal, the carrier is simultaneously modulated 40% by an audio wave.

SOLUTION

(a) Assume transmission is trough a 1 ohm resistance. For the unmodulated carrier with carrier frequency ω_0:

$$v_0(t) = A\cos\omega_0 t$$

$$P = \int v_0(t) = \frac{A^2}{2} = 9 \text{ kW, since:}$$

$$P = \frac{1}{2\pi}\int_0^{2} \frac{v^2(t)}{R}dt = \frac{A^2}{2\pi}\int_0^{2\pi} \cos^2 \omega_0 t dt$$

$$= \frac{A^2}{2\pi}\int_0^{2} \frac{\cos 2\omega_0 t + 1}{2}dt = \frac{A^2}{2\pi}\cdot\frac{2\pi}{2} = \frac{A^2}{2}$$

Then, for the modulated carrier with modulating frequency ω_1, and modulating coefficient "m", we obtain:

$$v_0(t) = A(1 + m\cos\omega_1 t)\cos\omega_0 t$$

$$= A\cos\omega_0 t + \frac{Am}{2}\cos(\omega_0 + \omega_1)t + \frac{Am}{2}\cos(\omega_0 - \omega_1)t$$

since a basic trigonometric relation states:

$\cos\alpha \cos\beta = 1/2 \cos(\alpha + \beta) + 1/2 \cos(\alpha - \beta)$

Therefore: $P = \dfrac{A^2 + \left(\dfrac{Am}{2}\right)^2 + \left(\dfrac{Am}{2}\right)^2}{2} = 9\left[\left(1 + 2\dfrac{m}{2}\right)^2\right] = 10.125$

$\dfrac{m}{2} = \sqrt{\dfrac{1.125}{18}} = 0.25$ and m = 0.5 , or 50% ANS.

(b) Denoting the audio wave frequency with ω_2, we obtain:

$$v_0(t) = A(1 + 0.5\cos\omega_1 t + 0.4\cos\omega_2 t)\cos\omega_0 t$$

$$= A\cos\omega_0 t + \frac{0.5A}{2}\cos(\omega_0 + \omega_1)t + \frac{0.5A}{2}\cos(\omega_0 - \omega_1)t$$

$$+ \frac{0.4A}{2}\cos(\omega_0 + \omega_2)t + \frac{0.4A}{2}\cos(\omega_0 - \omega_2)t$$

Then the effective radiated power is:

$$P = \frac{A^2 + 2\left(\frac{0.5A}{2}\right)^2 + \left(\frac{0.4A}{2}\right)^2}{2} = 10.125 + 9\left[2\left(\frac{0.4}{2}\right)^2\right]$$

$= 10.125 + 0.72 = 10.845$ kW ANS.

The frequency spectrum is as follows:

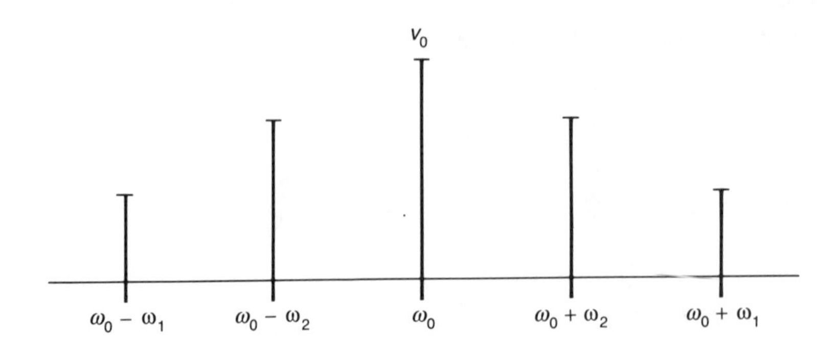

INFORMATION THEORY PROB. 9.9

SITUATION

In order to specify the communications link for a closed circuit television system, the bit rate must be known.

The monochrome television picture signal of this system requires 10 distinct levels brightness for good resolution. This television system also includes the following parameters:

(1) Frame rate, 15 frames per second
(2) Lines per frame, 1200
(3) Discrete picture elements, 100 per line

REQUIRED

Determine the channel capacity in bits per second required to transmit the above signal with all levels equally probable and with all elements assumed to vary independently.

List any assumptions that you make.

SOLUTION

The number of different possible pictures is:

$$P = 10 \times 10 \ldots \times 10 = 10^{1200 \times 100} = 10^{1.2 \times 10^5}$$

Probability of each element or picture is:

$$= \frac{1}{10^{1.2 \times 10^5}}$$

The Channel Capacity is defined as:

$$C = \lim_{T \to \infty} \frac{1}{T} \log_2 M(T)$$ where M(T) is the total number of messages in T seconds.

$$= S \log_2 P \quad \text{bit / sec}$$ where S is the signalling speed.

$$S = 15$$

$$C = 15 \log_2 10^{1.2 \times 10^5}$$

$$= 1.8 \times 10^6 \log_2 10$$

$$= 6.0 \times 10^6 \text{ bit / sec}$$

Assumption: It is assumed that the signal to noise ratio is large or the error probability is small so the channel is deemed noiseless.

COMMUNICATIONS PROB. 9.10

SITUATION

A series of remote stations is being planned to feed data to a large computer. The data are to be sent by the remote stations and recorded on a tape recording unit and then fed to the computer as necessary.

The data are expected to arrive from the remote stations with a Poisson distribution at an average rate of 10 transmissions from remote stations per hour. The recording time of the data varies exponentially, with a mean time of four minutes.

REQUIRED

(a) What is the average waiting time for a remote station before the data will begin to record?

(b) A second tape unit including automatic switching equipment is available at a cost of $2.50 per hour. The telephone lines cost 4 cents a minute per line when used. Is the second unit economically warranted? Show sufficient calculations to justify your answer.

SOLUTION

(a) This is the single stage, single server queuing model with Poisson arrivals and exponential service. Any basic operations research text gives the desired queue equations (e.g. Sasieni, Yaspan, and Friedman: <u>Operations Research - Methods and</u>

Problems, John Wiley, p126-138).

mean arrival rate λ = 10 transmissions/hour
mean service time = 1/μ = 4 minutes = 1/15 hour
mean service rate μ = 15 recordings/hour.

Average waiting time of an arrival

$$E(w) = \frac{\lambda}{\mu(\mu-\lambda)} = \frac{10}{15(15-10)} = \frac{10}{75} \text{ hour } = 8 \text{ minutes}$$

(b) This is the single stage, two server queuing situation. The various expectations for this model may also be obtained from a basic operations research (or queuing) text.

Before proceeding, a quick check can be made. We know a second recorder will substantially reduce (but not eliminate) waiting time. If the elimination of waiting time would not justify the second recorder then we need not bother to make the exact computation. Instead, we could simply conclude the second recorder is not economically warranted.

Hourly saving (assuming elimination of waiting time)
= mean arrival rate x mean waiting time reduction x line charge
= $\lambda[E(w_1) - E(w_2)]$ x 0.04 = (10)(8-0)(0.04) = \$3.20

Hourly cost = \$2.50

Thus we have been unable to show that the second recorder is uneconomical at zero waiting time. We must, therefore, proceed to compute the expectation of average waiting time.

$$P_0 = \frac{1}{\left[\sum_{n=0}^{k-1} \frac{1}{n!}\left(\frac{\lambda}{\mu}\right)^n\right] + \frac{1}{k!}\left(\frac{\lambda}{\mu}\right)^k \frac{k\mu}{k\mu - 1}}$$

For two service facilities:

$$k = 2 \qquad \lambda = 10 \qquad \mu = 15$$

$$P_0 = \frac{1}{\frac{\lambda}{\mu} + \frac{1}{2}\left(\frac{\lambda}{\mu}\right)^2 \frac{2\mu}{2\mu - \lambda}} = \frac{1}{\frac{10}{15} + \frac{100}{450} \cdot \frac{30}{20}} = \frac{1}{1} = 1$$

Average waiting time of an arrival

$$E(w) = \frac{\mu\left(\frac{\lambda}{\mu}\right)^k}{(k-1)!(k\mu-\lambda)^2} P_0 = \frac{15\cdot\left(\frac{10}{15}\right)^2}{(30-10)^2}(1) = \frac{1}{60} \text{ hour}$$

$= 1$ minute.

hourly saving of second recorder $= \lambda\left[\mathrm{E}(\mathrm{w}_1) - E(w_2)\right] \times 0.04$

$= (10)(8.0 - 1.0)(0.04) = \$2.80$

Thus the second recorder is economically warranted, for the hourly savings exceeds the hourly cost.

This page intentionally left blank.

PROBLEMS AND SOLUTIONS

BIOMEDICAL PROB. 10.1.

SITUATION

The output of a pair of Electrocardiograph skin electrodes is to be amplified with the diagramed differential instrumentation amplifier. The minimum peak differential signal amplitude expected from the electrodes is 0.5mV. The maximum peak differential signal amplitude expected from the electrodes is 4.0 mV. The minimum peak signal amplitude desired from the differential amplifier output is 0.5V.

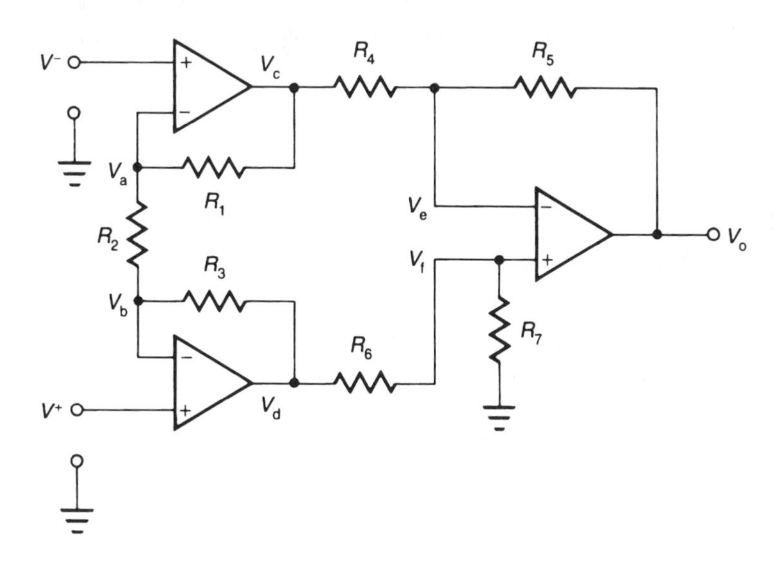

The gain of the output amplifier stage from V_c and V_d to V_o is desired to be 40 times the gain of the input amplifier state from V_- and V_+ to V_c and V_d. The differential gain of the entire amplifier (G_D) is the differential gain of the output stage (G_{Dout}) times the differential gain of the input stage (G_{Din}), or $G_D = G_{Dout} \bullet G_{Din}$.

REQUIRED

(a) If R_2 is 100kΩ, R_4 is 10kΩ, and R_6 is 10kΩ, what are the values of R_1, R_3, R_5, and R_7? Assume ideal op-amps.

(b) If the op-amps are capable of producing an output voltage within 1 V of the power supply voltage, what should the power supply voltage be so that the maximum peak signal amplitude from the differential amplifier output is not clipped:

SOLUTION

(a) $G_D = G_{Dout} \bullet G_{Din} = 1000$, $G_{Dout} = 40 \bullet G_{Din} = 5$ and $G_{Dout} = 200$. Thus, R_1 =200kΩ, R_3 = 200kΩ, R_5 = 2MΩ, and R_7 = 2MΩ.

(b) The maximum peak signal amplitude from the differential amplifier output is 4.0V. The minimum power supply voltage should be 5.0V.

BIOMEDICAL PROB. 10.2

SITUATION

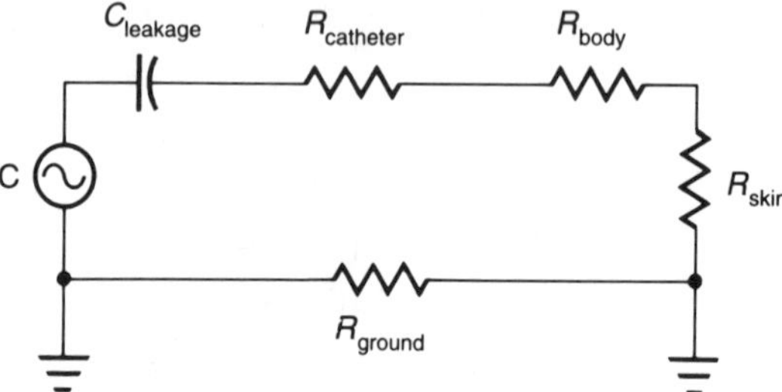

For the differential instrumentation amplifier diagramed in the problem 10.1 above:

REQUIRED

(a)What is the best CMRR that can be achieved using ±10% tolerance resistors, if all seven resistors were desired to be identical? Assume ideal op-amps.
(b): What is the minimum resistor tolerance necessary to achieve a CMRR of at least 100 dB? Assume ideal op-amps.

SOLUTION

(a)With R1, R3, R4, and R7 too low by 10%, and R2, R5, and R6 too high by 10%, the worst case CMRR is about 14.5 =23.2 dB.

(b) With R1, R3, R4, and R7 too low by 0.0015 %, and R2, R5, and R6 too high by 0.0015%, the CMRR is about 100,000 = 100 dB.

BIOMEDICAL PROB.10.3.

SITUATION

For the microshock circuit diagramed in problem 10.2 above:

REQUIRED

(a) If an isolation amplifier is inserted between the pressure transducer and the power supply, what is the minimum value of the series isolation resistance necessary to reduce the leakage current to one tenth of the maximum safe current limit?
(b) The engineer installing the isolation amplifier is concerned about saline solution from the catheter being spilled upon the isolation amplifier circuitry and thus bridging the isolation protection with a low impedance leakage path. The isolation

amplifier is 10 mm wide, and 20 mm long from the input terminals to the output terminals.

A column of saline solution has a cross-sectional resistance per length of about 200 Ω mm ^{2}/mm for each mm of column length per each mm^2 of column surface area. Thus, the total resistance of a saline column is (200 Ω mm)• (saline column length in mm)/(saline column cross-sectional area in mm^2). How deep would a layer of saline solution spanning the isolation amplifier have to be for the leakage current to reach the maximum safe microshock current limit?

(c) The engineer installing the isolation amplifier decides to encapsulate the amplifier with a plastic that has a primarily dielectric insulation characteristic. The leakage current through the effective encapsulation capacitor between the input and output of the isolation amplifier is to be no greater that the leakage current through the isolation amplifier. What is the maximum value of the encapsulation capacitor?

SOLUTION

(a) About 120 MΩ of isolation resistance will reduce the microshock current to about 120 V/120 MΩ = 1.0μA.

(b) The total leakage current through the saline leakage path in parallel with the isolation amplifier would be 10 μA. The leakage current through the isolation amplifier is 1.0 μA. The leakage current through the saline would be 9.0 μA. The total resistance of the saline leakage path would be 120 V/ 9.0 μA ≈ 13.3 MΩ . The total resistance of the saline leakage path would be 13.3 MΩ = (200 Ω mm)•(saline column length in mm)/(saline column cross-sectional area in mm2). The length of the saline column is 20 mm. The cross-sectional area of the saline column would be (200 Ω mm) • (20 mm)/ (13.3 MΩ) = 300 •10^{-6} mm^2. The width of the saline column is 10 mm. The depth of the saline column would be 300•10^{-6} mm^2 / 10 mm = 30 •10^{-6} mm = 30 nm.

(c) The minimum impedance of the encapsulation capacitor is 120 MΩ = 1/(2•π•60 Hz•C), so C≈ 22 pF.

Additional Problems:

In the past, most of the biomedical systems problems on the examination have been some form of an electronics problem that dealt with some aspect of biomedical measurements. Thus, for more study problems, one could take several problems from chapter 5 and add an op-amp detecting circuit of some kind, and proceed to the solution.

For example: from the table on page 10-3, a particular voltage measurement level could be determined, then adapt the op-amp portion of problem 5.1 to form a question that might include a voltage, current, or power output. The solution then would be straight forward as demonstrated (for a particular resistive load) in the solution for problem 5.1. The same thing may be said for problems 5.3, 5.7, 5.8, and 5.10.

PRINCIPLES AND PRACTICE OF ENGINEERING (PE)

Electrical Engineering Sample Exam

James H. Bentley, P.E.

INTRODUCTION

Principles and Practice of Engineering (PE) examinations are prepared by the National Council of Examiners for Engineering and Surveying (NCEES). All fifty states and five jurisdictions are represented by membership on the NCEES. Two times a year (spring and fall) the NCEES provides the state and jurisdiction licensing boards with examinations for their administration. All candidates for licensure as professional engineers must pass this uniform examination as one of the requirements for registration. Most jurisdictions also screen candidates based upon education, experience and other professional aspects. The board in the candidate's jurisdiction will provide information on requirements for professional registration as well as date and location for the next examination. The candidate should allow ample time in processing the application.

Problems, solutions and other information contained in this publication are intended to assist the candidate in preparing for the Electrical Engineering PE examination. Electrical engineering discipline and style of problems contained herein parallel those of recent exams. Be aware that the PE exam is different each time and the NCEES makes no guarantee that the current style and problem content will carry forward to future exams. The style has gradually and continually evolved into a more structured exam that leaves less latitude in problem solution. This makes the exam fairer and easier to grade. Recent exams more clearly test the knowledge of the candidate.

PE exams are prepared by committees comprised of professional engineers representing a variety of backgrounds -- private consulting, industry, government and education. While based on an understanding of basic engineering fundamentals, the problems require application of professional judgement and practical experience. Problems will be of such range as to require a variety of approaches and methods including analysis, design, application, economic considerations and operations.

EXAM STYLE

Currently, the eight-hour PE exam contains essay or free-response type questions in the four-hour a.m. session and multiple choice questions in the four-hour p.m. session. Multiple-part problems are designed so that the solution of one part does not depend upon the correct solution of a previous part. Four of twelve problems must be worked during each session.

Essay problems are presented as a "Situation" and "Requirements". The situation describes the environment in which an engineer might find himself when presented with a practical problem soon after becoming registered. The requirements define what is expected in response to the problem -- usually calculations and/or conclusions which demonstrate good engineering judgement. Partial credit is given to essay type problems. For selected problems the grader is instructed within narrow guidelines by a scoring plan as to how partial credit is to be awarded.

Multiple-choice problems consist of ten questions, each having five possible answers. An engineering scenario is presented at the start of each multiple-choice problem.

Probably the most difficult aspect of preparing and administering the PE exam is establishing a passing score which reflects a standard of minimum competency. Each jurisdiction is free to establish its minimum competency level for licensure. As an aid, the NCEES defines minimum competency for the examination component of the licensing process as follows:

> "The lowest level of knowledge at which a person can practice professional engineering in such a manner that will safeguard life, health and property and promote the public welfare."

EXAM SCORING

The PE exam is worth a maximum score of 80 points (eight problems at ten points each). By a scoring plan for essay problems, partial credit is given that results in an even-numbered score between 0 and 10 points. The problems and scoring are designed with priority given to defining what type of response will receive six points. The scoring plan follows this point format: 10 points (exceptional competence -- not necessarily perfect solution); 8 points (more than minimum but less than exceptional competence); 6 points (minimum competence -- incomplete solution); 4 points (more than rudimentary knowledge but insufficient to demonstrate competence); 2 points (errors in answers due to lack of understanding); 0 points (no indication of any knowledge of the problem or how to solve it).

Each of the ten questions in a multiple-choice problem is worth one point, with no penalty for wrong answers.

The NCEES recommends a minimum passing score of 48 points, or an average of six points per problem. Problems yielding a score of less than six may be compensated for by higher scores on other problems. Ultimately, the authority for deciding on licensure of a particular candidate rests with the individual registration board. The NCEES can only recommend a passing score based upon what it considers to be minimum competency.

EXAM PROCESS

The PE exam is administered in conformance with a rigid set of rules. The proctor will explain the process before the morning session begins. Briefly, the examinee can expect to be confronted with the following:

- The examination booklet will include a solution pamphlet for the a.m. essay questions and a machine-scoreable answer sheet for the p.m. multiple-questions.

- Solutions to the essay questions must be recorded in the solution pamphlet. All work to be considered for credit must be entered on right-facing pages. Left-facing pages are to be used for scratch work and will not be graded. The upper right-hand corner of each right-facing page contains space for Board Code, Examinee ID No., Problem No. and number of pages used in the solution.

- The front page of the solution pamphlet contains space for the examinee's name and the numbers of the problems that are to be scored. The candidate's solution pamphlet may contain more than four problem solutions, or partial solutions. Only the four to be scored must be identified on the front cover.

- The multiple-choice answer sheet is machine scored and requires the use of a #2 pencil, so several of these must be taken to the exam. If an answer is to be changed, the first answer must be completely erased so as not to be incorrectly graded. A good eraser is also a necessity.

- If the examinee completes the exam with more than 30 minutes remaining, s/he is free to leave. If there are fewer than 30 minutes remaining, s/he must stay until the end to avoid disrupting those who are still there. In any case, one may not leave without permission of the proctor.

- Before the day of the exam, the board will identify exactly what references and type of calculators may be used. In general, text books, handbooks, bound reference materials and battery-operated, silent, non-printing calculators may be brought to the exam. Writing tablets, loose papers and published problem solution manuals are generally not allowed.

- Copying problems for future reference is not permitted.

- An examinee having a disability should make this known to the board well in advance of the exam so that necessary arrangements can be made.

FUTURE EXAMS

It is anticipated that within five years, the examination will be computerized so that the examinee may answer just enough questions to accumulate a passing score, and then stop. In this type of exam, questions will start out at a high level of difficulty and gradually become easier. Thus, if a person can accumulate enough points early on, it is probable s/he will be able to answer subsequent problems with ease. To continue would merely be an unnecessary exercise, since his/her competence would already have been demonstrated. In any case, the content (if not the style) of the examination will not be unlike that of today's exam.

As technology evolves, there will be a gradual and steady change in the electrical engineering disciplines covered in the exam.

EXAM SPECS

The eight-hour exam contains a total of 24 problems that fall into 18 different categories. The boundary between categories is sometimes fuzzy. To one person, a problem appearing in one category might seem to another person to better fit into another category. However, all the problems here are similar to those that have appeared in recent past exams, so they are representative.

Following is a list of the 18 categories, with the quantity (1 or 2) of problems in that subject that will appear in the exam. The problem categories are defined and assigned by the NCEES. Each of the 24 problems in the 18 categories will appear either in the a.m. or p.m. session, with a total of 12 problems each session. Where there are two problems of the same category indicated, one will appear in the a.m. and the other in the p.m. In the chart below, "am/pm" means the problem will be in either a.m. OR p.m., with a.m. the preferred session; "pm/am" means p.m. is the preferred session. "Both" means that category is represented by a problem in both the a.m. and p.m. sessions.

In the actual PE exam, problems in the examination booklet for electrical engineering are numbered as follows:

for a.m., 130 thru 133, 290 thru 297
for p.m., 430 thru 433, 590 thru 597

This is the numbering system used in the sample exam to follow.

CAT.	DESCRIPTION	QTY	SES'N
1.	GENERATION SYSTEMS, FUNDAMENTAL DESIGN large scale power plants	1	am/pm
2.	GENERATION SYSTEMS, FINAL DESIGN AND APPLICATIONS same as #1 above	1	pm/am
3.	TRANSMISSION/DISTRIB. SYSTEMS, FUNDAMENTAL DESIGN transformers, protection, safety, overhead and underground lines, metering, batteries, relays, substations, circuit protectors, intrinsic safety devices	1	am/pm
4.	TRANSMISSION/DISTRIBUTION SYSTEMS, FINAL DESIGN AND APPLICATIONS same as #3 above	2	both
5.	ROTATING MACHINES, FINAL DESIGN motors, generators, transformers	1	pm/am
6.	INSTRUMENTATION, FINAL DESIGN instrument transformers, metering systems, measurement systems, transducers, test procedures	1	am/pm
7.	LIGHTING PROTECTION AND GROUNDING, FINAL DESIGN lightning protection, physical grounding of equipment and structures	1	pm/am
8.	CONTROL SYSTEMS DESIGN industrial process and operations control, feedback controls, transient theory	2	both
9.	ELECTRONIC DEVICES, DESIGN digital storage devices, integrated circuit components, op amps	2	both
10.	ELECTRONIC DEVICES, APPLICATIONS same as #9 above	1	am/pm
11.	INSTRUMENTATION, DESIGN same as #6 above	1	pm/am
12.	INSTRUMENTATION, APPLICATIONS same as #6 above	1	am/pm
13.	DIGITAL SYSTEMS, DESIGN digital systems including interfaces, protocols, standards	2	both
14.	COMPUTER SYSTEMS, DESIGN digital and analog systems	2	both
15.	DIGITAL SYSTEMS, APPLICATIONS	1	am/pm

	same as #14 above		
16.	COMMUNICATIONS SYSTEMS, DESIGN broadcast/voice/data communications, fiber optics antennas, microwave systems, HF transmission lines	2	both
17.	COMMUNICATIONS SYSTEMS, APPLICATIONS same as #16 above	1	pm/am
18.	BIOMEDICAL SYSTEMS DESIGN electrical applications in living systems	1	pm/am

SAMPLE PROBLEMS AND SOLUTIONS

FOR THE MORNING SESSION

OF THE EXAMINATION IN

ELECTRICAL ENGINEERING

BEGIN ON THE NEXT PAGE

130 (Category 1)

SITUATION:

A power distribution system consists of a 4800-volt, 3-phase bus supplied from an electric utility system having a 3-phase short-circuit capacity of 20MVA through a transformer whose reactance is 12% on a 5MVA base.

REQUIREMENTS:

In your solution pamphlet, neatly sketch a diagram of the distribution system. Calculate the percent voltage rise obtained by switching in a 3-phase bank of 1000 kVAR shunt capacitors.

SOLUTION:

A simplified drawing of the distribution is shown below.

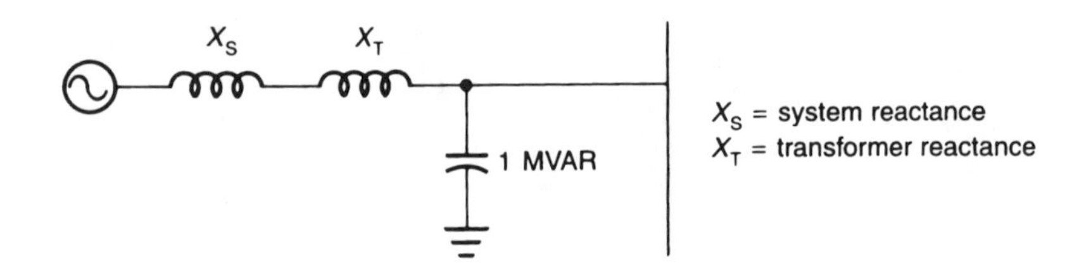

Choose short circuit capacity as BASE MVA = 20 MVA

$$\% X_S = \frac{\text{BASE MVA}}{\text{S.C. MVA}} \times 100\% = \frac{20}{20} \times 100 = 100\%$$

$$\% X_T = \frac{\text{BASE MVA}}{\text{XFMR MVA BASE}} \times 12\% = \frac{20}{5} \times 12\% = 48\%$$

$$\% X_{total} = 100\% + 48\% = 148\%$$

$$\text{Voltage Rise} = (\text{pu I})(\%X_{total})$$

$$\text{pu I} = \frac{\text{LOAD MVA}}{\text{BASE MVA}} = \frac{1}{20} = .05$$

•ANSWER Voltage Rise = (.05)(148%) = **7.4%**

131 (Category 3)

SITUATION:

A factory wishes to improve the power factor of its existing 2000kW load while adding more motor capacity. If the added load is obtained from overexcited synchronous motors, it can increase the power factor while at the same time increasing the factory motor capacity.

The existing load (induction motors and lighting) is 2000kW at a lagging power factor of .83. It is desired to improve the power factor to .94 with the added synchronous motors.

REQUIREMENTS:

(a) Calculate the kVA input to the synchronous motors if they have a leading power factor of .8.
(b) Calculate the resulting final load kVA.
(c) Calculate synchronous motor power output if efficiency is 90%.

SOLUTION:

Use subscripts L for existing load, M for added synchronous motors and F for resultant final load. Use P for real power (kW), Q for quadrature (apparent or reactive) power (kVAR) and S for apparent power (kVA). Power triangles for the existing load, the added motor load and the final resulting load are shown below:

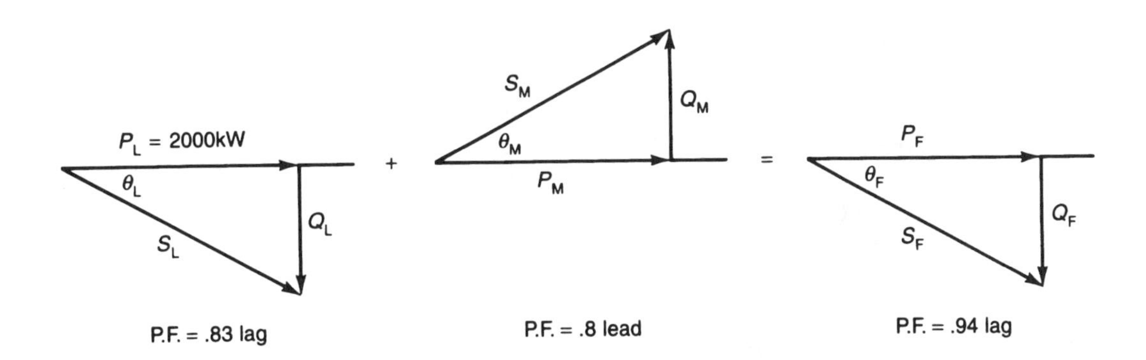

The problem is to calculate S_M, synchronous motor apparent power. Writing two equations, one for reals and one for imaginaries:

(1) $2000 + P_M = P_F$
(2) $Q_L - Q_M = Q_F$

131 (continued)

SOLUTION (cont.):

Calculating the existing load Θ_L, S_L and Q_L:

Θ_L = arc cos .83 = 33.9°
S_L = 2000/.83 = 2409.64 kVA
Q_L = 2409.64 sin 33.9° = 1344 kVAR

Calculating the synchronous motor and final load phase angles:

Θ_M = arc cos .8 = 36.87°
Θ_F = arc cos .94 = 19.95°

$\tan \Theta_M = Q_M/P_M$, $Q_M = P_M \tan 36.87° = 0.75P_P$
$\tan \Theta_F = Q_F/P_F$, $Q_F = P_F \tan 19.95° = 0.36P_F$

Substituting into equations (1) and (2):

(1) $2000 + P_M = P_F$
(2) $1344 - .75P_M = .36P_F$

Multiplying (1) by .75 and adding to (2):

$$
\begin{aligned}
1500 + .75P_M &= .75P_F \\
1344 - .75P_M &= .36P_F \\
\hline
2844 &= 1.11P_F
\end{aligned}
$$

P_F = 2844/1.11 = 2562 kW

From (1):

P_M - 2000 = 2562 - 2000 = 562 kW

Solving for Q_M and Q_F:

$Q_M = .75P_M$ = 422 kVAR
$Q_F = .36P_F$ = 922 kVAR

(a) and (b) Solving for S_M and S_F:

•ANSWER (a) $S_M = P_M/\cos \Theta_M$ = 562/.8 = **703 kVA**

•ANSWER (b) $S_F = P_F/\cos \Theta_F$ = 2562/.94 = **2726 kVA**

(c) Solving for synchronous motor output:

•ANSWER (c) P_{Mout} = 562 x .9 = **506 kW**

132 (Category 4)

SITUATION:

A partial one-line diagram of a power distribution system is shown below. Impedance values of the power system components are given for first cycle (subtransient) interrupting duty and for long term transient (eight cycles) interrupting duty for a three-phase fault at breaker F.

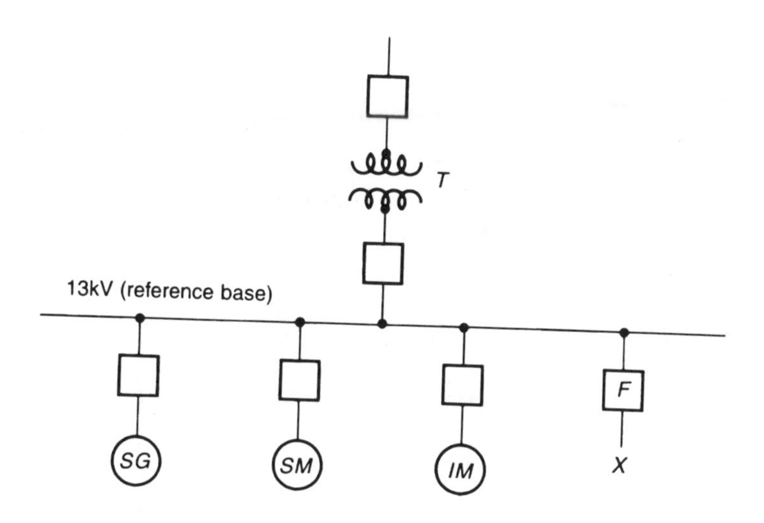

Assumptions:

- Ignore quadrature reactances.
- Motor efficiencies and p.f. are such that 1hp = 1kVA.
- Terminal voltage is 13kV on 13.9kV bus.
- Utility available short circuit is 3000MVA.
- A balanced 3-phase system with a 3-phase fault.
- Synchronous motor acts as synchronous generator into fault while it loses speed.
- kVA base = 10,000kVA = BASE

System component characteristics are:

transformer, T:	40,000kVA	X_T = 12%
sync gen., SG:	15,000kVA	X''_{dsg} = 10%, X'_{dsg} = 18%
sync motor, SM:	10,000hp	X''_{dsm} = 17%, X'_{dsm} = 30%
ind. motor, IM:	15,000hp	X'_{dim} = 28%

This problem is similar to a problem presented in ***Sample Problems and Solutions in Electrical Engineering,*** National Council of Examiners for Engineering and Surveying, 1996.

132 (continued)

REQUIREMENTS:

(a) For a three-phase fault at location F, draw two impedance diagrams, one for first cycle interrupting duty and one for eight-cycle interrupting duty.
(b) Determine the minimum momentary duty (rms amperes) for breaker F.
(c) Determine the minimum fault duty rating for breaker F.

SOLUTION:

Significant reactances during the initial fault condition are direct (subscript d), where line current lags voltage by 90°. Quadrature (subscript q) reactances are not important here. Reactances increase with increasing fault current from initial time zero. Primed values represent momentary initial conditions just after the fault (double primes), followed by the interrupting transient period (single primes).

(a) Calculating circuit values for impedance diagrams:

For available short circuit utility (3000MVA),

$$\text{pu } X_u = \text{BASE/sc MVA} = 10^4/3\text{x}10^6 = .0033$$

Transformer impedance,

$$\text{pu } X_T = (\text{BASE}/4\text{x}10^4) \text{ x } .12 = .03$$

1. Momentary generator/motor impedances are:

$$\text{pu } X''_{dsg} = (\text{BASE/kVA}) \text{ x } .1 = (10^4/1.5\text{x}10^4) \text{ x } .1 = .0667$$

$$\text{pu } X''_{dsm} = (\text{BASE/kVA}) \text{ x } .17 = (10^4/10^4) \text{ x } .17 = .17$$

$$\text{pu } X'_{dim} = (\text{BASE/kVA}) \text{ x } .28 = (10^4/1.5\text{x}10^4) \text{ x } .28 = .1867$$

Using these calculated circuit values, the momentary duty impedance diagram is **(ANSWER a1)**:

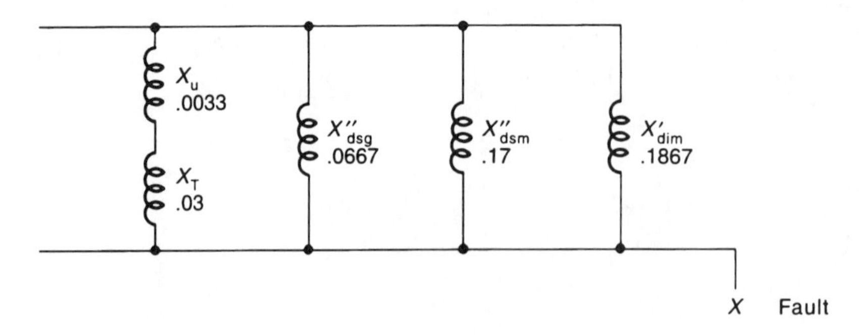

132 (continued)

SOLUTION (cont.):

2. Interrupting duty impedances are (the induction motor is no longer involved):

$$\text{pu } X'_{dsg} = (\text{BASE/kVA}) \times .18 = (10^4/1.5\times10^4) \times .18 = .12$$

$$\text{pu } X'_{dsm} = (\text{BASE/kVA}) \times .3 = (10^4/10^4) \times .3 = .3$$

Using the circuit values calculated above, the interrupting duty impedance diagram is **(ANSWER a2)**:

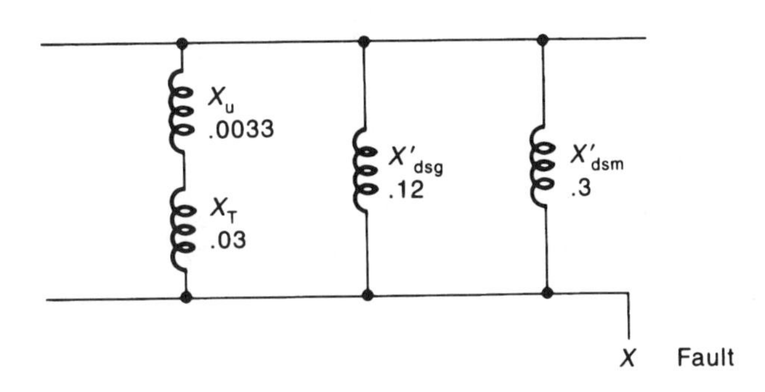

(b) Calculating the minimum momentary duty for breaker F:

The equivalent circuit impedance is the reciprocal of the sum of the reciprocals (parallel):

$$1/X = 1/(.0033+.03) + 1/.0667 + 1/.17 + 1/.1867 = 64.66$$

$$\text{pu } X = .0155, \quad \text{pu } I = 64.66\text{A}$$

$$I_{lbase} = \text{BASE VA}/\sqrt{3}\,V_l = 10^7/(\sqrt{3} \times 13.9\times10^3) = 415.35\text{A}$$

•ANSWER (b) $I_F = 415.36 \times 64.66 =$ **26.86kA**

(c) Calculating the minimum fault duty for breaker F:

$$1/X = 1/(.0033+.03) + 1/.12 + 1/.3 = 41.67$$

$$\text{pu } I = 41.67$$

•ANSWER (c) $I_F = 415.36 \times 41.67 =$ **17.31kA**

133 (Category 6)

SITUATION:

Two single-phase wattmeters are connected to a balanced three-phase, Y-connected load, as shown below. The three load impedances are, $Z = 4 + j3$, and the line voltage is 440 volts. Phase sequence is a-b-c as indicated in the phasor diagram below.

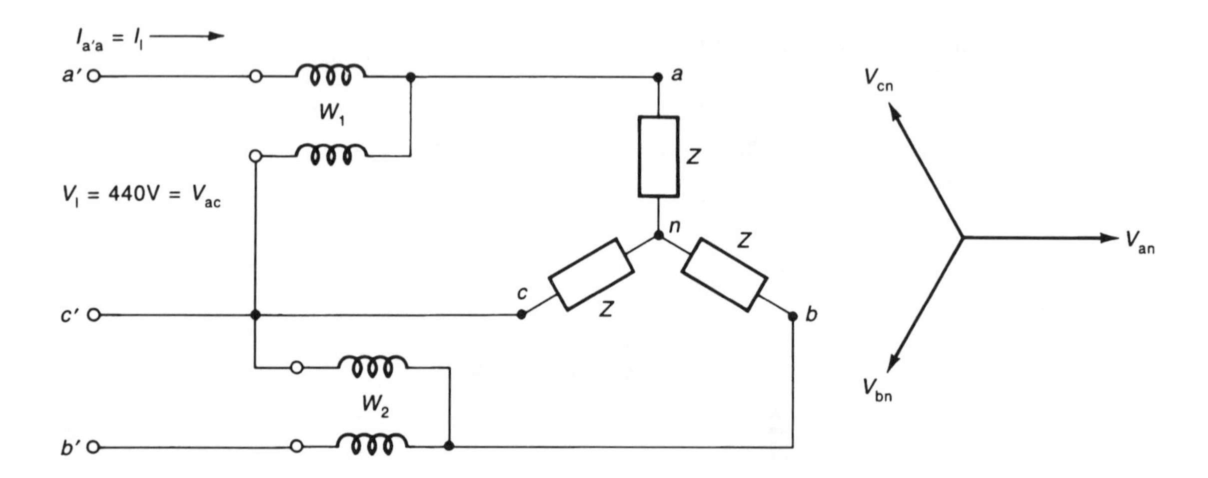

REQUIREMENTS:

(a) Calculate the line-to-neutral (phase) voltages.
(b) Calculate the line currents.
(c) Calculate total power dissipation of the load by two methods.
(d) Calculate the two wattmeter readings. Compare the sum of the wattmeter readings with the results of step (c).

SOLUTION:

(a) Calculating phase voltage, V_ϕ,

$$V_\phi = V_l / \sqrt{3} = 440 / \sqrt{3} = 254 \text{ volts}$$

•ANSWER (a) $\mathbf{V_\phi = 254\ V}$

(b) Calculating the line currents, I_l,

$$I_l = I_\phi = V_\phi / Z = 254/5\underline{/36.87°} = 50.8 \text{ amps}$$

•ANSWER (b) $\mathbf{I_l = 50.8\ A}$

133 (continued)

SOLUTION (cont.):

(c) Calculating total power dissipation of the load, P_{tot},

$P_{\phi} = I^2R = 50.8^2 \text{ x } 4 = 10{,}322.6$ watts

$P_{tot} = 3P_{\phi}$

OR an alternative way is by the formula:

$P_{tot} = \sqrt{3}\, V_l I_l \cos\Theta$

●ANSWERS (c)

$P_{tot} = 3 \text{ x } 10{,}322.6 =$ **30,968 W**

OR

$P_{tot} = 1.732 \text{ x } 440 \text{ x } 5.8 \text{ x } .8 =$ **30,972 W**

(d) Calculating the two individual wattmeter readings, the following equation may be used to represent power measured by wattmeters, W_1 and W_2:

$W_1 = W_{ac\text{-}a'a} = V_{ac}I_{a'a} \cos\Theta$,

$W_2 = W_{bc\text{-}b'b} = V_{bc}I_{b'b} \cos\Theta$,

where Θ is the angle between the line voltage and line current.

$V_{ac} = V_{an} + V_{nc} = 254\underline{/0°} + 254\underline{/-60°} = 440\underline{/-30°}$ volts

$I_{a'a} = V_{an}/Z_{an} = 254\underline{/0°}/5\underline{/36.87°} = 50.8\underline{/-36.87°}$ amps

$V_{bc} = V_{bn} + V_{nc} = 254\underline{/-120°} + 254\underline{/-60°} = 440\underline{/-90°}$ volts

$I_{b'b} = V_{bn}/Z_{bn} = 254\underline{/-120°}/5\underline{/36.87°} = 50.8\underline{/-156.87°}$ amps

● ANSWERS (d)

$W_1 = 440 \text{ x } 50.8 \cos 6.87° =$ **22,191.5 W**

$W_2 = 440 \text{ x } 50.8 \cos 66.87° =$ **8780.3 W**

CHECK:

$P_{tot} = W_1 + W_2 =$ **30,971.8 W**,

which compares closely with the previous calculations.

290 (Category 8)

SITUATION:

A servomechanism for controlling angular position by means of differentially connected potentiometers consists of a servo amplifier, a servomotor, and a feedback potentiometer whose transfer functions are as follows:

$G_a = 1/(s+3), \quad G_m = 1/(s+2), \quad H = K$

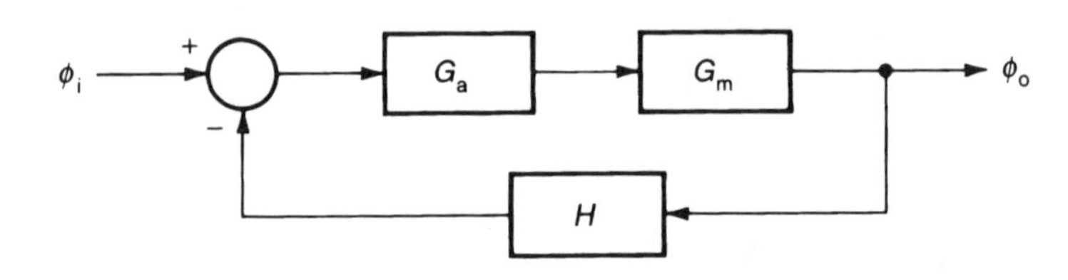

REQUIREMENTS:

Determine:
(a) The open loop transfer function.
(b) The system transfer function.
(c) The range of K for a stable system.

SOLUTION:

(a) Open loop transfer function = GH

•ANSWER (a) GH = **K/(s+3)(s+2)**

(b) System transfer function = G/(1+GH)

$$= \frac{\frac{1}{(s+3)(s+2)}}{1+\frac{K}{(s+3)(s+2)}} = \frac{1}{K+(s+3)(s+2)}$$

•ANSWER (b) $= \mathbf{\dfrac{1}{s^2+5s+(6+K)}}$

(c) For the system to be stable there must be no negative factors in the denominator. Therefore, (6+K) must always be positive. Hence,

$$(6+K) > 0$$

•ANSWER (c) **K > -6**

291 (Category 9)

SITUATION:

It is desired to buffer and amplify a low-level ac signal by using a common-emitter NPN transistor amplifier. The amplifier circuit is shown below, together with its small signal ac parameters.

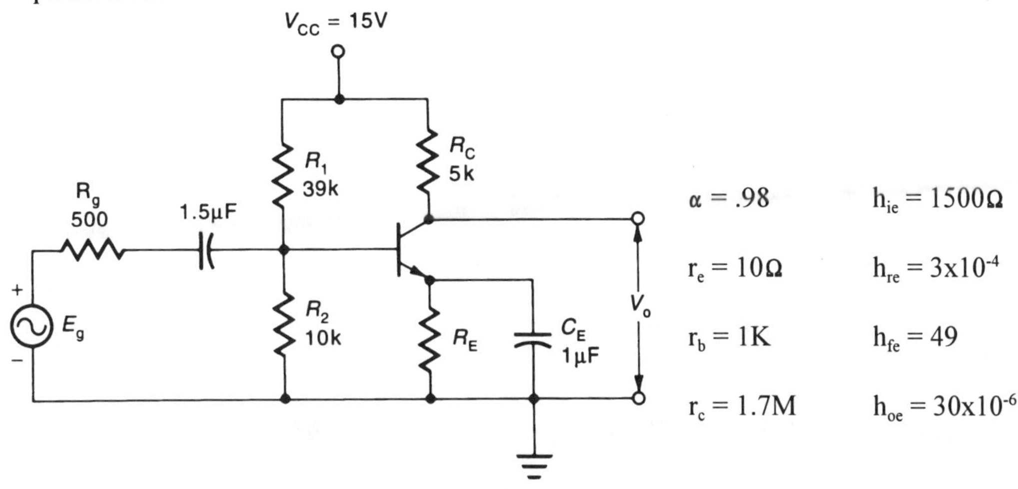

$\alpha = .98$ $\quad h_{ie} = 1500\Omega$

$r_e = 10\Omega$ $\quad h_{re} = 3\text{x}10^{-4}$

$r_b = 1\text{K}$ $\quad h_{fe} = 49$

$r_c = 1.7\text{M}$ $\quad h_{oe} = 30\text{x}10^{-6}$

REQUIREMENT:

Calculate the mid-frequency voltage gain, A_V, of the amplifier.

SOLUTION:

Convert the above common-emitter circuit to a hybrid-equivalent:

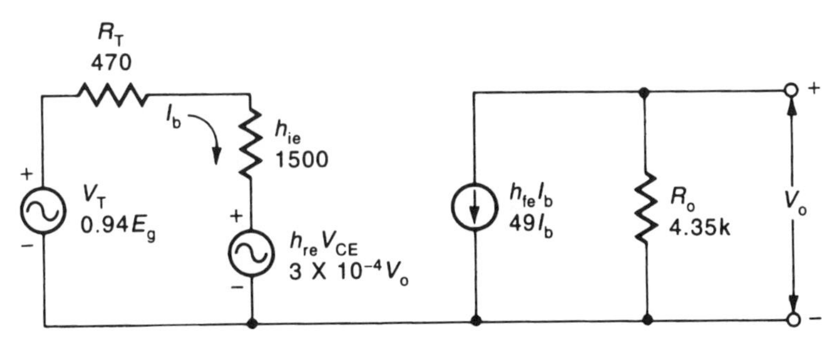

$$R_o = (1/h_{oe})\|R_C = (33\text{K})(5\text{K})/38\text{K} = 4.35\ \text{K}$$

$$V_T = \frac{R_1\|R_2E_g}{R_g + R_1\|R_2} = \frac{7959E_g}{8459} = .94E_g$$

$$R_T = (500)(7959)/8459 = 470\Omega$$

291 (continued)

SOLUTION (cont.)

From Kirchhoff's Laws:

$$V_o = -(4350)(49I_b)$$

$$.94E_g = (470+1500)I_b + 3x10^{-4}\ V_o$$

Solving for I_b:

$$.94E_g = 1970I_b - 3x10^{-4}\ V_o\ x\ 4350\ x\ 49I_b$$

$$= I_b(1970 - 64) = 1906I_b$$

$$I_b = (.94/1906)E_g$$

•ANSWER $A_V = V_o/E_g = -(4350)(49)/E_g\ x\ (.94E_g/1906) =$ **-105**

292 (Category 10)

SITUATION:

The schematic for a band-pass amplifier circuit using an NPN transistor is shown below. This circuit represents one stage of an R-C coupled amplifier.

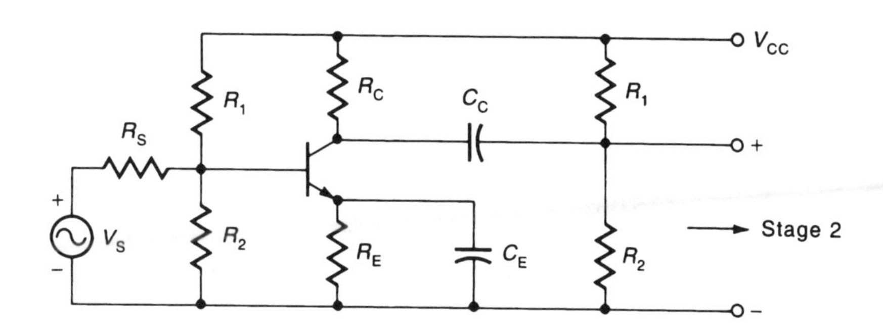

$h_{ie} = 1800\Omega$ $\quad R_B = R_1 \| R_2 = 4.7K$

$h_{re} = 3x10^{-4}$ $\quad R_E = 50\Omega$

$h_{fe} = 50$ $\quad C_p = 620pF$ (effective parallel capacitance)

$h_{oe} = 5x10^{-6}\mho$ $\quad R_s = 2200\Omega$

beta cutoff frequency, $f_B = 2.5MHz$

REQUIREMENTS:

(a) Calculate R_C to yield an upper cutoff frequency, f_2, of 300KHz.
(b) Determine the mid-frequency current gain.
(c) Determine the values of C_C and C_E to yield a lower cutoff frequency, f_1, of 200Hz.

SOLUTION:

(a) Since f_B is well above f_2, the output resistance becomes:

$$R_o = \frac{1}{2\pi f_2 C_p} = \frac{1}{2\pi \text{ x } 300 \text{ x } 10^3 \text{ x } 620 \text{ x } 10^{-12}} = 856\Omega$$

292 (continued)

SOLUTION (cont.)

$$G_o = \frac{1}{R_o} = \frac{1}{856} = h_{oe} + \frac{1}{R_C} + \frac{1}{R_{B2}} + \frac{1}{h_{ie2}}$$

$$= \frac{1}{R_C} + 5x10^{-6} + \frac{10^{-3}}{4.7} + \frac{1}{1800} = \frac{1}{R_C} + 7.73x10^{-4}$$

$$\frac{1}{R_C} = \frac{1}{856} - 7.73x10^{-4} = 3.95x10^{-4}$$

●ANSWER (a) R_C = **2530Ω**

(b) Determining mid-frequency current gain,

$$A_{Imid} = -h_{fe}R_o/h_{ie} = -50x856/1800$$

●ANSWER (b) A_{Imid} = **-23.78**

(c) Determining C_C and C_E,

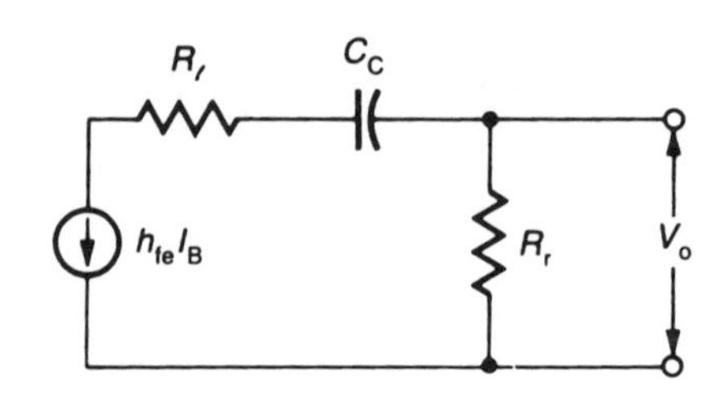

$$R_\ell = \frac{\frac{R_c}{h_{oe}}}{R_c + \frac{1}{h_{oe}}} = \frac{\frac{2530}{5} \times 10^{-6}}{2530 + \frac{1}{5 \times 10^{-5}}} = 2498.4\Omega$$

$$R_r = \frac{R_B h_{ie}}{R_B + h_{ie}} = \frac{4.7 \times 10^3 \times 1800}{4.7 \times 10^3 + 1800} = 1301.5\Omega$$

$$C_c = \frac{1}{2\pi f_1 (R_l + R_r)} = \frac{1}{2\pi \times 200(2498.4 + 1301.5)} = .21\mu F$$

●ANSWERS (c)

$$C_E = \frac{1 + h_{fe}}{2\pi f_1 R_s} = \frac{1 + 50}{2\pi \times 200 \times 2200} = 18.4\mu F$$

293 (Category 12)

SITUATION:

The value of an unknown capacitor is to be measured using a Schering bridge, shown below.

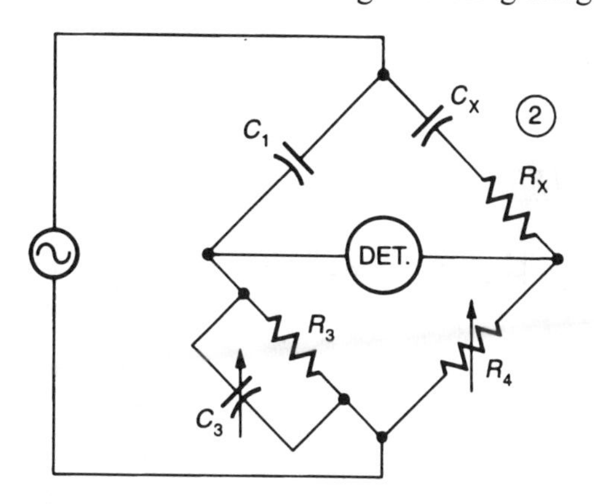

REQUIREMENTS:

The bridge capacitors and resistors are adjusted to the values delineated below so that the bridge is balanced (zero current through the detector). Calculate the value of the unknown capacitor and its series resistance.

$C_1 = 1000pF$ $C_3 = 10pF$ $R_3 = 100K\Omega$ $R_4 = 100\Omega$

SOLUTION:

For balanced conditions, $Z_1/Z_3 = Z_x/Z_4$ or $Z_x\ Z_4\ Y_3$

$Z_1 = \frac{-j}{\omega C_1}$ $Z_4 = R_4$ $Y_3 = \frac{1}{R_3} + j\omega C_3$

$Z_x = R_x - j/\omega C_x = [-j/\omega C_1][R_4][1/R_3 + j\omega C_3]$

Separating reals and imaginaries:

$R_x - j/\omega C_x = C_3R_4/C_1 - jR_4/\omega C_1R_3$

Evaluating reals and imaginaries:

$1/\omega C_x = R_4/\omega C_1R_3$ OR $C_x = C_1R_3/R_4 = 1\mu F$

$R_x = C_{3nR}4/C_1 = 1\Omega$

• ANSWER **$C_x = 1\mu F$ with internal series resistance of 1Ω.**

294 (Category 13)

SITUATION:

Because the microprocessor controlling a logic device is too slow, it is necessary to nest a high speed sequencer within the microprocessor loop.

REQUIREMENTS:

The sequencer has eight states. Two switches (numbered 1 and 2) can be disabled to cause jumps in the sequence. The state diagram is shown below.

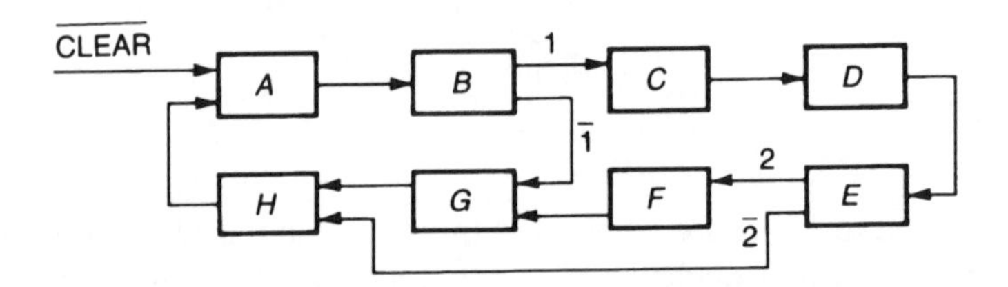

Assure that no timing race conditions can occur which would cause the state machine to fall out of sequence. The system is to be synchronously clocked, can be stopped at any state by turning off the clock and can be asynchronously cleared to state A at any time. Design the machine using J-K flip-flops (74107), a 3:8 line decoder (74138) and other TTL gates. Draw a Karnaugh map, develop the equations for the inputs to the J-K flip-flops and draw a logic diagram of the final design.

SOLUTION:

The Karnaugh map is drawn below with the eight states assigned so that only one flip-flop changes state at a time. Luckily, the two pairs of jump states are adjacent on a three-term Karnaugh map so that only three flip-flops are needed.

X \ YZ	00	01	11	10
0	A	B	C	D
1	H	G	F	E

294 (continued)

SOLUTION (cont.):

Next, a partial logic diagram is drawn with the three J-K flip flops labeled X (MSB), Y and Z (LSB).

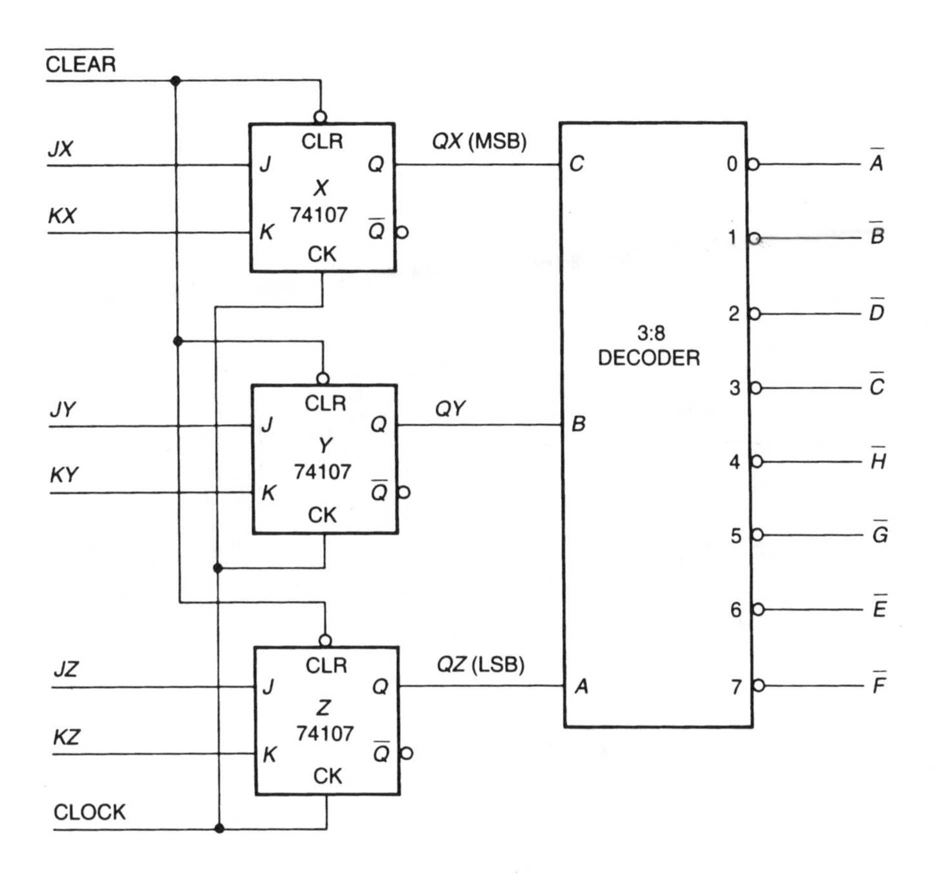

Now the six J-K input equations can be written. Each flip-flop term contains its present state if it is to change state at the next clock. A ONE on J causes the output to go HIGH and a ONE on K causes the output to go LOW, at the next clock time. An active LOW at the output of the decoder determines the state of the sequencer.

$$JX = D + B \bullet \overline{1} \qquad KX = H$$

$$JY = B \bullet 1 \qquad KY = F + E \bullet \overline{2}$$

$$JZ = A + E \bullet 2 \qquad KZ = C + G$$

As a check, count the number of terms, above (nine). This equals the number of arrows in the state diagram (not counting CLEAR).

294 (continued)

SOLUTION (cont.):

Finally, gating logic for the flip-flop inputs can be drawn. Gate inputs are obtained from the decoder outputs and the two switches.

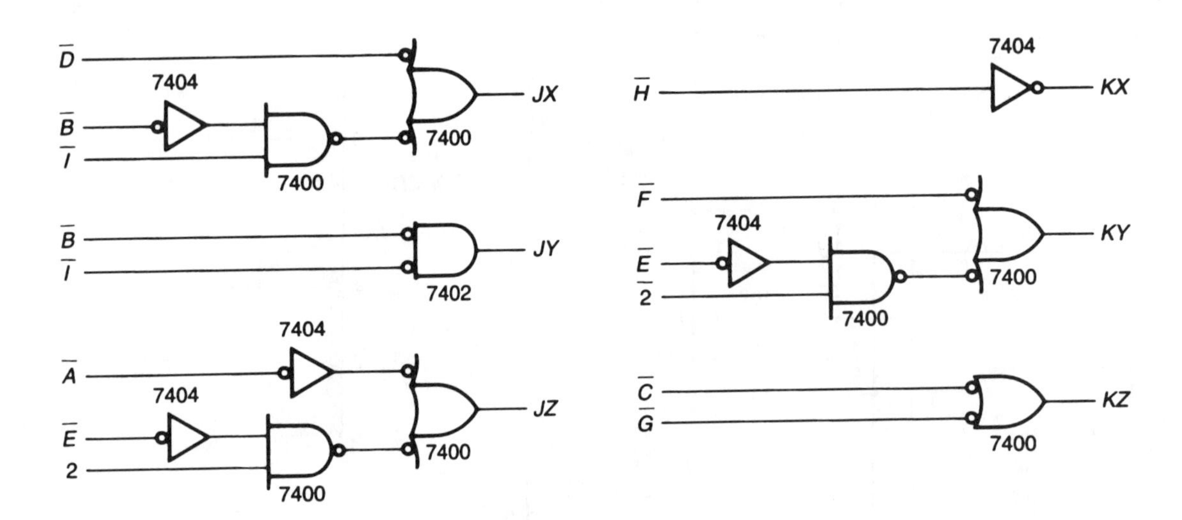

This completes the design.

295 (Category 14)

SITUATION:

For board space and power considerations, a logic designer wishes to use a minimum number of gates in a digital control system. He needs an exclusive OR function and has four spare NAND gates remaining from four 7400-type ICs scattered across the circuit board. Timing is not a consideration.

REQUIREMENTS:

Design an exclusive OR function from the four available spare NAND gates. Show a truth table, draw a Karnaugh map, write a Boolean equation, and draw a logic diagram of the circuit implementation.

SOLUTION:

The truth table and Karnaugh map for an XOR function are shown below:

A	B	Y
0	0	0
0	1	1
1	0	1
1	1	0

A \ B	0	1
0	0	1
1	1	0

From the Karnaugh map, one obvious equation is, $Y = \overline{A}B + A\overline{B}$, and its implementation could be,

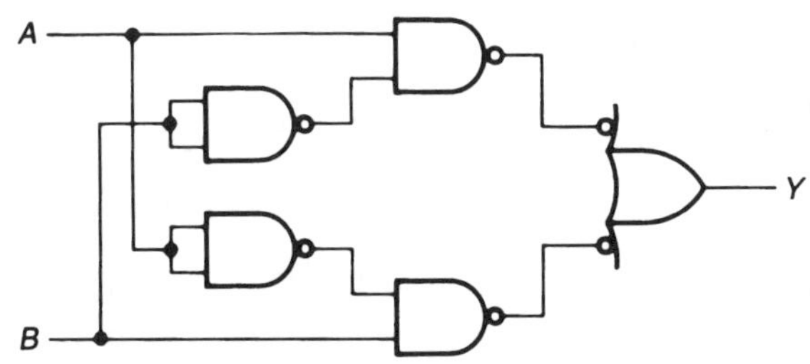

but this implementation requires the use of five NAND gates, and there are only four available. Let's try a Boolean algebra technique by adding $A\overline{A} + B\overline{B}$ to the equation, as follows:

$$\overline{A}B + A\overline{B} + A\overline{A} + B\overline{B} = A(\overline{A}+\overline{B}) + B(\overline{A}+\overline{B})$$

•ANSWER It's implementation is:

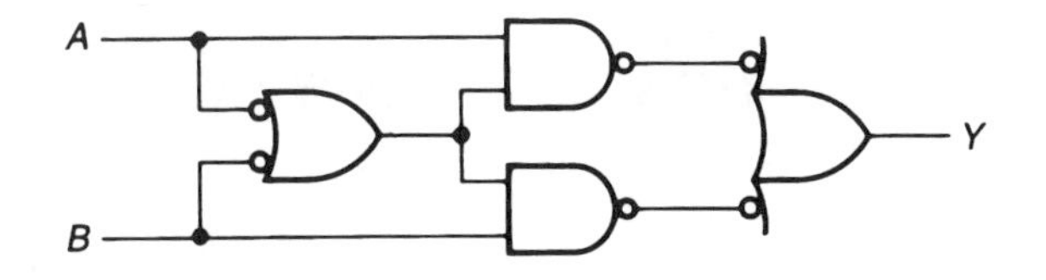

296 (Category 16)

SITUATION:

A newly installed vertical antenna at an AM broadcast station transmitter facility has an effective half-power radiation cross section as shown below:

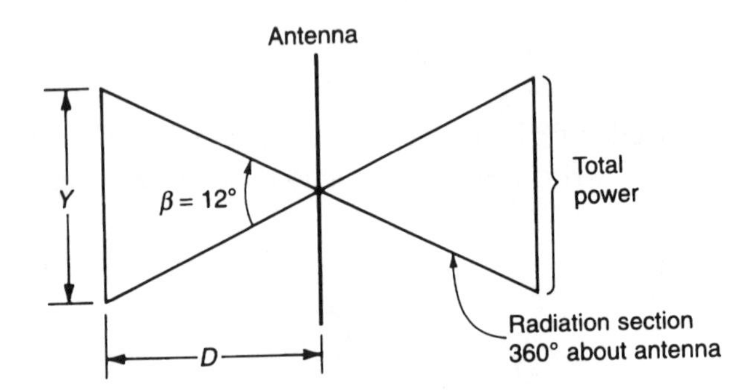

At 80 meters from the antenna, a field strength meter indicates the maximum rms value of field strength is 30 Volts-per-meter.

REQUIREMENTS:

Calculate the total value of time-average power that the antenna radiates under these conditions.

SOLUTION:

At 80 meters from the antenna, $E_{rms\ max} = 30V/m$

At half-power points, $E_{rms} = 30/\sqrt{2}\ V/m = E$

$$P = E^2/377 = [30/\sqrt{2}]^2/377 = 1.19W/m^2$$

where 377Ω is the impedance of free space

$\tan ß/2 = .5(Y/80)$, $Y = 160 \tan 6° = 16.82m$

Total area of the cylinder of revolution = $2\pi DY$
$= 2\pi 80 x 16.82 = 8453m^2$

•ANSWER P_{Total} = area x power = 8453 x 1.19 = **10,059W**

297 (Category 15)

SITUATION:

A portion of a digital system is to control an analog type device that requires an input voltage ranging from -10V to +10V. The output of the analog device will be drive an electromechanical device whose shaft position will be detected by a digital shaft encoder and fed back to a PIA port. A system block diagram is shown below.

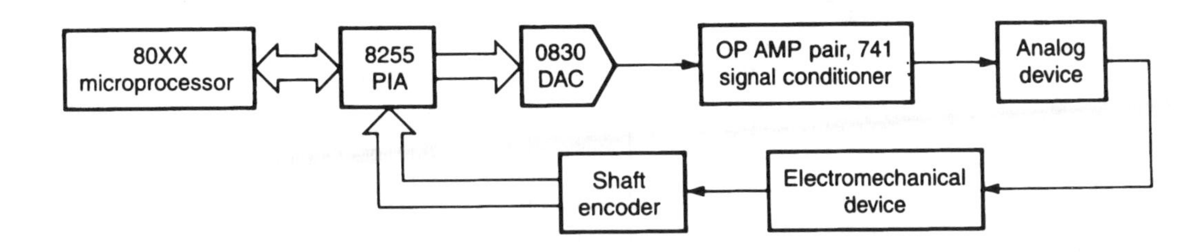

Assume that an 8-bit microprocessor is used in the digital system and that the Programmable Peripheral Interface Adapter (PIA) is an Intel 8255 IC. The analog device requires a resolution step size of not more than .1V. The D/A converter is a standard DAC 0830 (National).

REQUIREMENTS:

(a) Determine the resolution and full scale-output of the DAC for a 10-volt reference.

(b) If Port B of the PIA is interfaced to the DAC, what is the control word format (in hex) for proper data output operation of the PIA? Assume the simplest type of data output operation, using a MODE SET FLAG=1, and disregard any handshaking or input to the digital system for this portion of the operation.

(c) Use an offset technique in the signal conditioner circuit such that the data to the DAC is a decimal value of:

255 corresponding to an output of +10V,
128 corresponding to an output of 0 V,
0 corresponding to an output of -10V.

Show a possible circuit arrangement using the op amp(s) for the signal conditioner. (No need to show pin connections or control lines.)

297 (continued)

The pertinent characteristics of the various devices are:

8255 PIA:

Control Word (for mode set flag ACTIVE):

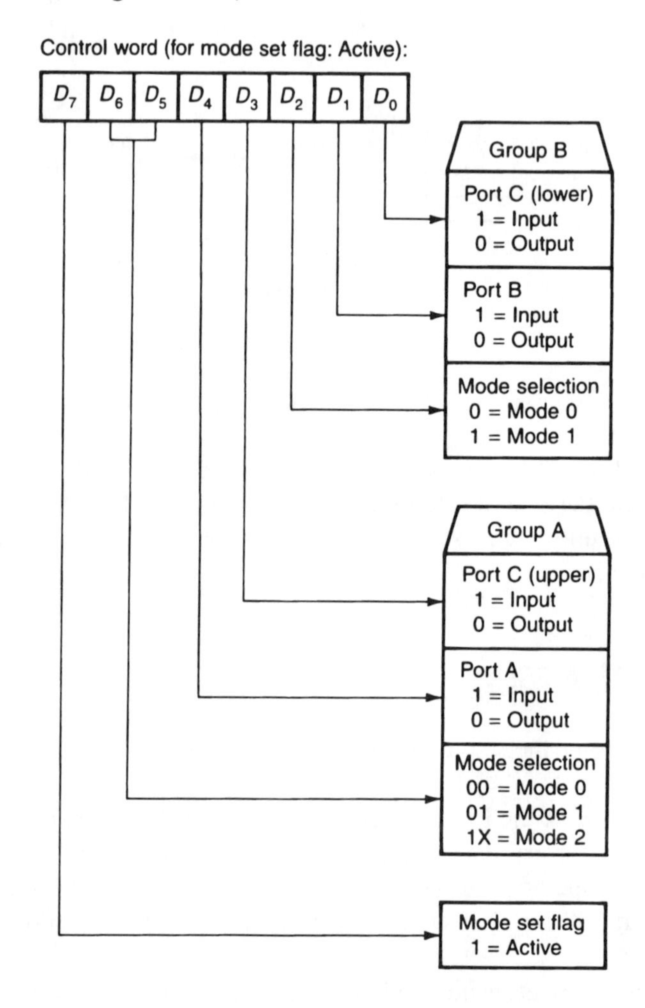

DAC 0830:

The 0830 DAC is one of many 8-bit multiplying analog-to-digital converters that have an on-chip feedback resistor for use as the shunt feedback resistor for an external op-amp (provides for an output voltage for the DAC). The resistor matches the on-chip resistors for the internal R-2R ladder network (R nominally is 15K).

OP AMP 741:

The typical 741 op amp has a high output capability and has at least a full ±10V swing into a 1K load with a unity gain frequency response of 1MHz, with an open loop gain greater than 50,000 (200,000 being typical).

297 (continued)

SOLUTION:

(a) The voltage resolution of the DAC (with direct connection of the first operational amplifier and measured at its output) is the reference voltage corresponding to the least significant bit,

$$\text{Resolution} = E_R/2^n = 10/2^8 = .039 \text{ volt},$$

and the full scale voltage output is,

●ANSWER (a1) $FSV = E_R(1-½^n) = 10(1-1/256) =$ **9.96 volts.**

Resolution of the second operational amplifier voltage output will be double these amounts since its voltage swing is to be from -10V to +10V.

●ANSWER (a2) The resolution of the offsetting op amp is, 2 x .039 = **.078V.**

This is well within the design requirement of .1V.

(b) Since the output of the signal conditioner drives an electro-mechanical device, the speed of operation and handshaking is unimportant. Therefore, the output port arrangement may be as simple as possible. The 8255 PIA may be programmed simply in Mode 0 (since a MODE SET FLAG being set to ONE was assumed) with Port B being chosen as the output latched port to the DAC.

●ANSWER (b) Thus, a control word such as binary 10001001 or $\mathbf{89_H}$ will give one possible correct configuration. In general, any code that is compatible with binary **1XXXX00X** is acceptable.

(c) The signal conditioner circuit is to give a +10V output (actually 9.96V) when the DAC data input is all ONEs and a -10V output when the DAC input is all ZEROs. Also, the output should be approximately 0V for a DAC binary input of 10000000 (or 128_D). This requirement may be achieved by using the second op amp for both multiplying by two and offsetting by 10V.

●ANSWER (c) See the circuit arrangement below,

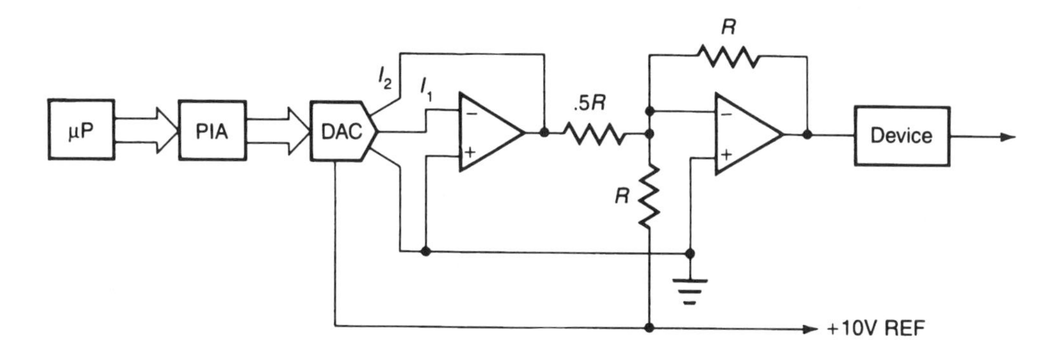

SAMPLE PROBLEMS AND SOLUTIONS

FOR THE AFTERNOON SESSION

OF THE EXAMINATION IN

ELECTRICAL ENGINEERING

BEGIN ON THE NEXT PAGE

430 (Category 2)

A simplified one-line diagram representing a 3-phase, 60 Hz distribution facility is shown below, together with the two transformer specifications.

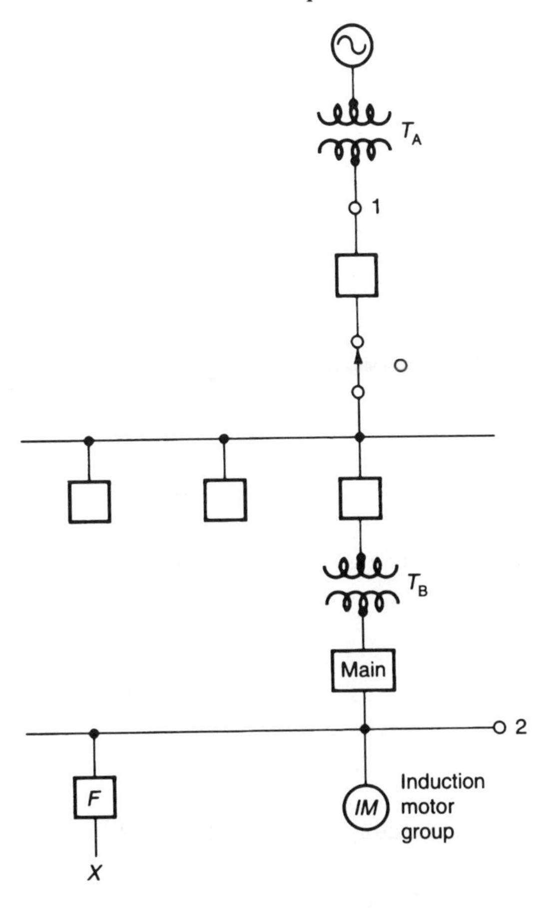

T_A: 15MVA, 7% Z, 15MVA base, 115kVΔ -- 12.47kVY/7.2kV

T_B: 2MVA, 6% Z, 2MVA base, 12.47kVΔ -- 480Y/277

The induction motor group has a power output of 1200 hp. Assume 1hp = 1kW. Use 1000 kVA base in calculations.

1. What are the per unit impedances (ignore resistance) of transformers T_A and T_B?

(A) j.0021 and j.075
(B) j.0033 and j.02
(C) j.0047 and j.03
(D) j.0068 and j.86
(E) j.075 and j.125

2. At point 2, what are the base current (in amps) at 12.47kV and the base impedance (in ohms)?

(A) 46 and .23
(B) 32 and .13
(C) 74 and .94
(D) 92 and .02
(E) 27 and 1.68

This problem is similar to a problem presented in ***Sample Problems and Solutions in Electrical Engineering,*** National Council of Examiners for Engineering and Surveying, 1996.

430 (continued)

3. Assuming that four times the full load current of the motors feeds back into a fault, how much current (in amps) would the motors contribute to a fault immediately after the fault occurs?
(A) 6742
(B) 5774
(C) 3728
(D) 2537
(E) 9413

4. What would be the resistance (in ohms) of a neutral grounding resistor at T_B used to limit phase-to-ground faults to 12 amps?
(A) 932
(B) 138
(C) 12
(D) 47
(E) 23

5. Assume that 20 short circuit MVA are available at point 2. What would be the minimum required amperes interrupting rating for a symmetrical fault?
(A) 13,470
(B) 94,860
(C) 28,870
(D) 33,340
(E) 124,000

6. What would be the result of connecting an overexcited 480V, 700kVA synchronous generator at point 2?
(A) Decreases reactive component of line current.
(B) Increases power factor.
(C) Improves bus voltage regulation.
(D) Reduces I^2R losses in the power distribution line.
(E) All of the above.

7. The main circuit breaker on the secondary side of transformer B is rated to interrupt fault current having a power factor as low as 15%. If a fault occurs having a power factor of 12%, which of the following would be the most appropriate conclusion one would reach?
(A) The breaker will fail.
(B) The breaker design should be upgraded.
(C) Breaker interrupting rating is 3% below the fault.
(D) Accounting for rating tolerance, fault is within spec.
(E) Breaker application should be re-evaluated.

430 (continued)

8. The voltage at point 2 normally runs too low. What would be the most expeditious way to correct this low voltage situation?
(A) Increase the MVA rating at the source.
(B) Lower T_B transformer primary tap by 5%.
(C) Upgrade the cable capacity between the two transformers.
(D) Install cooling fans on transformer T_B.
(E) Raise T_B transformer primary tap by 5%.

9. The synchronous generator of question 6 was purchased under the assumption that its annual operating and maintenance costs would be $1500 less than actual experience. Assuming the added costs occur at the end of each year, a life of 20 years and cost of money being 8% annually, what will be the approximate additional cost of the generator over its lifetime?
(A) $30000
(B) $15000
(C) $60000
(D) $70000
(E) $90000

10. The induction motor group had an initial cost of $100,000. Its anticipated life is 20 years, a salvage value of 15,000, and an annual maintenance cost of $4000. Assuming the cost of money is 8% on an annual basis, what if the approximate equivalent uniform annual cost of the motors?
(A) $12,570
(B) $14,450
(C) $15,210
(D) $11,760
(E) $13,860

430 (continued)

SOLUTION:

1. P.U. Z of T_A is, $Z = j(.07) \times 1000/15000 = j.0047$

   P.U. Z of T_B is, $Z = j(.06) \times 1000/2000 = j.03$

• THE CORRECT ANSWER IS (C), j.0047 and j.03.

2.

$$I_{base} = \frac{1000 \text{ base kVA}}{\sqrt{3} \times 12.47\text{kV}} = 46.3 \text{ amps}$$

$$Z_{base} = \frac{\text{kV}^2}{\text{kVA}_{base}} = \frac{480^2}{1000\text{x}1000} = \frac{230400}{1{,}000{,}000} = .2304\Omega$$

• THE CORRECT ANSWER IS (A), 46 and .23.

3.

$$I_{rated} = \frac{1200\text{kVA}}{\sqrt{3} \times 480} = 1.433\text{kVA} = 1{,}433 \text{ amps}$$

$$I_{fault} = 4 \times I_{rated} = 5{,}774 \text{ amps}$$

• THE CORRECT ANSWER IS (B), 5774.

4. $R = V_\phi / I, \quad V = V_L / \sqrt{3}, \quad I = 12 \text{ amps}$

$$R = 480/ \sqrt{3} \times 12 = 23.09\ \Omega$$

• THE CORRECT ANSWER IS (E), 23.

5. $I_{IR} = I_{fpu} \times I_{base} = 20 \times 1433 = 28{,}870$ A

• THE CORRECT ANSWER IS (C), 28,870.

6. Adding an overexcited synchronous generator has the same effect as adding a bank of capacitors for power factor correction. This also decreases the reactive current which improves voltage regulation.

• THE CORRECT ANSWER IS (E), All of the above.

7. Since the lower power factor is beyond the present design limit, the first course of action should be to further evaluate the breaker for this application.

• THE CORRECT ANSWER IS (E), Breaker application should be re-evaluated.

430 (continued)

SOLUTION (cont.):

8. The least expensive and effective way to achieve the desired result is to lower the tap on the T_B transformer primary.

● **THE CORRECT ANSWER IS (B), Lower T_B transformer tap by 5%.**

9. The Future Value of Annuity table in a set of interest tables will show that for 20 years at 8% interest, the multiplication factor is 45.76. The following formula can be used to calculate the total future additional cost of the generator,

F = A [FVA @20 years, 8%] = $1500 x 45.76 = $68640

♦ **THE CORRECT ANSWER IS (D), $70000.**

10. The equivalent uniform annual cost of the motors may be calculated as follows:

A = $100,000/[FVA @20 years, 8%] - 1500/[PVA @20 years, 8%] + 4000

= $100,000/9.818 - $15000/45.75 + $4000 = $13,857.57

● **THE CORRECT ANSWER IS (E), $13,860.**

431 (Category 4)

A small factory is being established that will have the following loads:

55 KVA,240 volt, single-phase, unity power factor
14 KW, 240 volt, three-phase, .82 power factor

Primary power to the factory is 7200 volts, three-phase, 60 Hz. Phase sequence is a-b-c. Since the single-phase load is so much larger than the three-phase load, it has been decided to distribute power using two single-phase transformers in the configuration shown below. Also shown is a phasor diagram of the three secondary voltages.

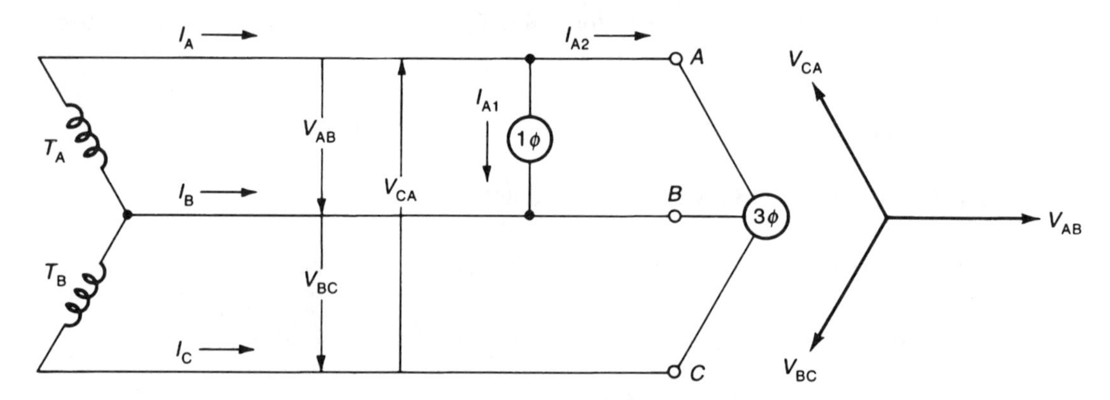

1. What is a name for this type of configuration?
(A) Open delta.
(B) V-V.
(C) V-connection.
(D) All of the above.
(E) None of the above.

2. Which best describes this configuration?
(A) Used in case one phase fails.
(B) Used with greatly unbalanced loads to save money.
(C) Current out of phase with the voltage.
(D) Has about 58% capacity of full delta connection.
(E) All of the above.

3. What is the magnitude of current I_{A1} in amps?
(A) 229
(B) 254
(C) 337
(D) 548
(E) 150

431 (continued)

4. What is the phase angle of I_{A1} with respect to reference V_{ab}?
(A) 0°
(B) -30°
(C) +30°
(D) -150°
(E) +120°

5. What is the magnitude of current I_{A2} in amps?
(A) 19.4
(B) 36.2
(C) 41.1
(D) 52.0
(E) 26.4

6. What is the phase angle of I_{A2} with respect to reference V_{ab}?
(A) -60°
(B) -64.9°
(C) -34.9°
(D) -55.1°
(E) -61.8°

7. What is the magnitude of current I_C in amps?
(A) 41.1
(B) 37.0
(C) 45.2
(D) 34.9
(E) 43.2

8. What is the phase angle of I_C with respect to reference V_{ab}?
(A) 49.6°
(B) 55.1°
(C) 52.3°
(D) 77.1°
(E) 41.1°

9. What is the KVA rating of transformer T_A?
(A) 10
(B) 50
(C) 40
(D) 60
(E) 20

10. What is the KVA rating of transformer T_B?
(A) 10
(B) 30
(C) 60
(D) 40
(E) 50

431 (continued)

SOLUTION:

1. The most popular name is open delta, but V-V and V-connection are also used. Thus, choice (D) is the best answer.

• **THE CORRECT ANSWER IS (D), All of the above.**

2. (A) through (D) are correct statements for this configuration.

• **THE CORRECT ANSWER IS (E), All of the above.**

3. $I_{A1} = KVA/V_{ab} = 55{,}000/240\underline{/0°} = 229.17\underline{/0°}$ amps.

• **THE CORRECT ANSWER IS (A), 229.**

4. Since the load in branch 1 is unity power factor, the current is in phase with the line voltage and the phase angle of the current is 0°.

• **THE CORRECT ANSWER IS (A), 0°.**

5. $$I_{A2} = \frac{P}{\sqrt{3}V_L \cos\Theta} \angle -(\ +30\ ) = \frac{14{,}000 \quad \angle -64.92°}{\sqrt{3}x240x.82}$$

$= 41.07\underline{/-64.92°}$ amps (in polar coordinates), OR

$= 19.41 - j36.19$ amps (in rectangular coordinates)

• **THE CORRECT ANSWER IS (C), 41.1.**

6. In an open delta configuration, current lags line voltage by 30°. In addition, power factor is .82 so Θ is arc COS .82 = 34.92°. Therefore, the total phase angle between current, I_{A2} and voltage, V_{AB} is 0° - (30°+34.92°) = -64.92°.

• **THE CORRECT ANSWER IS (B), -64.9°.**

7. The line current in branch C has the same magnitude as the other two 3-phase load currents, namely 41.07 amps.

• **THE CORRECT ANSWER IS (A), 41.1.**

8. The current, I_C, lags line voltage, V_{CA}, by 30° plus the power factor angle. Therefore, the phase angle of line current, I_C, with respect to reference voltage, V_{AB}, is 120° - (30°+34.92°) = 55.08°.

• **THE CORRECT ANSWER IS (B), 55.1°.**

431 (continued)

SOLUTION (cont.):

9. The KVA rating of transformer A is equal to the product of line voltage (240V) and line current. Calculating line current,

$I_A = I_{A1} + I_{A2} = 229.17 + 19.41 - j36.19 = 248.58 - j36.19$

$= 251.2\underline{/-8.28°}$ amps.

Therefore, $KVA_A = .24 \times 251.2 = 60$ KVA.

● **THE CORRECT ANSWER IS (D), 60.**

10. The KVA rating of transformer B is similarly calculated as follows:

$KVA_B = .24 \times 41.07 = 9.86$ KVA.

Therefore, a 10KVA transformer should be selected.

● **THE CORRECT ANSWER IS (A), 10.**

432 (Category 5)

The figure below represents the equivalent circuit for a 4 hp, 4-pole, 60 Hz, 3-phase induction motor having a wound rotor.

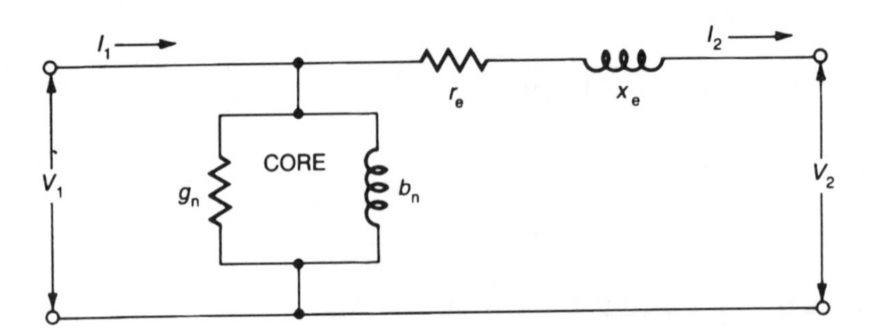

While drawing 3670 watts, the motor delivers 3.95 hp. Losses determined by locked rotor, no-load and other tests and calculations include:

primary winding loss	350 watts
secondary winding loss	123 watts
core loss (attributed to secondary)	200 watts
friction/windage stray load loss	50 watts

1. What losses does a locked rotor test yield?
(A) Primary winding loss.
(B) Secondary winding loss.
(C) Core loss.
(D) Friction/windage, stray load loss.
(E) All of the above.

2. What losses does a no-load test yield?
(A) Primary winding loss.
(B) Secondary winding loss.
(C) Core loss.
(D) Friction/windage, stray load loss.
(E) (C) & (D), above.

3. What losses are not determined by motor tests?
(A) Primary winding loss.
(B) Secondary winding loss.
(C) Core loss.
(D) Friction/windage, stray load loss.
(E) None of the above.

432 (continued)

4. What is the output power, in watts?
(A) 2358
(B) 2652
(C) 2947
(D) 3242
(E) 3536

5. What are the total losses of the motor, in watts?
(A) 823
(B) 523
(C) 600
(D) 450
(E) 723

6. What is the motor "synchronous" speed, n_1, in rpm?
(A) 1200
(B) 1633
(C) 1800
(D) 2400
(E) 3600

7. What is the value of motor slip, s?.
(A) .0297
(B) .0371
(C) .0425
(D) .0541
(E) NONE OF THE ABOVE

8. What is motor speed (rpm), n_2, under the given load?
(A) 1559
(B) 1646
(C) 1681
(D) 1733
(E) 1768

9. What is the motor efficiency under the given load condition?
(A) .803
(B) .776
(C) .901
(D) .824
(E) .863

10. What is line current (in amps) if line voltage is 220 V? Assume power factor = 1.
(A) 10.11
(B) 9.10
(C) 8.54
(D) 7.99
(E) 9.63

432 (continued)

SOLUTION:

1. The locked rotor test is a full load test that draws full line current through the primary. Thus, the majority of the motor losses occur in the primary winding.

• **THE CORRECT ANSWER IS (A), Primary winding loss.**

2. The no-load test causes only light current to flow. Thus, the only measurable losses occur due to core loss, friction and windage and stray load losses.

• **THE CORRECT ANSWER IS (E), (C) & (D).**

3. Secondary winding losses must be calculated. It is the difference between total losses and measurable losses. Total losses are the difference between motor input and output power, with the motor under load.

• **THE CORRECT ANSWER IS (B), Secondary winding loss.**

4. Motor output power is given as 3.95 hp. Multiplying this figure by 746 watts/hp yields 2947 watts. An alternate method is to subtract the total losses from the input power (3670 - 723 = 2947).

• **THE CORRECT ANSWER IS (C), 2947.**

5. Total losses are the sum of the losses given in the problem statement (723 watts), or the difference between input power and output power (3670 - 2947 = 723).

• **THE CORRECT ANSWER IS (E), 723.**

6. If an induction motor could operate at synchronous speed, its speed, n_1 would be 120f/p rpm; or the line frequency (f) multiplied by 60 min/sec (to convert to rpm), and divided by the number of pole pairs. Thus, $n_1 = 120 \times 60/4 = 1800$ rpm.

• **THE CORRECT ANSWER IS (C), 1800.**

432 (continued)

SOLUTION (cont.):

7. Slip, s, is the difference between synchronous speed and actual speed divided by synchronous speed, n_1-n_2/n_1. It also can be calculated from the formula:

secondary winding loss/{P_{in} - (primary winding loss + core loss)}

In this case, core loss is stated as being attributed entirely to the secondary side so it does not enter into the calculation here. Therefore,

$$s = 123/(3670 - 350) = .03714$$

● **THE CORRECT ANSWER IS (B), .0371.**

8. The actual rpm of the motor, n_2, depends upon its load. It can also be expressed by the formula:

$$n_2 = n_1(1\text{-}s) = 1800(1\text{-}.0371) = 1733 \text{ rpm}$$

● **THE CORRECT ANSWER IS (D), 1733.**

9. Efficiency, Eff, is output divided by input, or input minus losses divided by input. One way to calculate it here is,

$$Eff = P_{out}/P_{in} = 2947/3670 = .803$$

● **THE CORRECT ANSWER IS (A), .803.**

10. The formula for 3-phase power is, $P = \sqrt{3}VI\cos\Theta$. In this case, $\cos\Theta = 1$. Therefore, the line current becomes,

$$I_1 = P / \sqrt{3}V = 3670/(1.732 \text{ x } 220) = 9.63 \text{ amps}$$

● **THE CORRECT ANSWER IS (E), 9.63.**

433 (Category 7)

Four induction motors are to be fed from a 3ϕ, 60Hz, 460 V source. Two of the motors are squirrel cage type (with full voltage starting) having code markings of type B. The other two are wound-rotor types (with no code markings).

The figure below shows how the motors are connected. A late edition (1993 or later) of the National Electrical Code© is needed.

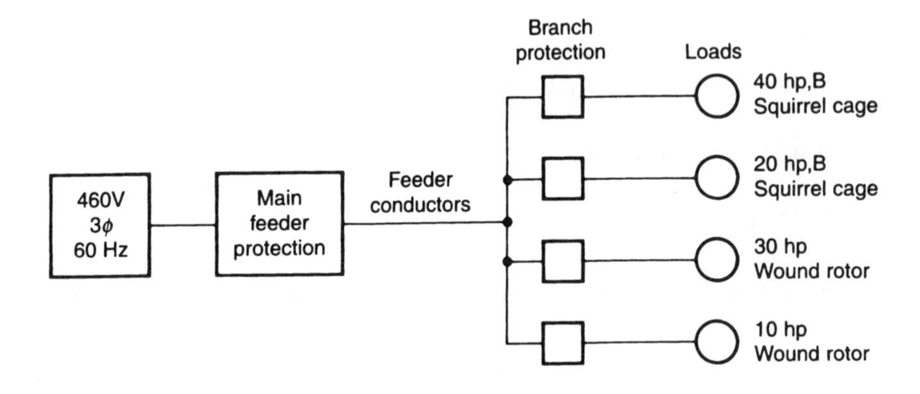

1. Determine squirrel cage motors branch feeder wire sizes.
(A) #3 and #4
(B) #4 and #6
(C) #4 and #8
(D) #6 and #12
(E) #6 and #14

2. Determine the wound rotor motors branch feeder wire sizes.
(A) #3 and #4
(B) #4 and #6
(C) #4 and #8
(D) #6 and #12
(E) #6 and #14

3. Assume the wires to the squirrel cage motor will be carried in a common trade size metal conduit. Also, assume the conductor sizes for these two motors are three #8 wires and three #10 wires along with one common #10 neutral wire (all are TW insulated copper wires). Determine the minimum size conduit acceptable (use the 40% fill relationship and do not round down).
(A) ½"
(B) ¾"
(C) 1"
(D) 1¼"
(E) 1½"

433 (continued)

4. Determine the branch circuit current protection using nontime delay fuses for the squirrel cage motors.
(A) 125 A and 60 A
(B) 150 A and 70 A
(C) 175 A and 100 A
(D) 40 A and 15 A
(E) 150 A and 60 A

5. Determine the branch circuit current protection using nontime delay fuses for the wound rotor motors.
(A) 125 A and 50 A
(B) 60 A and 20 A
(C) 52 A and 20 A
(D) 40 A and 14 A
(E) 156 A and 60 A

6. Determine the main feeder current protection using a nontime delay fuse. Assume that rounding up cannot be justified. Use these assumed full-load current for the motors (even though they may not agree with the previous calculated values):

40 hp --> 50 A; 20 hp --> 30 A; 30 hp --> 40 A; 10 hp --> 16 A

(A) 160 A
(B) 175 A
(C) 200 A
(D) 250 A
(E) 300 A

7. The nontime delay fuse of the previous question is to be replaced with an inverse time breaker. Determine current rating for this device (use the assumed motor current values as given in question 6).
(A) 150 A
(B) 175 A
(C) 200 A
(D) 225 A
(E) 250 A

8. When the two original squirrel cage motors that were ordered arrived, the requested code markings were missing. To decide whether the existing branch feeder protective nontime delay fuses need new values, recompute the two motor current ratings of the protective devices (without code marking on the machines).
(A) 125 A and 50 A
(B) 60 A and 20 A
(C) 175 A and 100 A
(D) 40 A and 15 A
(E) 150 A and 80 A

433 (continued)

9. Consider the original 30 hp wound rotor motor circuit. The motor is known to be 90% efficient and has a power factor of .707 (lagging) at full load. Rather than using the current rating as given by the NEC handbook, calculate the value of current.
(A) 38 A
(B) 40 A
(C) 45 A
(D) 48 A
(E) 52 A

10. Consider the previous problem. The power factor remains at .707; but with a different efficiency, the full load current is 40 A. Now add a balanced lighting load of 20 kW at the terminals of the 30 hp motor. What size conductors (TW type) are needed for this modified branch circuit to carry the combined loads (assume the illumination load is continuous)?
(A) #10
(B) #8
(C) #6
(D) #4
(E) #3

433 (continued)

SOLUTION:

1. Using a 1993 or later National Electrical Code manual, determine the motor current ratings (Table 430-150) and conductor size (Table 310-16).

40 hp motor --> 52 A; 1.25 x 52 = 65 A --> #4 type TW wire

20 hp motor --> 27 A; 1.25 x 27 = 33.75 A --> #8 type TW wire

• **THE CORRECT ANSWER IS (C), #4 and #8.**

2. Repeat question 1 for the wound rotor motors.

30 hp motor --> 40 A; 1.25 x 40 = 50 A --> #6 type TW wire

10 hp motor --> 14 A; 1.25 x 14 = 17.5 A --> #12 type TW wire

Footnote to 310-16 on #14 wire requires any overcurrent protection not to exceed 15A; therefore, #12 wire size will be specified.

• **THE CORRECT ANSWER IS (D), #6 and #12.**

3. For conduit sizes, refer to Table 5, Chapter 9 in the National Electrical Code manual. Here the 40% fill code is applicable.

3 each #10 TW wire area --> .0224 x 3 = .0672 sq. inches

3 each #8 TW wire area --> .0471 x 3 = .1413 sq. inches

1 each #10 TW wire area --> .0224 x 1 =.0224 sq. inches

======

Total Area .2309 sq. inches

• **THE CORRECT ANSWER IS (C), 1".**

4. Branch circuit protection for code B squirrel cage motors using a nontime delay fuse is 250% of full load current (see Table 430-152 of the National Electrical Code).

40 hp motor --> 52 A; 2.5 x 52 = 135 A --> 125 A

20 hp motor --> 27 A; 2.5 x 27 = 67.5 A --> 60 A

The 1993 code change requires rounding down, unless the lower rating or setting is inadequate to carry the starting current.

• **THE CORRECT ANSWER IS (A), 125 A AND 60 A.**

433 (continued)

SOLUTION (cont.):

5. Branch circuit protection for the wound rotor motors using nontime delay fuses is now changed to 150% of full load current (again, refer to Table 430-152).

30 hp motor --> 40 A; 1.5 x 40 = 60 A --> 60 A

10 hp motor --> 14 A; 1.5 x 14 = 21 A --> 20 A (round down)

• **THE CORRECT ANSWER IS (B), 60 A and 20 A.**

6. The current rating for the main feeder wire protector must not be greater than the largest branch circuit protection plus the sum of the full load currents of the other branches.

2.5x50 + 30 + 40 + 16 = 211 A --> 200 A (rounded down)

• **THE CORRECT ANSWER IS (C), 211 A.**

7. The inverse time breaker rating is found the same way as in answer 6, except that the rating is now 200% for the largest code B squirrel cage motor.

2.0x50 + 30 + 40 + 16 = 186 A --> 175 A (rounded down)

• **THE CORRECT ANSWER IS (B), 175 A.**

8. The two motor current ratings are found the same way as in answer 1. However, for no code markings, the full load currents are not increased by 300%.

40 hp motor --> 52 A; 3 x 52 = 156 A --> 150 A (rounded down)

20 hp motor --> 27 A; 3 x 27 = 81 A --> 80 A (rounded down)

• **THE CORRECT ANSWER IS (E), 150 A and 80 A.**

9. The full load power output is 30 x 746 = 22,380 watts. The power input is found to be 22,380/.9 = 24,870 watts. Thus, input current, I, is,

$$I = P_{in} / (\sqrt{3}\, V \cos \Theta) = 24870 / (1.732 \text{x} 460 \text{x} .707) = 44.1 \text{ A} \dashrightarrow 45 \text{ A}$$

• **THE CORRECT ANSWER IS (C), 45 A.**

433 (continued)

SOLUTION (cont.):

10. The balanced lighting current is 20,000/(1.732x460) = 25.1 A. Section 220-10 of the code requires the continuous load to be increased by 125%. The total current for computing the wire size (without regard to phase) is given as,

40 + 1.25x25.1 = 40 + 31.4 = 71.4 A --> #3 wire.

• **THE CORRECT ANSWER IS (E), #3.**

590 (Category 8)

A control system is shown in the figure below. Transfer functions for the forward and feedback blocks are defined below the figure.

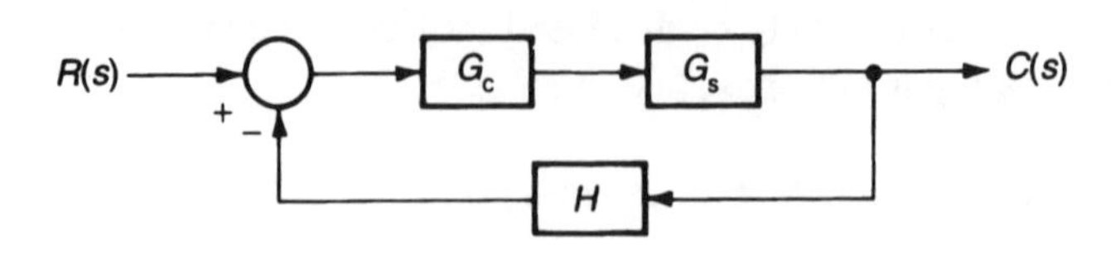

$G_s(s) = \frac{2}{s+2}$, $G_c(s) = \frac{45}{(s+3)(s+5)}$, H = 1

1. What is the open loop transfer function?
(A) $\frac{45}{(s+2)(s+3)(s+5)}$
(B) $\frac{K}{(s+2)(s+3)(s+5)}$
(C) $\frac{2x45}{(s+2)(s+3)(s+5)}$
(D) (s+3)(s+5)
(E) None of the above.

2. What equation defines the closed loop transfer function?
(A) G/(1+GH)
(B) ±KGH
(C) 1/GB
(D) 1-KGH
(E) R(s)-C(s)

3. What are the poles of the system(for K= 45 being replaced with K→0)?
(A) -4.22,3,-2
(B) -2,-3,-5
(C) -4.22,-2.45
(D) 2,3,5
(E) -31,-10,-6

4. What are the zeros of the system?
(A) 2,3,5
(B) 0
(C) 3
(D) 5
(E) None

590 (continued)

5. What is the third term in the first column of the Routh table?

(A) 31

(B) $\frac{31 \times 10 - 120}{10}$

(C) 10

(D) $\frac{31 \times 10 - 90}{10}$

(E) 0

6. What is the characteristic equation of the system?

(A) (s+2)(s+3)(s+5)
(B) $s^3+10s^2+31s+120$
(C) (s+2)(s+3)(s+5)+90
(D) $s^3+10s^2+31s+90$
(E) None of the above.

7. Which best describes the stability of the system?

(A) Stable
(B) Unstable
(C) Marginally stable
(D) All of the above
(E) Not enough information

8. What number is the system "type"?
(A) 0
(B) 1
(C) 2
(D) 3
(E) 4

9. Where do the asymptotes of the root locus intersect the real axis of the s-plane?
(A) -4.2
(B) -2.45
(C) -3
(D) -5
(E) -10/3

10. Where does the root locus intersect the real axis?
(A) -3
(B) -5
(C) -2
(D) -2.45
(E) -4.2

590 (continued)

SOLUTION:

1. The open loop transfer function is GH. Thus,

$$GH = \frac{2x45}{(s+2)(s+3)(s+5)}$$

- **THE CORRECT ANSWER IS (C),** $GH = \frac{2x45}{(s+2)(s+3)(s+5)}$

2. The general equation for the closed loop transfer function is,

$$G/(1+GH)$$

- **THE CORRECT ANSWER IS (A), G/1+GH.**

3. The poles of the system are the roots of the denominator of the open loop transfer function, namely -2, -3 and -5.

- **THE CORRECT ANSWER IS (B), -2,-3,-5.**

4. The zeros of the system are the roots of the numerator. Since there are no roots, there are no zeros.

- **THE CORRECT ANSWER IS (E), None.**

5. The Routh table must first be constructed, as follows:

1	31
10	120
$\frac{31x10-120}{10}$	0
120	

Thus, it is seen that the third term in the first column is:

$$\frac{31x10-120}{10}$$

- **THE CORRECT ANSWER IS (B),** $\frac{31x10-120}{10}$

590 (continued)

<u>*SOLUTION (cont.):*</u>

6. The characteristic equation of the system, using the open loop transfer function, is given by,

$\text{Denom.}_{GH} + \text{Num.}_{GH} = 0$, $(s+2)(s+3)(s+5)+90 = 0$,

$s^3+10s^2+31s+120 = 0$

● **THE CORRECT ANSWER IS (B), $s^3+10s^2+31s+120 = 0$.**

7. The system is seen to be stable from at least two standpoints. There are no positive poles, and there is no difference in sign for the first column terms in the Routh table.

● **THE CORRECT ANSWER IS (A), Stable.**

8. The system "type" is equal to the difference between the number of the s-term exponent in the numerator and the number of the s-term exponent in the denominator of the open loop transfer function. Since there are no s-terms in either the numerator or denominator of the characteristic equation, the system type is 0.

● **THE CORRECT ANSWER IS (A), 0.**

9. The location, σ_c, of the intersection of the asymptotes of the root locus with the real axis of the s-plane is determined as follows:

The intersection is equal to the difference between the sum of poles less the sum of zeros, divided by the difference between the number of poles and zeros. Thus,

$\sigma_c = -\ [(2-3-5)-0]/(3-0) = 10/3$

Since the intersection is in the LHP, the correct result is -10/3.

● **THE CORRECT ANSWER IS (E), -10/3.**

590 (continued)

SOLUTION (cont.):

10. The location, σ_b, of the intersection of the root locus itself with the real axis of the s-plane is determined as follows:

$$1/(\sigma_b+2) + 1/(\sigma_b+3) + 1/(\sigma_b+5) = 0$$

$$(\sigma_b+3)(\sigma_b+5) + (\sigma_b+2)(\sigma_b+5) + (\sigma_b+2)(\sigma_b+3) = 0$$

$$3\sigma_b^2 + 20\sigma_b + 31 = 0$$

Solving for the roots of the quadratic equation yields, -4.22 and
-2.45. We know the intersection must occur between the polesat -2 and -3; therefore, the intersection occurs at -2.45.

• **THE CORRECT ANSWER IS (D), -2.45.**

591 (Category 9)

Shown below is a 2-stage transistor amplifier circuit. Parameters of the circuit are:

$h_{ie} = 1500\Omega$ $h_{re} = 2.5x10^{-4}$ $h_{fe} = 49$ $h_{oe} = 2.5x10^{-5}$mho

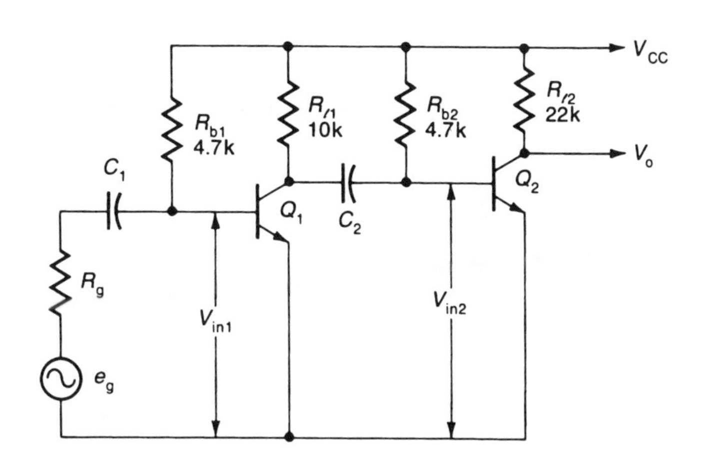

1. What is the configuration of each of the two transistor amplifiers?
(A) NPN Common Emitter
(B) NPN Common Collector
(C) NPN Common Base
(D) NPN Emitter Follower
(E) PNP Common Emitter

2. What is the second stage voltage gain, A_{V2}?
(A) -31
(B) -610
(C) -275
(D) -525
(E) -379

3. What is the input impedance, Z_{in2}, to the second stage?
(A) 1176Ω
(B) 1326Ω
(C) 2189Ω
(D) 4703Ω
(E) 1547Ω

4. What is the first stage voltage gain, A_{V1}?
(A) -530
(B) -260
(C) -65
(D) -15
(E) -30

591 (continued)

5. What is the input impedance, Z_{in1}, to the first stage?
(A) 1519Ω
(B) 2239Ω
(C) 1489Ω
(D) 1327Ω
(E) 3906Ω

6. What is the input power, P_{in}, (in watts) to the first stage, in terms of input voltage to the first stage?

(A) $V_i^2/1131$

(B) $V_i^2/1327$

(C) $V_i^2/1519$

(D) $V_i^2/2339$

(E) none of the above

7. What is the output power, P_{out}, (in watts) into the 22K load resistor, in terms of output voltage?

(A) $V_o^2/4700$

(B) $V_o^2/22000$

(C) $V_o^2/1500$

(D) $V_o^2/10000$

(E) $V_o^2/47$

8. What is total power gain, in dB?

(A) 47

(B) 63

(C) 94

(D) 33

(E) 71

591 (continued)

9. What is the interstage power, P_{IL}, (in watts) lost in resistors R_{l1} and R_{b2}, in terms of V_{in1}?

(A) $.14V_{in1}^2$

(B) $.28V_{in1}^2$

(C) $139V_{in1}^2$

(D) $9.2V_{in1}^2$

(E) $2.8V_{in1}^2$

10. What is the pre-first-stage power, P_{PFSL}, (in watts) lost in resistor R_{b1}, in terms of V_{in1}?

(A) $3x10^{-3}\ V_{in1}^2$

(B) 40

(C) 2.7

(D) $2x10^{-4}\ V_{in1}^2$

(E) $4x10^{-6}\ V_{in1}^2$

591 (continued)

<u>*SOLUTION:*</u>

1. Each transistor is an NPN and they are both connected in the common emitter configuration.

● **THE CORRECT ANSWER IS (A), NPN Common Emitter.**

2. The formula for voltage gain is,

$$A_V = -h_{fe}/[h_{ie}G_l + h_{ie}h_{oe} - h_{re}h_{fe}]$$

Calculating voltage gain for the second stage,

$$A_{V2} = V_o/V_{in2} = -49/[1500/22000 + 1500x2.5x10^{-5} - 2.5x10^{-4}x49]$$

$$= -525.47$$

● **THE CORRECT ANSWER IS (D), -525.**

3. The formula for input impedance is,

$$Z_{in} = h_{ie} - h_{re}h_{fe}/(h_{oe}+G_l)$$

Calculating the input impedance to the second stage,

$$Z_{in2} = 1500 - 2.5x10^{-4}x49/[2.5x10^{-4} + 1/22000]$$

$$= 1500 - 174 = 1326\Omega$$

● **THE CORRECT ANSWER IS (B), 1326Ω.**

4. The load conductance of the first stage must first be calculated,

$$G_{l1} = 1/R_{l1} + 1/R_{b2} + 1/Z_{in2} = 1/10000 + 1/4700 + 1/1326$$

$$= 1.067x10^{-3} \text{ mho.}$$

Calculating the voltage gain of the first stage,

$$A_{V1} = -49/[1500x1.067x10^{-3} + 1500x2.5x10^{-5} - 2.5x10^{-4}x49]$$

$$= -49/1.626 = -30.14.$$

● **THE CORRECT ANSWER IS (E), -30.**

591 (continued)

SOLUTION (cont.):

5. Input impedance to the first stage is,

$Z_{in1} = 1500 - [2.5x10^{-4}x49]/[2.5x10^{-5} + 1.067x10^{-3}] = 1500 - 11$

$= 1489\Omega$

• THE CORRECT ANSWER IS (C), 1489Ω.

6. $P_{in} = V_i^2/[R_{b1} \| Z_{in1}]$

Calculating the parallel resistance, $\frac{(4700)(1489)}{4700 + 1489} = 1131\Omega$

$P_{in} = V_{in}^2/1131$ watts

• THE CORRECT ANSWER IS (A), $V_{in}^2/1131$

7. $P_{out} = V_o^2/R_{l2} = V_o^2/22000.$

• THE CORRECT ANSWER IS (B), $V_o^2/22000$.

8. $G = P_{out}/P_{in} = [V_o^2/22000][1130.76/v_i^2] = 5.14x10^{-2} A_V^2$

$A_v = A_{v1} A_{v2} = (-30.14)(-525.47) = 15837$

$G = 5.14x10^{-2} (15827)^2 = 1.289x10^7$

In dB, $G_{dB} = 10\log G = 10(7+.11) = 71.1$ dB.

• THE CORRECT ANSWER IS (E), 71.

9. Interstage power loss is that power lost in the parallel combination of R_{l1} and R_{b2}. The parallel combination is,

$\frac{10K \times 4.7K}{10K + 4.7K} = 3197\Omega$

Thus, $P_{IL} = V_{in1}^2/3197 = (30.14V_{in1})^2/3197 = .28V_{in1}^2$ watts.

• THE CORRECT ANSWER IS (B), $.28V_{in1}^2$.

10. Pre first stage power loss is that power lost in R_{b1}.
Thus, $P_{PFSL} = V_{in1}^2/R_{b1} = V_{in1}^2/4700 = 2.13x10^{-4} V_{in1}^2$ watts.

• THE CORRECT ANSWER IS (D), $2x10^{-4} V_{in1}^2$.

592 (Category 11)

Consider the LH0045 Two Wire Transmitter, an instrumentation amplifier for use with thermocouples, strain gauges and thermistors. The circuit below shows its connection with external circuitry for this application. Also defined below are circuit parameters needed for the problem calculations.

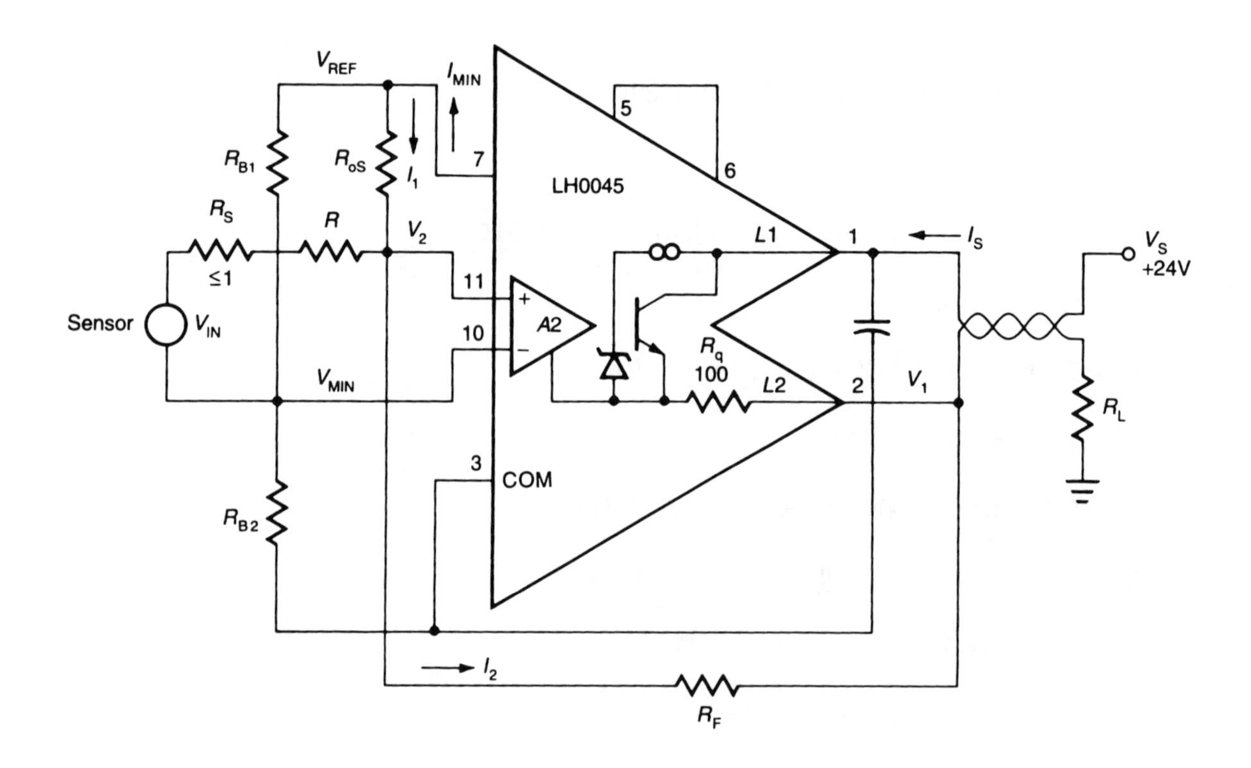

Circuit Parameters:

Internal characteristics are:

Internal current set resistor, $R_9 = 100\Omega$

Output characteristics are:

Supply voltage, $V_S = +24V$.

I_S span is 16ma [null = 4mA; full scale = 20mA]

$I_{2span} = 1.0\mu A$

$V_{1span} = -1.6V$ [-(16ma)(R_9) = -1.6V],

or null = -.4V; full scale = -2.0V

592 (continued)

Input (sensor) characteristics are:

V_{IN} span is 100mV [zero scale = 0mV; full scale = 100mV]

Source impedance, R_5, $\leq$ 1.0Ω

V_{MIN} = 1.0V [minimum input common mode voltage at pin 10]

I_{SOL} = 3.0mA [maximum open loop supply current]

V_{REF} = 5.1V nominal, when pins 5 and 6 are shorted together

1. What problem(s) can occur when differential signals are sent through long cables?
(A) Noise is picked up
(B) Signal bandwidth is reduced
(C) AC common mode rejection is degraded
(D) Induced noise and ground noise are amplified
(E) All of the above.

2. What is the value in ohms of resistor R_F?
(A) 1.0M
(B) 1.6M
(C) 1.5M
(D) 800K
(E) 2.0M

3. What is the minimum value in ohms of resistor R_{B2}?
(A) 2K
(B) 10K
(C) 4.7K
(D) 2.2K
(E) 1K

4. What is the value in ohms of resistor R_{B1}?
(A) 4.1K
(B) 4.7K
(C) 3.3K
(D) 2.2K
(E) 5.6K

5. As an alternative, what may be used to bias the signal amplifier, A_2?
(A) A reference (zener) diode
(B) An op amp
(C) A differentiator
(D) A and B
(E) None of the above.

592 (continued)

6. What should be the nominal value in ohms of resistor R_{OS} to provide an I_S null current of 4.0mA?
(A) 5.6K
(B) 1K
(C) 4.7M
(D) 3.3K
(E) 6.8K

7. What is the value in ohms of R, the total impedance in signal path between pins 10 and 11?
(A) 200K
(B) 50K
(C) 100K
(D) 150K
(E) 330K

8. Which factor(s) can cause a change in V_{REF}?
(A) Ambient temperature changes
(B) Supply voltage variations
(C) Self-heating due to power dissipation variations
(D) (B) and (C)
(E) All of the above

9. How can V_{OS} be caused to drift?

(A) Supply voltage variations
(B) Too much heat sinking
(C) Thermal resistance variations
(D) Ambient temperature changes
(E) All of the above

10. What factors can cause R_9 to effect system errors?
(A) Temperature coefficient (TCR) of R_9
(B) Self-heating of the device
(C) Ambient temperature changes
(D) Inadequate heat sinking
(E) All of the above

592 (continued)

<u>*SOLUTION:*</u>

1. When differential signals are sent through long cables, three problems can occur: (a) noise (both common-mode and differential) is picked up; (b) signal bandwidth is reduced by the RC low-pass filter formed by the source impedance and the cable capacitance; (c) when these RC time constants are not identical (unbalanced source impedance and/or unbalance capacitance), AC common-mode rejection is degraded, amplifying both induced noise and "ground" noise.

• **THE CORRECT ANSWER IS (E), All of the above.**

2. R_F is a function of V_{1span} and I_{2span} as follows:

$$R_2 = V_{1span}/I_{2span} = 1.6V/1.0\mu A = 1.6M$$

• **THE CORRECT ANSWER IS (B), 1.6M.**

3. $R_{B2} \geq V_{MIN}/I_{MIN}$,

I_{MIN} is the difference between I_S at null and I_{SOL} max. = 4 - 3 = 1mA.

Therefore, R_{B2} = 1V/1mA = 1K, min.

• **THE CORRECT ANSWER IS (E), 1K.**

4. R_{B1} can be solved from the equation,

$$V_{REF}R_{B2}/(R_{B1} + R_{B2}) = V_{MIN} = 1.0V$$

Solving for R_{B1},

$$R_{B1} = R_{B2}(V_{REF} - 1.0V)/1.0V = 1K(5.1 - 1.0) = 4.1K$$

• **THE CORRECT ANSWER IS (A), 4.1K.**

5. A 1.22V reference diode (e.g., LM113) or an op amp (e.g., LM108) may be used to bias amplifier A_2, as shown in the figures below. These techniques have the advantage of lowering the impedance seen at pin 10.

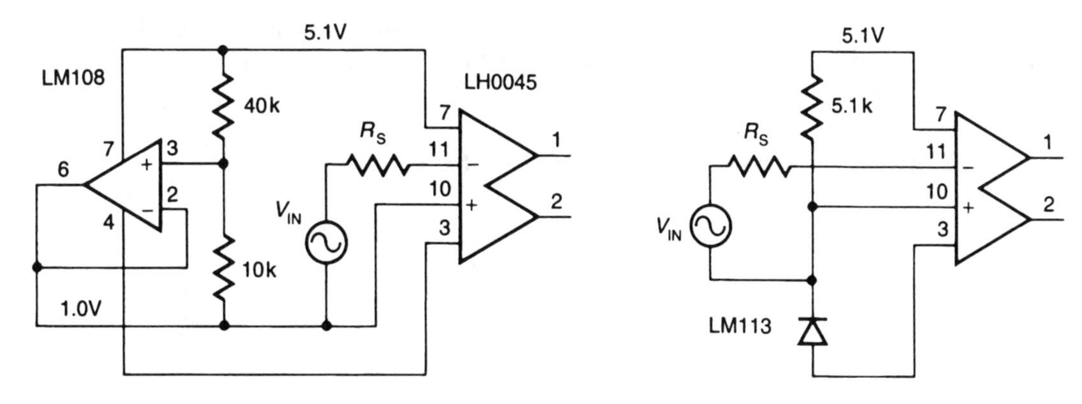

THE CORRECT ANSWER IS (D), A and B.

592 (continued)

SOLUTION (cont.):

6. R_{OS} (for offset) is selected to provide the 4.0mA I_S null current of 4.0mA such that V_{1null} = 4.0mA x R_9 = 0.4V. The voltage V_2 at pin 11 is,

$$V_2 = V_{MIN} + V_{OS} = V_{MIN} = 1.0V, \quad \text{for } V_{IN} = 0V.$$

Hence, the current required to generate the null voltage, I_{2null} is,

$$I_{2null} = (V_{MIN} - V_{1null})/R_F = \frac{1.0 - (-0.4)}{1.6M\Omega} = 0.875\mu A$$

This current must be provided by R_{OS} from V_{REF}. Therefore,

$$R_{OS} = (V_{REF} - V_{min})/I_{2null}$$

The nominal value for V_{REF} is 5.1V; hence, the nominal value for R_{OS} is,

$$R_{OS} = \frac{5.1V - 1.0V}{0.875\mu A} = 4.7M\Omega$$

• **THE CORRECT ANSWER IN OHMS IS (C), 4.6M.**

7. From feedback theory and the gain equation, the following relationship holds:

I_S span = 16ma = $V_{IN}R_F/(R \times R_9)$, where

feedback resistor is assumed to be 1.6K
V_{IN} = full scale input voltage = 100mV

Thus, $R = \frac{100mV \times 1.6M\Omega}{16mA \times 100\Omega} = 100K$

• **THE CORRECT ANSWER IN OHMS IS (C), 100K.**

8. Factors that can cause V_{REF} to change are thermal and supply voltage effects. These are defined in (A), (B) and (C). Since (D) includes these first three choices, the first four choices are correct.

• **THE CORRECT ANSWER IS (E), All of the above.**

592 (continued)

SOLUTION (cont.):

9. Variations in V_{OS} can induce a significant error in I_S. Causes include:
   self heating of the device
   ambient temperature changes

Thermal resistance is considered to be a constant, and supply voltage variations have no effect. Therefore, the only correct choice is ambient temperature changes.

- **THE CORRECT ANSWER IS (D), Ambient temperature changes.**

10. Errors can occur due the changes in R_9. These changes can be caused by:
   temperature coefficient (TCR), causing output current errors
   self-heating of the device
   ambient temperature changes

These are included in the available choices.

- **THE CORRECT ANSWER IS (E), All of the above.**

593 (Category 18)

Consider the circuit (shown below) for use in a biomedical system application.

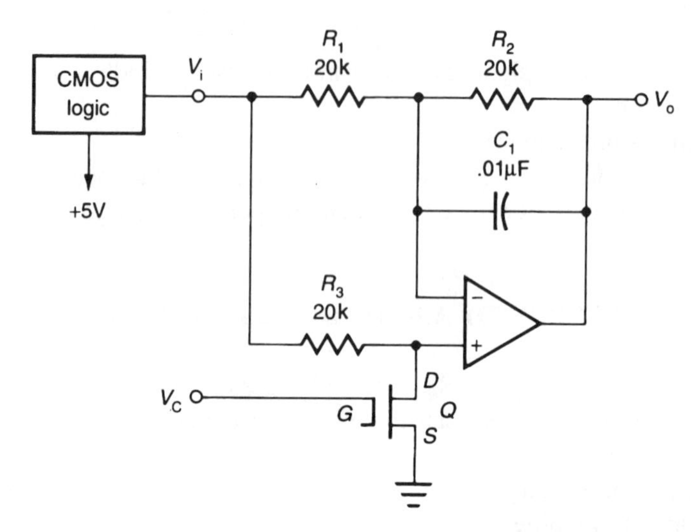

Assume an ideal op amp. The FET has an on-resistance of 1Ω and it cuts off at a gate voltage below -6V. FET gate reverse current is approximately 1nA. Input V_i is controlled by CMOS logic (switches rail-to-rail between 0V and +5V). For this problem, V_i is active high, logic ONE.

For questions 1 through 5, $V_C = 0V$.

1. What is the output voltage (in Volts)?

(A) -10
(B) -5
(C) 0
(D) 5
(E) 10

2. What is the current (in μA) through resistor R_1?

(A) 0
(B) 100
(C) 250
(D) 500
(E) 1000

3. What is the FET static drain current (in μA)?

(A) 0
(B) 250
(C) 500
(D) 1000
(E) 2500

4. The small signal transfer function is $V_o(s)/V_i(s)$. How many poles and zeros are there?

(A) None
(B) 1 zero
(C) 1 pole
(D) 1 pole and 1 zero
(E) 2 zeros

This problem is similar to a problem presented in ***Sample Problems and Solutions in Electrical Engineering,*** National Council of Examiners for Engineering and Surveying, 1996.

593 (continued)

5. For a steady state ac input, at what frequency is the output voltage down 3dB (small signal transfer function = .707)?
(A) .8
(B) 1.6
(C) 3.2
(D) 16
(E) None of the above

For questions 6 through 10, V_C = -10V.

6. What is the output voltage (in Volts)?
(A) -10
(B) -5
(C) 0
(D) 5
(E) 10

7. What is the current (in μA) through resistor R_1?
(A) 0
(B) 100
(C) 250
(D) 500
(E) 1000

8. What is the FET static drain current (in μA)?
(A) 0
(B) 250
(C) 500
(D) 1000
(E) 2000

9. How many poles and zeros are there?
(A) None
(B) 1 zero
(C) 1 pole
(D) 1 pole and 1 zero
(E) 2 poles

10. For a steady state ac input, at what frequency is the output voltage down 3dB?
(A) .8
(B) 1.6
(C) 3.2
(D) 9.6
(E) A frequency does not exist.

593 (continued)

SOLUTION:

1. The FET is conducting when $V_C = 0V$. The on-resistance is low (1Ω) so that the positive input to the op amp is effectively grounded. Therefore, the amplifier acts as an inverter with a gain of one ($R_2/R_1 = 20K/20K = 1$).

$$V_o = -V_i\ (R_2/R_1) = -5V.$$

THE CORRECT ANSWER IS (B), -5.

2. The current through R_1 is,
$V_i/R_1 = 5/20{,}000 = 250\mu A$

THE CORRECT ANSWER IS (C), 250.

3. For a +5V input, the static drain current is,
$I_D = V_i/R_3 = 5/20{,}000 = 250\mu A.$

THE CORRECT ANSWER IS (B), 250.

4. T.F. = $-R_2/R_1[1/(R_2Cs + 1)]$. Thus, it is seen that there is

one pole at $s = -1/R_2C$.

THE CORRECT ANSWER IS (C), One pole.

5. By substituting in the circuit parameters, the absolute value of the transfer function (defined above) becomes,

$$1/(2x10^{-4}s + 1),\ \tau = 2 \times 10^{-4},\ \ \omega = 1/\tau$$

Therefore, $|V_o(j\omega)/V_i(j\omega)| = .707$ when $\omega = 10^4/2$,

or $f = \omega/2\pi = 0.796Hz$.

THE CORRECT ANSWER IS (E),none of these.

6. The FET is in cutoff at -10V. Thus,

$$V_o = V_i[-R_2/R_1) + (R_1 + R_2)/R_1] = 5[-1 + 2] = 5V.$$

THE CORRECT ANSWER IS (D), 5.

7. With the FET not conducting and the source essentially at +5V, there is no voltage drop across R_1. Thus, the current through R_1 is zero.

THE CORRECT ANSWER IS (A), 0.

593 (continued)

SOLUTION (cont.):

8. The static drain current through the FET is zero since it is in cutoff.

THE CORRECT ANSWER IS (A), 0.

9. The small signal transfer function is,

$$V_o(s)/V_i(s) = -Z_2(s)/R_1 + (R_1 + Z_2(s))/R_1 = R_1/R_1 = 1.$$

THE CORRECT ANSWER IS (A), None.

10. The transfer function is 1, as calculated above, which is independent of frequency. Thus, there is no frequency to satisfy the requirement that the output voltage be down 3dB.

THE CORRECT ANSWER IS (E), A frequency does not exist.

594 (Category 13)

The diagram below defines the sequence of a synchronous state machine. Three TTL flip-flops are used in its implementation, as shown in the partial logic diagram.

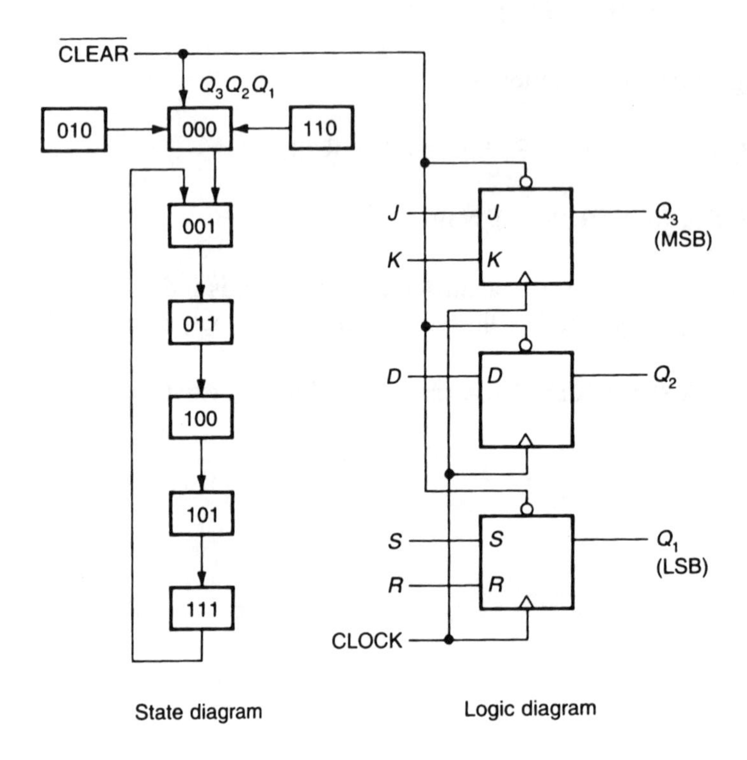

State diagram

Logic diagram

1. Since the two invalid states are not used, what is the purpose of their connection as shown?
(A) It provides an escape route back to the main loop.
(B) It serves no useful purpose.
(C) It could cause problems and should be deleted.
(D) The states can be used to simplify the logic equations.
(E) None of the above.

2. What sort of problem can occur with this implementation?
(A) Three different flip-flops are used.
(B) Gray code is not used.
(C) No problem since TTL logic is used.
(D) A decoding glitch can send the sequencer to a wrong state.
(E) None of the above.

594 (continued)

3. How can faithful operation be better assured?
(A) Use a two phase clock -- one to clock the flip-flops and the other for decoding to set up the flip-flop inputs.
(B) Use only J-K flip-flops.
(C) Change the state codes to a gray code sequence.
(D) (A) and (C).
(E) All of the above.

4. If standard 7400 series logic is used, what is the maximum safe clock rate?
(A) 15MHz
(B) 25MHz
(C) 40MHz
(D) 50MHz
(E) 75MHz

5. What is the simplified input equation for the J input?

(A) $\overline{Q_3}\, Q_2$

(B) $Q_3\, Q_2\, Q_1$

(C) $\overline{Q_3}\, Q_2\, Q_1$

(D) $Q_2\, \overline{Q_1}$

(E) Q_1

6. What is the simplified input equation for the K input?

(A) $Q_3\, Q_2\, Q_1$

(B) $Q_3\, Q_2$

(C) $Q_3\, Q_2\, \overline{Q_1}$

(D) $\overline{Q_3}\, Q_2$

(E) $\overline{Q_2}\, Q_1$

594 (continued)

7. What is the simplified input equation for the D input?

(A) $\overline{Q}_3 \overline{Q}_2 Q_1$

(B) $Q_3 \overline{Q}_2 Q_1$

(C) $\overline{Q}_3 Q_2$

(D) $\overline{Q}_2 Q_1$

(E) $Q_3 Q_2 Q_1$

8. What is the simplified input equation for the S input?

(A) $\overline{Q}_2$

(B) $\overline{Q}_3 \overline{Q}_2 \overline{Q}_1$

(C) $Q_3 \overline{Q}_2 \overline{Q}_1$

(D) $\overline{Q}_2 \overline{Q}_1$

(E) $Q_3 \overline{Q}_2 \overline{Q}_1$

9. What is the simplified input equation for the R input?

(A) $\overline{Q}_3 \overline{Q}_2$

(B) $\overline{Q}_3 Q_2 Q_1$

(C) $\overline{Q}_3 Q_2 \overline{Q}_1$

(D) $Q_3 Q_2 \overline{Q}_1$

(E) $\overline{Q}_3 Q_2$

10. If one of the invalid states were inserted into the main sequencer loop to eliminate the condition in which more than one flip-flop changes at a clock time, what would be the effect on the system?

(A) One clock period would be added to the loop time.
(B) The cycle time of the sequencer loop would increase.
(C) The sequencer would cycle through the loop using gray code.
(D) The added state would serve as an "idle" transition state.
(E) All of the above.

594 (continued)

SOLUTION:

1. By chance, if the sequencer finds itself in one of these invalid states it can find its way back to the main loop at the next clock pulse. Otherwise, the sequencer would be stuck in the invalid state forever.

• **THE CORRECT ANSWER IS (A), It provides an escape route back to the main loop.**

2. In several cases, gray code is not implemented; thus, more than one bit changes when going from one state to the next. It is possible that a decoding glitch can send the sequencer to an incorrect next state.

• **THE CORRECT ANSWER IS (D), A decoding glitch can send the sequencer to a wrong state.**

3. Since gray code is not used, it is possible for the sequencer to fall out of sequence if the system uses only one clock signal to clock the flip-flops and to do decoding between the flip-flop outputs and inputs. This problem can be eliminated by use of a two-phase clock and/or arranging the states so that only one state changes at a time.

• **THE CORRECT ANSWER IS (D), (A) and (C).**

4. Referring to a handbook for standard 7400 logic will lead to a maximum safe clock frequency of typically 25MHz.

• **THE CORRECT ANSWER IS (B), 25MHz.**

5. The J input must be set to a ONE prior to the next clock. There is only one state where the MSB goes from ZERO to ONE. Thus,

$$J = \overline{Q}_3 Q_2 Q_1$$

THE CORRECT ANSWER IS (C), $\overline{Q}_3 Q_2 Q_1$.

6. The K input must be set to a ZERO prior to the next clock. There are two states (one is an invalid state) where the MSB goes from a ONE to a ZERO. Thus,

$$K = Q_3 Q_2 Q_1 + Q_3 Q_2 \overline{Q}_1$$

Simplifying, $K = Q_3 Q_2$.

• **THE CORRECT ANSWER IS (B), $Q_3 Q_2$.**

594 (continued)

SOLUTION (cont.):

7. The output of the D flip-flop reflects the input after the clock pulse. The D input must be set to a ONE each time the next state is a ONE, even if the current state is already a ONE. All other times, the D input must be a ZERO. There are two states where the middle bit becomes a ONE (the invalid states don't count). Thus,

$$D = \overline{Q}_3 \overline{Q}_2 Q_1 + Q_3 \overline{Q}_2 Q_1$$

Simplifying, $D = \overline{Q}_2 Q_1$.

• **THE CORRECT ANSWER IS (D), $Q_2 Q_1$.**

8. The SR flip-flop can be implemented using two cross-coupled NAND gates. Each gate is fed by a NAND gate, one of whose inputs is the clock and the other the S or the R input. A ONE on the S input sets the output to a ONE. There are two states (in [ ] below) where the LSB changes from a ZERO to a ONE. There are also three more states where the output remains at a ONE (don't cares) that can be used to simplify the logic. Thus,

$$S = [\overline{Q}_3 \overline{Q}_2 \overline{Q}_1 + Q_3 \overline{Q}_2 \overline{Q}_1] + \overline{Q}_3 \overline{Q}_2 Q_1 + Q_3 \overline{Q}_2 Q_1 + Q_3 Q_2 Q_1$$

Simplifying, $S = \overline{Q}_2$.

• THE CORRECT ANSWER IS (A), $\overline{Q}_2$.

9. Similar to 8 above, the SR flip-flop output is set to a ZERO on the next clock pulse when the R input is a ONE. In addition to one transition from ONE to ZERO the two invalid states (don't cares) can be used to simplify the logic. Thus,

$$R = [\overline{Q}_3 Q_2 Q_1] + \overline{Q}_3 Q_2 \overline{Q}_1 + Q_3 Q_2 \overline{Q}_1$$

Simplifying, $S = \overline{Q}_3 Q_2$.

• **THE CORRECT ANSWER IS (E),** $\overline{Q}_3 Q_2$.

10. Each one of the choices is correct.

• **THE CORRECT ANSWER IS (E), All of the above.**

595 (Category 14)

A Johnson counter can be used as a multiphase clock generator. It generates non-overlapping pulses that provide for glitch-free decoding. A simplified block diagram based on a 74194 4-bit bidirectional universal shift register is shown below. After the shift register is initially cleared, a series of four ONEs will be right-shifted, followed by four ZEROs. This sequence repeats at the input clock rate. The sequence of ONEs and ZEROs can be decoded into eight separate clock phases as shown below.

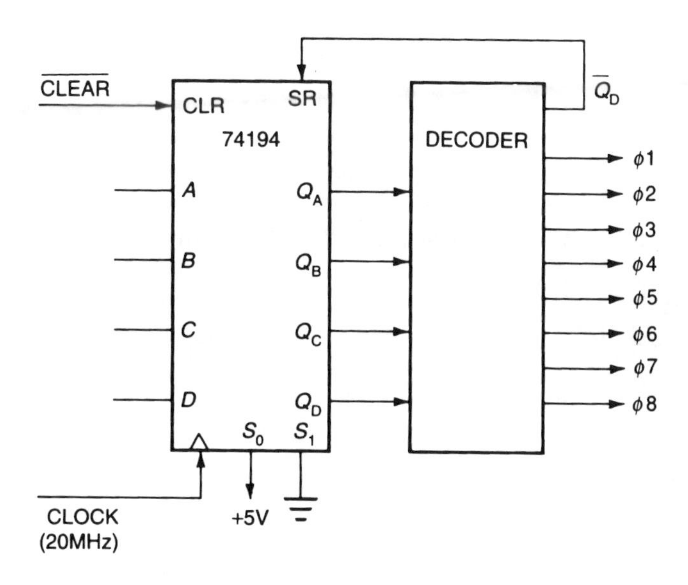

Each clock phase is a function of two shift register outputs. For example,

$\phi 1 = \overline{Q}_A \overline{Q}_D$, $\phi 2 = Q_A \overline{Q}_B$, $\phi 3 = Q_B \overline{Q}_C$, $\phi 4 = Q_C \overline{Q}_D$, $\phi 5 = Q_A Q_D$, etc.

Drawing a detailed logic diagram of the decoder and a timing diagram showing the relation of the input clock, the shift register outputs and the eight output phases will help in answering some of the following questions.

1. What is the maximum number of clock phases that a three-bit Johnson counter can generate?
   (A) two
   (B) four
   (C) six
   (D) eight
   (E) ten

595 (continued)

2. What is the active (high) pulse width of each of the output clock phases if the input clock frequency is 20MHz (assume 50% duty cycle)?
(A) 25ns
(B) 50ns
(C) 100ns
(D) 150ns
(E) 200ns

3. What is the frequency of each of the output clock phases?
(A) 40MHz
(B) 20MHz
(C) 10MHz
(D) 5MHz
(E) 2.5MHz

4. What is a maximum reliable frequency at which the 74194 shift register can be clocked?
(A) 10MHz
(B) 20MHz
(C) 35MHz
(D) 50MHz
(E) 100MHz

5. What is the logic equation for $\phi 6$?

(A) $\overline{Q_A}\ Q_B$

(B) $\overline{Q_B}\ Q_C$

(C) $\overline{Q_C}\ Q_D$

(D) $Q_A\ \overline{Q_D}$

(E)None of the above.

6. What is the logic equation for $\phi 7$?

(A) $\overline{Q_A}\ Q_B$

(B) $\overline{Q_B}\ Q_C$

(C) $\overline{Q_C}\ Q_D$

(D) $Q_A\ \overline{Q_D}$

(E)None of the above.

595 (continued)

7. What is the logic equation for $\phi 8$?

(A) $\overline{Q_A}\ Q_B$

(B) $\overline{Q_B}\ Q_C$

(C) $\overline{Q_C}\ Q_D$

(D) $Q_A\ \overline{Q_D}$

(E)None of the above.

8. How can the frequency of each of the eight output phases be increased?
(A) Increase the input clock frequency.
(B) Reduce the duty cycle of the input clock.
(C) Use a shift register having fewer bits.
(D) Use three, rather than two, shift register outputs to decode each clock phase.
(E) All of the above.

9. What would be the advantages/disadvantages of using LS logic rather than standard TTL?
(A) Higher speed capability.
(B) Lower power consumption.
(C) Higher operating temperature.
(D) Greater fanout.
(E) All of the above.

10. How can the shift register outputs be cleared to all ZEROs?
(A) Temporarily set Clear low.
(B) Set S_0, S_1 and input lines A through D to ONEs followed by a clock pulse.
(C) Temporarily power the chip OFF and then ON.
(D) (A) and (B).
(E) None of the above.

595 (continued)

SOLUTION:

1. A Johnson counter can generate 2n phases, where n is the length of the shift register. Since this is a 4-bit shift register, the maximum number of phases that can be generated is eight.

• **THE CORRECT ANSWER IS (C), eight.**

2. Each phase requires eight clock pulses. It has a 50% duty cycle. Therefore, the active portion is four clock pulses wide, or 200ns.

• **THE CORRECT ANSWER IS (E), 200 ns.**

3. Each phase is one eight the frequency of the input clock, or 2.5MHz.

• **THE CORRECT ANSWER IS (E), 2.5MHz.**

4. Referring to a TTL handbook for the 74194, the maximum frequency given is typically 36MHz. The closest answer is 35MHz.

• **THE CORRECT ANSWER IS (C), 35MHz.**

5. A timing diagram showing the input clock, the four shift register outputs and the eight clock phases would show that the logic equation for $\phi 6$ is $\overline{Q}_A Q_B$.

• **THE CORRECT ANSWER IS (A),** $\overline{Q}_A Q_B$.

6. Similar to question 5, the logic equation for $\phi 7$ is $\overline{Q}_B Q_C$.

• **THE CORRECT ANSWER IS (B),** $\overline{Q}_B Q_C$.

7. Similar to question 5, the logic equation for $\phi 8$ is $\overline{Q}_C Q_D$

• **THE CORRECT ANSWER IS (C),** $\overline{Q}_C Q_D$

8. The frequency of the output phases can be increased by raising the frequency of the input clock. This is the only valid choice.

• **THE CORRECT ANSWER IS (A), Increase the input clock frequency.**

9. LS logic uses less power and runs at about the same speed. The only valid choice given is lower power consumption.

• **THE CORRECT ANSWER IS (B), Lower power consumption.**

595 (continued)

SOLUTION (cont.):

10. Choices (A) and (B) are two ways that the shift register may be cleared to all ZEROs.

• **THE CORRECT ANSWER IS (D), (A) and (B).**

596 (Category 16)

This problem concerns high frequency transmission line parameters and attenuators. Some questions refer to the equivalent T-section transmission line lump and the attenuator configurations, below.

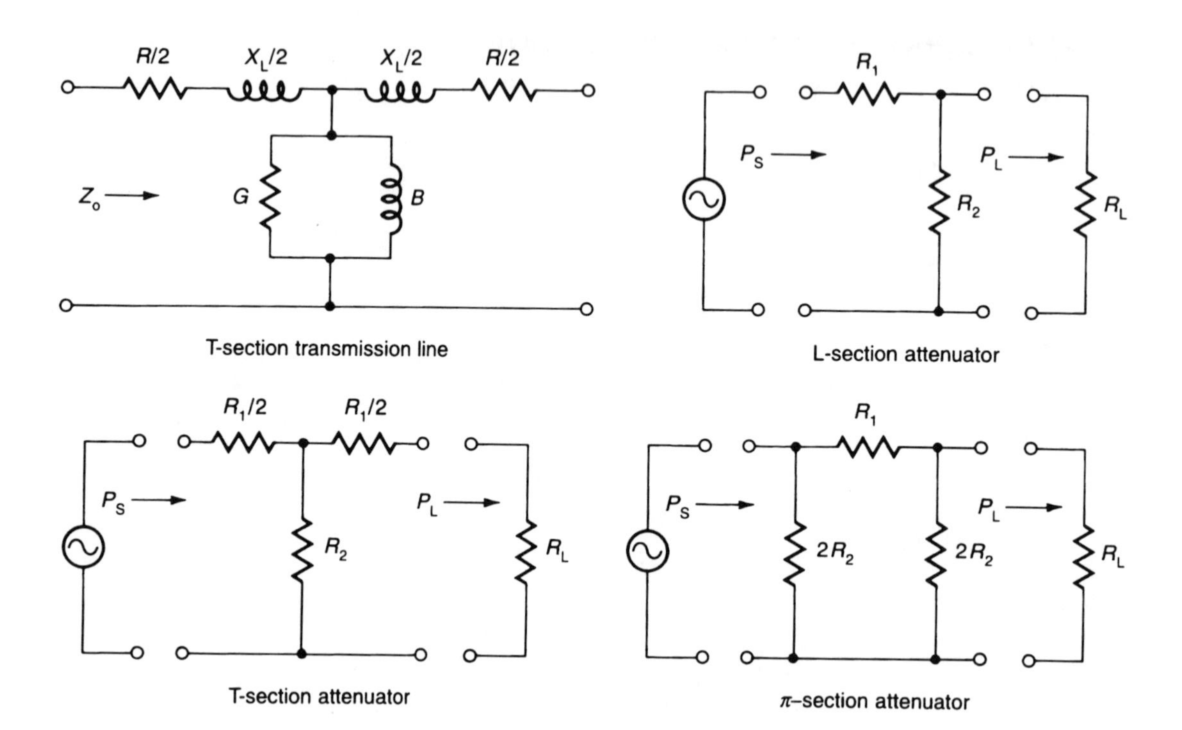

1. The equation for the characteristic impedance, Z_o, of a transmission line is , $\sqrt{Z/Y}$ where $Z = R + jX_L$ (series component) and $Y = G + jB$ (shunt component). At high frequencies certain assumptions can be made that lead to the following:
(A) Resistive elements of the transmission line can be ignored.
(B) Characteristic impedance becomes $\sqrt{L/C}$.
(C) $Z_o = R_o$.
(D) $Z = jX_L$.
(E) All of the above.

2. If the transmission line is a coaxial cable, the dielectric constant of the insulating material between the center conductor and outer shield, ϵ, becomes important. In this case, which choice for velocity is correct (c is the speed of light)?

(A) $v = c/\sqrt{\varepsilon}$
(B) $v = c/\ \varepsilon$
(C) $v = c\ \sqrt{\varepsilon}$
(D) $v = c\varepsilon$
(E) None of the above

596 (continued)

3. Which is correct for the propagation constant, γ, of a transmission line?

(A) $= \sqrt{ZY}$

(B) γ is made up of attenuation and phase shift.
(C) The attenuation factor is insignificant at high frequencies.
(D) At high frequencies, $\gamma = j\omega LC$
(E) All of the above.

4. Which passive attenuator characteristic is correct?
(A) An attenuator should not change the apparent load impedance.
(B) Power from the source should not change.
(C) Power to the load should be less.
(D) The attenuator will absorb some power.
(E) All of the above.

5. For an L-section 10dB passive attenuator, if load resistance, R_L, is 500Ω, what is the value of R_2?

(A) 168Ω
(B) 207Ω
(C) 231Ω
(D) 249Ω
(E) 347Ω

6. For an L-section 10dB passive attenuator, if load resistance, R_L, is 500Ω and R_2 is assumed to be 250Ω, what is the value of R_1?
(A) 351Ω
(B) 664Ω
(C) 333Ω
(D) 175Ω
(E) 212Ω

7. For a T-section 12dB passive attenuator, if load resistance, R_L, is 72Ω, what is the value of R_2?
(A) 30Ω
(B) 60Ω
(C) 45Ω
(D) 65Ω
(E) 87Ω

8. For a T-section 12dB passive attenuator, if load resistance, R_L, is 72Ω, what is the value of R_1?
(A) 72Ω
(B) 144Ω
(C) 265Ω
(D) 134Ω
(E) 179Ω

596 (continued)

9. For a π-section 12dB passive attenuator, if load resistance, R_L, is 72Ω, what is the value of R_2?

(A) 87Ω
(B) 72Ω
(C) 65Ω
(D) 45Ω
(E) 60Ω

10. For a π-section 12dB passive attenuator, if load resistance, R_L, is 72Ω, what is the value of R_1?

(A) 134Ω
(B) 179Ω
(C) 144Ω
(D) 72Ω
(E) 265Ω

596 (continued)

SOLUTION:

1. All choices are correct.

• **THE CORRECT ANSWER IS (E), All of the above.**

2. Velocity is inversely proportional to the square root of the dielectric constant.

• **THE CORRECT ANSWER IS (A),** $v = \frac{c}{\sqrt{\varepsilon}}$.

3. All choices are correct.

• **THE CORRECT ANSWER IS (E), All of the above.**

4. All choices are correct.

• **THE CORRECT ANSWER IS (E), All of the above.**

5. $R_2 = R_L/[10^{dB/20}-1] = 500/[10^{½}-1] = 231.24Ω$.

• **THE CORRECT ANSWER IS (C), 231Ω.**

6. $R_1 = R_L^2/[R_2 + R_L] = 500^2/[250 + 500] = 333.33Ω$.

• **THE CORRECT ANSWER IS (C), 333Ω.**

7. $dB = 20\log E_1/E_2 = 20\log K$ (the reflection coefficient).

$12 = 20\log K$, $K = 3.981$, $2R_2 = R_L[(K+1)/K-1)] = 72(4.981)/(2.981) = 120.3Ω$
$R_2 = 60.15Ω$

• **THE CORRECT ANSWER IS (B), 60Ω**

8. $R_1 = R_L^2/[R_2-(R_L^2/4R_2)] = 72^2/[60.15-(72^2/240.6)] = 134.3Ω$.

• **THE CORRECT ANSWER IS (D), 134 Ω.**

9. R_2 is the same as was calculated in problem 7.

• **THE CORRECT ANSWER IS (E), 60 Ω.**

10. R_1 is the same as was calculated in problem 8.

• **THE CORRECT ANSWER IS (A), 134Ω.**

597 (Category 17)

The following questions relate to the transmission of digital signals over balanced and unbalanced transmission lines. Design requirements for this type of data transmission are defined by EIA Standards RS-422/432.

1. Which of the following is **NOT** a controlling factor in a voltage digital interface?
(A) Cable length.
(B) Temperature.
(C) Modulation rate.
(D) Interconnection cable characteristics.
(E) Signal rise time.

2. Which of the following choices is correct? Unbalanced (vs. balanced) interfaces,
(A) Are intended for lower modulation rates.
(B) Are more susceptible to noise that can corrupt data.
(C) Need more isolation between data and clock signals.
(D) Operate over shorter cable lengths.
(E) All of the above.

3. Which does **NOT** influence maximum cable length?
(A) Amount of signal distortion.
(B) Cable termination above 200 Kbaud or where rise time is four times the one way propagation delay time of the cable.
(C) Baud rate.
(D) Amount of common-mode noise.
(E) None of the above.

4. Which of the following is incorrect? Modulation rate for use on data, timing or control circuits,
(A) is speced below 100 Kbaud for unbalanced interfaces.
(B) is speced up to 10 Mbaud for balanced interfaces.
(C) must be lowest at 50% duty cycle where distortion is highest.
(D) may be so high (especially where the duty cycle greatly deviates from 50%) that the signal may not have a fast enough rise or fall time to achieve its binary logic threshold before the next binary bit time.
(E) need not be met over its entire specified range.

5. Which choice is correct? Distortion can be caused by or affected by:
(A) Signal rise time.
(B) Duty cycle.
(C) Cable attenuation.
(D) Cable delay.
(E) All of the above.

597 (continued)

6. Which of the following driver specs is correct?
(A) An unbalanced driver circuit should be a low impedance (50Ω or less) voltage source that will produce a voltage applied to the interconnecting cable in the range of 4V to 6V.
(B) The voltage across a 450Ω resistor connected between an unbalanced driver output and ground should not be less than 90% of the magnitude of either binary state.
(C) A balanced driver circuit should be a low impedance (100Ω or less) balanced voltage source that will produce a differential voltage applied to the interconnecting cable in the range of 2V to 6V.
(D) The voltage across a 100Ω resistor connected between balanced driver output terminals shall be not less than 2.0V.
(E) All of the above.

7. Which is incorrect for a binary signal?
(A) The frequency (reciprocal of the period) of the signal is the bit rate.
(B) The reciprocal of the period of the active portion of the signal is the baud rate.
(C) For NRZ coding, the baud rate is equal to the bit rate.
(D) NRZ coding is the most efficient telecommunication code.
(E) None of the above.

8. Which statement is **NOT** correct for unbalanced data transmission?
(A) Uses a single conductor with voltage referenced to a common signal ground.
(B) Only one line is switched.
(C) Operates reliably in noisy environments.
(D) For multiple channels, approximately half the number of conductors are needed as compared to balanced transmission.
(E) Cabling and connector costs can be lower than for balanced transmission.

9. Which statement is **NOT** correct for balanced data transmission?
(A) Requires two conductors per signal.
(B) Two lines are switched during data transmission.
(C) The binary states are referenced by the difference of potential between the lines and not with respect to ground.
(D) Differential transceivers do not work well in a noisy environment.
(E) A differential probe is handy for viewing differential signals on an oscilloscope.

597 (continued)

10. Of the coding schemes described below, which description is **NOT** correct?
(A) For NRZ, instead of discrete pulses for each data bit, the signal rises or falls only when a ZERO bit is followed by a ONE bit, or a ONE by a ZERO.
(B) For NRZ1, each ONE causes the signal to change state.
(C) For FM, every bit cell interval changes state at the start of the cell. ONEs are marked by an additional state change at the middle of the cell, thus doubling the frequency for a series of ONEs.
(D) For MFM, a state change occurs at the beginning of a cell only if no ONE exists in both the preceding and current bit cell. A ONE in a cell causes a transition in the middle of the cell.
(E) For RLL 2,7 code, each ONE is separated by either two or seven ZEROs.

597 (continued)

<u>*SOLUTION:*</u>

1. Within the thermal operating range defined by the components being used, temperature should not affect operation of the interface. All other factors have an effect on operation of the interface and must be considered in the design.

- **THE CORRECT ANSWER IS (B), Temperature.**

2. Unbalanced transmission lines are not as robust for long distance transmission. All of the choices are correct.

- **THE CORRECT ANSWER IS (E), All of the above.**

3. All choices have an influence on maximum cable length. Therefore, the answer is a double negative response.

- **THE CORRECT ANSWER IS (E), None of the above.**

4. Selection (E) is incorrect. Duty cycle of the transmitted signal contributes to distortion. At a duty cycle of 50%, distortion is lowest and can, thus, support the highest modulation rate.

- **THE CORRECT ANSWER IS (C), must be lowest at 50% duty cycle where distortion is highest.**

5. All choices are correct

- **THE CORRECT ANSWER IS (E), All of the above.**

6. All choices are correct.

- **THE CORRECT ANSWER IS (E), All of the above.**

7. Every choice is correct. Therefore, there is no incorrect answer.

- **THE CORRECT ANSWER IS (E), None of the above.**

8. One of the main disadvantages of unbalanced data transmission is that is shares a common ground which makes it inherently more susceptible to noise. Therefore, choice (C) is incorrect.

- **THE CORRECT ANSWER IS (C), Operates reliably in noisy environments.**

597 (continued)

SOLUTION (cont.):

9. Differential transceivers are ideal for use in a noisy environment. Therefore, choice (D) is incorrect.

• THE CORRECT ANSWER IS (D), Differential transceivers do not work well in a noisy environment.

10. All choices are correct except for (E). It should state that each ONE is separated by any number of ZEROs from two to seven.

• THE CORRECT ANSWER IS (E), For RLL 2,7 code, each ONE is separated by either two or seven ZEROs.

For additional information please contact:

National Council of Examiners for Engineering and Surveying (NCEES), P.O. Box 1686, Clemson, S.C. 29633-1686;
Telephone 864-654-6824, FAX 864-654-6033.

This page left intentionally blank.

Appendix A

Engineering Economics Problems and Solutions

Donald G. Newnan

This page left intentionally blank.

Problems and Solutions

A-1. A loan was made $2^{1}/_{2}$ years ago at 8% simple annual interest. The principal amount of the loan has just been repaid along with $600 of interest. The principal amount of the loan was closest to

(a) $300
(b) $3000
(c) $4000
(d) $5000
(e) $7500

Solution

$$F = P + Pin$$

$$600 + P = P + P(0.08)(2.50)$$

$$P = [600]/[0.08(2.50)] = \$3000$$

The answer is (b).

A-2. A $1000 loan was made at 10% simple annual interest. It will take how many years for the amount of the loan and interest to equal $1700?

(a) 6 years
(b) 7 years
(c) 8 years
(d) 9 years
(e) 10 years

Solution

$$F = P + Pin$$

$$1700 = 1000 + 1000(0.10)(n)$$

$$n = [700]/[1000(0.10)] = 7 \text{ years}$$

The answer is (b).

A-3. A retirement fund earns 8% interest, compounded quarterly. If $400 is deposited every three months for 25 years, the amount in the fund at the end of 25 years is nearest to

(a) $50,000
(b) $75,000
(c) $100,000
(d) $125,000
(e) $150,000

Solution

$$F = A(F/A,i\%,n) = 400(F/A,2\%,100)$$

$$= 400(312.23) = \$124,890$$

The answer is (d).

A-4. For some interest rate i, and some number of interest periods n, the uniform series capital recovery factor is 0.2091 and the sinking fund factor is 0.1941. The interest rate i must be closest to

(a) $1\frac{1}{2}\%$
(b) 2%
(c) 3%
(d) 4%
(e) 5%

Solution

The relationship between the capital recovery factor and the sinking fund factor is $(A/P,i\%,n) = (A/F,i\%,n) + i$. Substituting the values in the problem

$$0.2091 = 0.1941 + i$$
$$i = 0.2091 - 0.1941 = 0.015 = 1\tfrac{1}{2}\%$$

The answer is (a).

A-5. The repair costs for some handheld equipment is estimated to be \$120 the first year, increasing by \$30 per year in subsequent years. The amount a person will need to deposit into a bank account, paying 4% interest, to provide for the repair costs for the next five years is nearest to

(a) \$500
(b) \$600
(c) \$700
(d) \$800
(e) \$900

Solution

$$\begin{aligned} P &= A(P/A,i\%,n) + G(P/G,i\%,n) \\ &= 120(P/A,4\%,5) + 30(P/G,4\%,5) \\ &= 120(4.452) + 30(8.555) = \$791 \end{aligned}$$

The answer is (d).

A-6. An "annuity" is defined as the

(a) earned interest due at the end of each interest period
(b) cost of producing a product or rendering a service
(c) total annual overhead assigned to a unit of production
(d) amount of interest earned by a unit of principal in a unit of time
(e) series of equal payments occurring at equal periods of time

Solution

The answer is (e).

A-7. One thousand dollars is borrowed for one year at an interest rate of 1% per month. If this same sum of money is borrowed for the same period at an interest rate of 12% per year, the saving in interest charges is closest to

(a) \$0
(b) \$3
(c) \$5
(d) \$7
(e) \$14

Solution

At i = 1%/month: $F = 1000(1 + 0.01)^{12} = \$1126.83$

At i = 12%/year: $F = 1000(1 + 0.12)^{1} = 1120.00$

Saving in interest charges = 1126.83 – 1120.00 = \$6.83

The answer is (d).

A-8. How much should a person invest in a fund that will pay 9%, compounded continuously, if he wishes to have \$10,000 in the fund at the end of 10 years? The amount is nearest to

(a) \$4000
(b) \$5000
(c) \$6000
(d) \$7000
(e) \$8000

Solution

$$P = Fe^{-rn} = 10{,}000e^{-0.09(10)} = 4066$$

The answer is (a).

A-9. A store charges $1\frac{1}{2}$ % interest per month on credit purchases. This is equivalent to a nominal annual interest rate of

(a) 1.5%
(b) 15.0%
(c) 18.0%
(d) 19.6%
(e) 21.0%

Solution

The nominal interest rate is the annual interest rate ignoring the effect of any compounding. Nominal interest rate = $1\frac{1}{2}\% \times 12 = 18\%$. The answer is (c).

A-10. A small company borrowed \$10,000 to expand its business. The entire principal of \$10,000 will be repaid in two years, but quarterly interest of \$330 must be paid every three months. The nominal annual interest rate the company is paying is closest to

(a) 3.3%
(b) 5.0%
(c) 6.6%
(d) 10.0%
(e) 13.2%

Solution

The interest paid per year = 330 × 4 = 1320. The nominal annual interest rate = 1320/10,000 = 0.132 = 13.2%. The answer is (e).

A-11. A store policy is to charge 3% interest every two months on the unpaid balance in charge accounts. The effective interest rate is closest to

(a) 6%
(b) 12%
(c) 15%
(d) 18%
(e) 19%

Solution

$i_{\text{eff}} = (1 + r/m)^m - 1 = (1 + 0.03)^6 - 1 = 0.194 = 19.4\%$

The answer is (e).

A-12. The effective interest rate is 19.56%. If there are 12 compounding periods per year, the nominal interest rate is closest to

(a) 1.5%
(b) 4.5%
(c) 9.0%
(d) 18.0%
(e) 19.6%

Solution

$i_{\text{eff}} = (1 + r/m)^m - 1$

$r/m = (1 + i_{\text{eff}})^{1/m} - 1 = (1 + 0.1956)^{1/12} - 1 = 0.015$

$r = 0.015(m) = 0.015 \times 12 = 0.18 = 18\%$

The answer is (d).

A-13. A deposit of $300 was made one year ago into an account paying monthly interest. If the account now has $320.52, the effective annual interest rate is closest to

(a) 7%
(b) 10%
(c) 12%
(d) 15%
(e) 18%

Solution

$i_{\text{eff}} = 20.52/300 = 0.0684 = 6.84\%$

The answer is (a).

A-14. In a situation where the effective interest rate per year is 12%, based on monthly compounding, the nominal interest rate per year is closest to

(a) 8.5%
(b) 9.3%
(c) 10.0%
(d) 11.4%
(e) 12.0%

Solution

$i_{eff} = (1 + r/m)^m - 1$

$0.12 = (1 + r/12)^{12} - 1$

$(1.12)^{1/12} = (1 + r/12)$

$1.00949 = (1 + r/12)$

$r = 0.00949 \times 12 = 0.1138 = 11.38\%$

The answer is (d).

A-15. If 10% nominal annual interest is compounded daily, the effective annual interest rate is nearest to

(a) 10.00%
(b) 10.38%
(c) 10.50%
(d) 10.75%
(e) 18.00%

Solution

$i_{eff} = (1 + r/m)^m - 1 = (1 + 0.10/365)^{365} - 1 = 0.1052 = 10.52\%$

The answer is (c).

A-16. If 10% nominal annual interest is compounded continuously, the effective annual interest rate is nearest to

(a) 10.00%
(b) 10.38%
(c) 10.50%
(d) 10.75%
(e) 18.00%

Solution

$i_{eff} = e^r - 1$

where r = nominal annual interest rate

$i_{eff} = e^{0.10} - 1 = 0.10517 = 10.52\%$

The answer is (c).

A-17. If the quarterly effective interest rate is $5\frac{1}{2}\%$ with continuous compounding, the nominal interest rate is nearest to

(a) 5.5%
(b) 11.0%
(c) 16.5%
(d) 21.4%
(e) 22.0%

Solution

For 3 months: $i_{eff} = e^r - 1;\quad 0.055 = e^r - 1$

The rate per quarter year is $r = \log_e(1.055) = 0.05354;\quad r = 4 \times 0.05354 = 0.214 = 21.4\%$ per year

The answer is (d).

A-18. A continuously compounded loan has what effective interest rate if the nominal interest rate is 25%? Select one of the five choices.

(a) $e^{1.25}$
(b) $e^{0.25}$
(c) $\log_e(1.25)$
(d) $\log_e(0.25)$
(e) $e^{0.25} - 1$

Solution

$i_{eff} = e^r - 1 = e^{0.25} - 1$

The answer is (e).

A-19. A continuously compounded loan has what *nominal interest rate* if the *effective interest rate* is 25%? Select one of the five choices.

(a) $e^{1.25}$
(b) $e^{0.25}$
(c) $\log_e(1.25)$
(d) $\log_e(0.25)$
(e) $\log_{10}(1.25)$

Solution

$i_{eff} = e^r - 1 = 0.25 \quad e^r = 1.25$
$\log_e(e^r) = \log_e(1.25)$

$r = \log_e(1.25)$

The answer is (c).

A-20. An individual wishes to deposit a certain quantity of money now so that he will have $500 at the end of five years. With interest at 4% per year, compounded semiannually, the amount of the deposit is nearest to

(a) $340
(b) $400
(c) $410
(d) $416
(e) $608

Solution

$P = F(P/F,i\%,n) = 500(P/F,2\%,10) = 500(0.8203) = \$410$

The answer is (c).

A-21. A steam boiler is purchased on the basis of guaranteed performance. A test indicates that the operating cost will be $300 more per year than the manufacturer guaranteed. If the expected life of the boiler is 20 years, and money is worth 8%, the amount the purchaser should deduct from the purchase price to compensate for the extra operating cost is nearest to

(a) \$2950 (d) \$5520
(b) \$3320 (e) \$6000
(c) \$4100

Solution

$P = 300(P/A,8\%,20) = 300(9.818) = \$2945$

The answer is (a).

A-22. A consulting engineer bought a fax machine. There will be no maintenance cost the first year as it was sold with one year's free maintenance. In the second year the maintenance is estimated at \$20. In subsequent years the maintenance cost will increase \$20 per year (that is, 3rd year maintenance will be \$40, 4th year maintenance will be \$60, and so forth). The amount that must be set aside now at 6% interest to pay the maintenance costs on the fax machine for the first six years of ownership is nearest to

(a) \$101 (d) \$284
(b) \$164 (e) \$300
(c) \$229

Solution

Using single payment present worth factors:

$P = 20(P/F,6\%,2) + 40(P/F,6\%,3) + 60(P/F,6\%,4) + 80(P/F,6\%,5) + 100(P/F,6\%,6) = \$229$

Alternate solution using the gradient present worth factor:

$P = 20(P/G,6\%,6) = 20(11.459) = \$229$

The answer is (c).

A-23. An investor is considering buying a 20-year corporate bond. The bond has a face value of \$1000 and pays 6% interest per year in two semiannual payments. Thus the purchaser of the bond will receive \$30 every six months, and in addition he will receive \$1000 at the end of 20 years, along with the last \$30 interest payment. If the investor believes he should receive 8% annual interest, compounded semiannually, the amount he is willing to pay for the bond value is closest to

(a) \$500 (d) \$800
(b) \$600 (e) \$900
(c) \$700

Solution

$PW = 30(P/A,4\%,40) + 1000(P/F,4\%,40) = 30(19.793) + 1000(0.2083) = \$802$

The answer is (d).

A-24. Annual maintenance costs for a particular section of highway pavement are $2000. The placement of a new surface would reduce the annual maintenance cost to $500 per year for the first five years and to $1000 per year for the next five years. The annual maintenance after ten years would again be $2000. If maintenance costs are the only saving, the maximum investment that can be justified for the new surface, with interest at 4%, is closest to

(a) $5,500 (d) $10,340
(b) $7,170 (e) $12,500
(c) $10,000

Solution

Benefits are $1500 per year for the first five years and $1000 per year for the subsequent five years.

As Fig. A-24 indicates, the benefits may be considered as $1000 per year for ten years, plus an additional $500 benefit in each of the first five years.

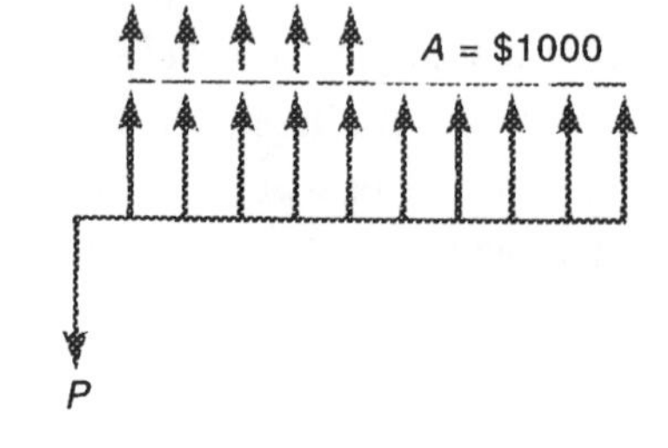

Fig. A-24

Maximum investment = Present worth of benefits

$$= 1000(P/A,4\%,10) + 500(P/A,4\%,5)$$
$$= 1000((8.111) + 500(4.452) = \$10,337$$

The answer is (d).

A-25. A project has an initial cost of $10,000, uniform annual benefits of $2400, and a salvage value of $3000 at the end of its 10-year useful life. At 12% interest the net present worth of the project is closest to

(a) $2500 (d) $5500
(b) $3500 (e) $6500
(c) $4500

Solution

$$\text{NPW} = \text{PW of benefits} - \text{PW of cost}$$
$$= 2400(P/A,12\%,10) + 3000(P/F,12\%,10) - 10{,}000 = \$4526$$

The answer is (c).

A-26. A person borrows $5000 at an interest rate of 18%, compounded monthly. Monthly payments of $167.10 are agreed upon. The length of the loan is closest to

(a) 12 months (d) 30 months
(b) 20 months (e) 40 months
(c) 24 months

Solution

$$\text{PW of benefits} = \text{PW of cost}$$
$$5000 = 167.10(P/A,1.5\%,n)$$
$$(P/A,1.5\%,n) = 5000/167.10 = 29.92$$

From the $1^1/_2\%$ interest table, $n = 40$. The answer is (e).

A-27. A machine costing $2000 to buy and $300 per year to operate will save labor expenses of $650 per year for eight years. The machine will be purchased if its salvage value at the end of eight years is sufficiently large to make the investment economically attractive. If an interest rate of 10% is used, the minimum salvage value must be closest to

(a) $100 (d) $400

(b) $200 (e) $500

(c) $300

Solution

$$\begin{aligned}\text{NPW} &= \text{PW of benefits} - \text{PW of cost} = 0\\ &= (650 - 300)(P/A,10\%,8) + S_8(P/F,10\%,8) - 2000 = 0\\ &= 350(5.335) + S_8(0.4665) - 2000 = 0\end{aligned}$$

$$S_8 = 132.75/0.4665 = \$285$$

The answer is (c).

A-28. The amount of money deposited 50 years ago at 8% interest that would now provide a perpetual payment of $10,000 per year is nearest to

(a) $3,000 (d) $70,000

(b) $8,000 (e) $90,000

(c) $50,000

Solution

The amount of money needed now to begin the perpetual payments is $P' = A/i = 10{,}000/0.08 = 125{,}000$. From this we can compute the amount of money, P, that would need to have been deposited 50 years ago:

$$P = 125{,}000(P/F,8\%,50) = 125{,}000(0.0213) = \$2663$$

The answer is (a).

A-29. An industrial firm must pay a local jurisdiction the cost to expand its sewage treatment plant. In addition, the firm must pay $12,000 annually toward the plant operating costs. The industrial firm will pay sufficient money into a fund, that earns 5% per year, to pay its share of the plant operating costs forever. The amount to be paid to the fund is nearest to

(a) $15,000 (d) $120,000

(b) $30,000 (e) $240,000

(c) $60,000

Solution

$$P = A/i = 12000/0.05 = \$240{,}000$$

The answer is (e).

A-30. At an interest rate of 2% per month, money will double in value in how many months?

(a) 20 months
(b) 22 months
(c) 24 months
(d) 30 months
(e) 35 months

Solution

$2 = 1(F/P,i\%,n)$

$(F/P,2\%,n) = 2$

From the 2% interest table, n = about 35 months. The answer is (e).

A-31. A woman deposited $10,000 into an account at her credit union. The money was left on deposit for 80 months. During the first 50 months the woman earned 12% interest, compounded monthly. The credit union then changed its interest policy so that the woman earned 8% interest compounded quarterly during the next 30 months. The amount of money in the account at the end of 80 months is nearest to

(a) $10,000
(b) $12,500
(c) $15,000
(d) $17,500
(e) $20,000

Solution

At end of 50 months

$$F = 10{,}000(F/P,1\%,50) = 10{,}000(1.645) = \$16{,}450$$

At end of 80 months

$$F = 16{,}450(F/P,2\%,10) = 16{,}450(1.219) = \$20{,}053$$

The answer is (e).

A-32. An engineer deposited $200 quarterly in her savings account for three years at 6% interest, compounded quarterly. Then for five years she made no deposits or withdrawals. The amount in the account after eight years is closest to

(a) $1200
(b) $1800
(c) $2400
(d) $3000
(e) $3600

Solution

$$\begin{aligned} \text{FW} &= 200(F/A,1.5\%,12)(F/P,1.5\%,20) \\ &= 200(13.041)(1.347) = \$3513 \end{aligned}$$

The answer is (e).

A-33. A sum of money, Q, will be received six years from now. At 6% annual interest the present worth now of Q is $60. At this same interest rate the value of Q ten years from now is closest to

(a) $60 (d) $107
(b) $77 (e) $120
(c) $90

Solution

The present sum $P = 60$ is equivalent to Q six years hence at 6% interest. The future sum F may be calculated by either of two methods:

$$F = Q(F/P,6\%,4) \text{ and } Q = 60\ (F/P,6\%,6) \quad (1)$$

$$F = P(F/P,6\%,10) \quad (2)$$

Since P is known, the second equation may be solved directly.

$$F = P(F/P,6\%,10) = 60(1.791) = \$107$$

The answer is (d).

A-34. If $200 is deposited in a savings account at the beginning of each of 15 years and the account earns interest at 6%, compounded annually, the value of the account at the end of 15 years will be most nearly

(a) $4500 (d) $5100
(b) $4700 (e) $5300
(c) $4900

Solution

$$F' = A(F/A,i\%,n) = 200(F/A,6\%,15) = 200(23.276) = \$4655.20$$

$$F = F'(F/P,i\%,n) = 4655.20(F/P,6\%,1) = 4655.20(1.06) = \$4935$$

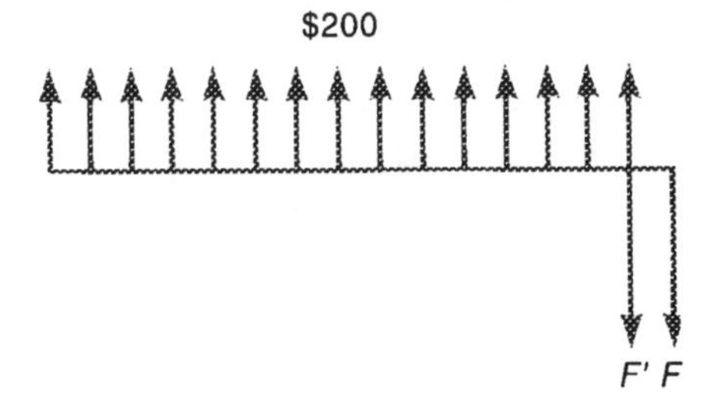

Fig. A-34

The answer is (c).

A-35. The maintenance expense on a piece of machinery is estimated as follows:

Year	1	2	3	4
Maintenance	\$150	\$300	\$450	\$600

If interest is 8%, the equivalent uniform annual maintenance cost is closest to

(a) \$250 (d) \$400
(b) \$300 (e) \$450
(c) \$350

Solution

$$\text{EUAC} = 150 + 150(A/G,8\%,4) = 150 + 150(1.404) = \$361$$

The answer is (c).

A-36. A payment of \$12,000 six years from now is equivalent, at 10% interest, to an annual payment for eight years starting at the end of this year. The annual payment is closest to

(a) \$1000 (d) \$1600
(b) \$1200 (e) \$1800
(c) \$1400

Solution

$$\begin{aligned}\text{Annual payment} &= 12{,}000(P/F,10\%,6)(A/P,10\%,8)\\ &= 12{,}000(0.5645)(0.1874) = \$1269\end{aligned}$$

The answer is (b).

A-37. A manufacturer purchased \$15,000 worth of equipment with a useful life of six years and a \$2000 salvage value at the end of the six years. Assuming a 12% interest rate, the equivalent uniform annual cost is nearest to

(a) \$1500 (d) \$4500
(b) \$2500 (e) \$5500
(c) \$3500

Solution

$$\begin{aligned}\text{EUAC} &= 15{,}000(A/P,12\%,6) - 2000(A/F,12\%,6)\\ &= 15{,}000(0.2432) - 2000(0.1232) = \$3402\end{aligned}$$

The answer is (c).

A-38. Consider a machine as follows:

Initial cost: \$80,000

End-of-useful life salvage value: \$20,000

Annual operating cost: \$18,000

Useful life: 20 years

Based on 10% interest, the equivalent uniform annual cost for the machine is closest to

(a) \$21,000 (d) \$27,000

(b) \$23,000 (e) \$29,000

(c) \$25,000

Solution

$$\begin{aligned}\text{EUAC} &= 80{,}000\,(A/P,10\%,20) - 20{,}000\,(A/F,10\%,20) + \text{annual operating cost}\\ &= 80{,}000\,(0.1175) - 20{,}000\,(0.0175) + 18{,}000\\ &= 9400 - 350 + 18{,}000 = \$27{,}050\end{aligned}$$

The answer is (d).

A-39. Consider a machine as follows:

Initial cost: \$80,000

Annual operating cost: \$18,000

Useful life: 20 years

What must be the salvage value of the machine at the end of 20 years for the machine to have an equivalent uniform annual cost of \$27,000? Assume a 10% interest rate. The salvage value is closest to

(a) \$10,000 (d) \$40,000

(b) \$20,000 (e) \$50,000

(c) \$30,000

Solution

$$\begin{aligned}\text{EUAC} &= \text{EUAB}\\ 27{,}000 &= 80{,}000(A/P,10\%,20) + 18{,}000 - S(A/F,10\%,20)\\ &= 80{,}000(0.1175) + 18{,}000 - S(0.0175)\end{aligned}$$

$$S = (27{,}400 - 27{,}000)/0.0175 = \$22{,}857$$

The answer is (b).

A-40. Twenty-five thousand dollars is deposited in a savings account that pays 5% interest, compounded semiannually. Equal annual withdrawals are to be made from the account beginning one year from now and continuing forever. The maximum amount of the equal annual withdrawals is closest to

(a) \$625 (d) \$1265

(b) \$1000 (e) \$1365

(c) \$1250

Solution

The general equation for an infinite life, $P = A/i$, must be used to solve the problem.

$i_{eff} = (1 + 0.025)^2 - 1 = 0.050625$

The maximum annual withdrawal will be $A = Pi = 25{,}000(0.050625) = \$1266$

The answer is (d).

A-41. An investor is considering the investment of $10,000 in a piece of land. The property taxes are $100 per year. The lowest selling price the investor must receive if she wishes to earn a 10% interest rate after keeping the land for 10 years is

(a) $20,000
(b) $21,000
(c) $23,000
(d) $25,000
(e) $27,000

Solution

$$\begin{aligned}\text{Minimum sale price} &= 10{,}000(F/P,10\%,10) + 100(F/A,10\%,10)\\ &= 10{,}000(2.594) + 100(15.937) = \$27{,}530\end{aligned}$$

The answer is (e).

A-42. The rate of return for a $10,000 investment that will yield $1000 per year for 20 years is closest to

(a) 1%
(b) 4%
(c) 8%
(d) 12%
(e) 18%

Solution

$\text{NPW} = 1000(P/A,i\%,20) - 10{,}000 = 0$

$(P/A,i\%,20) = 10{,}000/1000 = 10$

From interest tables: $6\% < i < 8\%$. The answer is (c).

A-43. An engineer invested $10,000 in a company. In return he received $600 per year for six years and his $10,000 investment back at the end of the six years. His rate of return on the investment was closest to

(a) 6%
(b) 10%
(c) 12%
(d) 15%
(e) 18%

Solution

The rate of return was = $600/10{,}000 = 0.06 = 6\%$

The answer is (a).

A-44. An engineer made ten annual end-of-year purchases of $1000 of common stock. At the end of the tenth year, just after the last purchase, the engineer sold all the stock for $12,000. The rate of return received on the investment is closest to

(a) 2%
(b) 4%
(c) 8%
(d) 10%
(e) 12%

Solution

$$F = A(F/A,i\%,n)$$
$$12{,}000 = 1000(F/A,i\%,10)$$
$$(F/A,i\%,10) = 12{,}000/1000 = 12$$

In the 4% interest table: $(F/A,4\%,10) = 12.006$, so $i = 4\%$. The answer is (b).

A-45. A company is considering buying a new piece of machinery.

Initial cost: $80,000

End-of-useful life salvage value: $20,000

Annual operating cost: $18,000

Useful life: 20 years

The machine will produce an annual saving in material of $25,700. What is the before-tax rate of return if the machine is installed? The rate of return is closest to

(a) 6%
(b) 8%
(c) 10%
(d) 15%
(e) 20%

Solution

PW of cost = PW of benefits

$$80{,}000 = (25{,}700 - 18{,}000)(P/A,i\%,20) + 20{,}000(P/F,i\%,20)$$

Try $i = 8\%$

$$80{,}000 = 7700(9.818) + 20{,}000(0.2145) = 79{,}889$$

Therefore, the rate of return is very close to 8%. The answer is (b).

A-46. Consider the following situation: Invest $100 now and receive two payments of $102.15—one at the end of Year 3, and one at the end of Year 6. The rate of return is nearest to

(a) 6%
(b) 8%
(c) 10%
(d) 12%
(e) 18%

Solution

PW of cost = PW of benefits

$$100 = 102.15(P/F,i\%,3) + 102.15(P/F,i\%,6)$$

Solve by trial and error:

Try $i = 12\%$

$$100 = 102.15(0.7118) + 102.15(0.5066) = 124.46$$

The PW of benefits exceeds the PW of cost. This indicates that the interest rate i is too low.

Try $i = 18\%$

$$100 = 102.15(0.6086) + 102.15(0.3704) = 100.00$$

Therefore, the rate of return is 18%. The answer is (e).

A-47. Two mutually exclusive alternatives are being considered:

Year	A	B
0	–$2500	–$6000
1	+746	+1664
2	+746	+1664
3	+746	+1664
4	+746	+1664
5	+746	+1664

The rate of return on the difference between the alternatives is closest to

(a) 6% (d) 12%

(b) 8% (e) 15%

(c) 10%

Solution

The difference between the alternatives:

Incremental cost = 6000 – 2500 = $3500

Incremental annual benefit = 1664 – 746 = $918

PW of cost = PW of benefits

$$3500 = 918(P/A,i\%,5)$$

$$(P/A,i\%,5) = 3500/918 = 3.81$$

From the interest tables, i is very close to 10%. The answer is (c).

A-48. A project will cost $50,000. The benefits at the end of the first year are estimated to be $10,000, increasing $1000 per year in subsequent years. Assuming a 12% interest rate, no salvage value, and an eight-year analysis period, the Benefit-Cost ratio is closest to

(a) 0.78 (d) 1.45

(b) 1.00 (e) 1.60

(c) 1.28

Solution

$$B/C = \frac{\text{PW of benefits}}{\text{PW of cost}} = \frac{10{,}000(P/A,12\%,8) + 1000(P/G,12\%,8)}{50{,}000}$$

$$= \frac{10{,}000(4.968) + 1000(14.471)}{50{,}000} = 1.28$$

The answer is (c).

A-49. Two alternatives are being considered.

	A	B
Initial cost:	$500	$800
Uniform annual benefit:	$140	$200
Useful life, years:	8	8

The Benefit-Cost ratio of the difference between the alternatives, based on a 12% interest rate, is closest to

(a) 0.60
(b) 0.80
(c) 1.00
(d) 1.20
(e) 1.40

Solution

$$B/C = \frac{\text{PW of benefits}}{\text{PW of cost}} = \frac{60(P/A,12\%,8)}{300} = \frac{60(4.968)}{300} = 0.99$$

Alternate Solution:

$$B/C = \frac{\text{EUAB}}{\text{EUAC}} = \frac{60}{300(A/P,12\%,8)} = \frac{60}{300(0.2013)} = 0.99$$

The answer is (c).

A-50. An engineer will invest in a mining project if the Benefit-Cost ratio is greater than one, based on an 18% interest rate. The project cost is $57,000. The net annual return is estimated at $14,000 for each of the next eight years. At the end of eight years the mining project will be worthless. The Benefit-Cost ratio is closest to

(a) 1.00
(b) 1.05
(c) 1.21
(d) 1.57
(e) 1.96

Solution

$$B/C = \frac{\text{PW of benefits}}{\text{PW of cost}} = \frac{14{,}000(P/A,18\%,8)}{57{,}000} = \frac{14{,}000(4.078)}{57{,}000} = 1.00$$

The answer is (a).

A-51. A city has retained your firm to do a Benefit-Cost analysis of the following project:

Project cost: $60,000,000

Gross income: $20,000,000 per year

Operating costs: $5,500,000 per year

Salvage value after 10 years: None

The project life is ten years. Use 8% interest in the analysis. The computed Benefit-Cost ratio is closest to

(a) 0.80
(b) 1.00
(c) 1.20
(d) 1.40
(e) 1.60

Solution

$$B/C = \frac{\text{EUAB}}{\text{EUAC}} = \frac{20{,}000{,}000 - 5{,}500{,}000}{60{,}000{,}000(A/P, 8\%, 10)} = 1.62$$

The answer is (e).

A-52. A piece of property is purchased for $10,000 and yields a $1000 yearly profit. If the property is sold after five years, the minimum price to break even, with interest at 6%, is closest to

(a) $5000
(b) $6500
(c) $7700
(d) $8300
(e) $9700

Solution

$$F = 10{,}000(F/P,6\%,5) - 1000(F/A,6\%,5)$$
$$= 10{,}000(1.338) - 1000(5.637) = \$7743$$

The answer is (c).

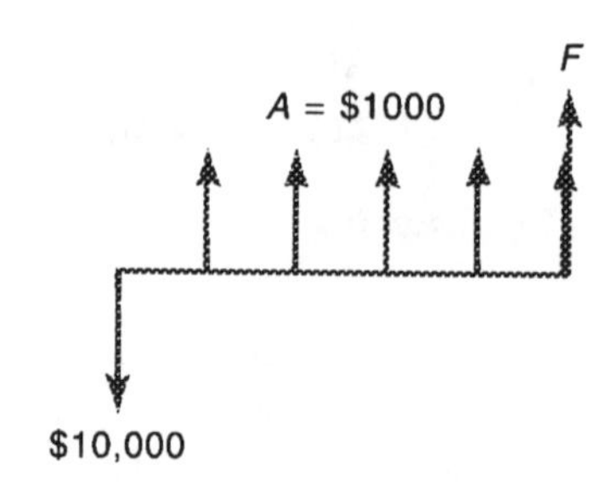

Fig. A-52

A-53. Given two machines:

	A	B
Initial cost:	$55,000	$75,000
Total annual costs:	$16,200	$12,450

With interest at 10% per year, at what service life do these two machines have the same equivalent uniform annual cost? The service life is closest to

(a) 4 years
(b) 5 years
(c) 6 years
(d) 7 years
(e) 8 years

Solution

$$\text{PW of cost}_A = \text{PW of cost}_B$$
$$55{,}000 + 16{,}200(P/A,10\%,n) = 75{,}000 + 12{,}450(P/A,10\%,n)$$
$$(P/A,10\%,n) = (75{,}000 - 55{,}000)/(16{,}200 - 12{,}450) = 5.33$$

From the 10% interest tables, $n = 8$ years. The answer is (e).

A-54. A machine part that is operating in a corrosive atmosphere is made of low-carbon steel. It costs $350 installed, and lasts six years. If the part is treated for corrosion resistance it will cost $700 installed. How long must the treated part last to be as economic as the untreated part, if money is worth 6%?

(a) 8 years
(b) 11 years
(c) 15 years
(d) 17 years
(e) 20 years

Solution

$$\text{EUAC}_{\text{untreated}} = \text{EUAC}_{\text{treated}}$$

$$350(A/P,6\%,6) = 700(A/P,6\%,n)$$

$$350(0.2034) = 700(A/P,6\%,n)$$

$$(A/P,6\%,n) = 71.19/700 = 0.1017$$

From the 6% interest table, n = 15+ years. The answer is (c).

A-55. A firm has determined the two best paints for its machinery are Tuff-Coat at $45 per gallon and Quick at $22 per gallon. The Quick paint is expected to prevent rust for five years. Both paints take $40 of labor per gallon to apply, and both cover the same area. If a 12% interest rate is used, how long must the Tuff-Coat paint prevent rust to justify its use?

(a) 5 years
(b) 6 years
(c) 7 years
(d) 8 years
(e) 9 years

Solution

$$\text{EUAC}_{\text{T-C}} = \text{EUAC}_{\text{Quick}}$$

$$(45 + 40)(A/P,12\%,n) = (22 + 40)(A/P,12\%,5)$$

$$(\text{A/P},12\%,n) = 17.20/85 = 0.202$$

From the 12% interest table, n = 8. The answer is (d).

A-56. Two alternatives are being considered:

	A	B
Cost:	$1000	$2000
Useful life in years:	10	10
End-of-useful-life salvage value:	100	400

The net annual benefit of A is \$150. If interest is 8%, what must be the net annual benefit of B for the two alternatives to be equally desirable? The net annual benefit of B must be closest to

(a) \$150
(b) \$200
(c) \$225
(d) \$275
(e) \$325

Solution

At breakeven,

$$\text{NPW}_A = \text{NPW}_B$$

$$150(P/A,8\%,10) + 100(P/F,8\%,10) - 1000 = \text{NAB}(P/A,8\%,10) + 400(P/F,8\%,10) - 2000$$

$$52.82 = 6.71(\text{NAB}) - 1814.72$$

Net annual benefit (NAB) = (1814.72 + 52.82)/6.71 = \$278. The answer is (d).

A-57. Which one of the following is *NOT* a method of depreciating plant equipment for accounting and engineering economics purposes?

(a) Double-entry method
(b) Modified accelerated cost recovery system
(c) Sum-of-years-digits method
(d) Straight-line method
(e) Sinking-fund method

Solution

"Double entry" probably is a reference to double entry accounting. It is not a method of depreciation. The answer is (a).

A-58. A machine costs \$80,000, has a 20-year useful life, and an estimated \$20,000 end-of-useful-life salvage value. Assuming Sum-Of-Years-Digits depreciation, the book value of the machine after two years is closest to

(a) \$21,000
(b) \$42,000
(c) \$59,000
(d) \$69,000
(e) \$79,000

Solution

Sum-Of-Years-Digits depreciation:

$$D_j = \frac{n-j+1}{\frac{n}{2}(n+1)}(C - S_n)$$

$$D_1 = \frac{20-1+1}{\frac{20}{2}(20+1)}(80{,}000 - 20{,}000) = 5714$$

$$D_2 = \frac{20-2+1}{\frac{20}{2}(20+1)}(80{,}000 - 20{,}000) = 5429$$

Total: \$11,143

$$\text{Book value} = \text{Cost} - \text{Depreciation to date}$$
$$= 80{,}000 - 11{,}143 = \$68{,}857$$

The answer is (d).

A-59. A machine costs $100,000. After its 25-year useful life, its estimated salvage value is $5,000. Based on Double-Declining-Balance depreciation, what will be the book value of the machine at the end of three years? The book value is closest to

(a) $16,000 (d) $78,000
(b) $22,000 (e) $83,000
(c) $58,000

Solution

Double-Declining-Balance depreciation:

$$BV_j = C\left(1 - \frac{2}{n}\right)^j$$

$$BV_3 = 100{,}000\left(1 - \frac{2}{25}\right)^3 = \$77{,}869$$

The answer is (d).

Questions A-60 to A-63

Special tools for the manufacture of finished plastic products cost $15,000 and have an estimated $1000 salvage value at the end of an estimated three-year useful life.

A-60. The third-year straight line depreciation is closest to

(a) $3000 (d) $4500
(b) $3500 (e) $5000
(c) $4000

Solution

$D_3 = (C - S)/n = (15{,}000 - 1000)/3 = \$4666$

The answer is (d).

A-61. The first year Modified-Accelerated-Cost-Recovery-System (MACRS) depreciation is closest to

(a) $3000 (d) $4500
(b) $3500 (e) $5000
(c) $4000

Solution

The half-year convention applies here; Double-declining balance must be used with an assumed salvage value of zero. In general,

$$D_j = \frac{2C}{n}\left(1 - \frac{2}{n}\right)^{j-1}$$

For a half-year in Year 1:

$$D_1 = \frac{1}{2} \times \frac{2 \times 15{,}000}{3}\left(1 - \frac{2}{3}\right)^{1-1} = \$5000$$

The answer is (e).

A-62. The second-year Sum-Of-Years-Digits (SOYD) depreciation is closest to

(a) \$3000 (d) \$4500
(b) \$3500 (e) \$5000
(c) \$4000

Solution

$$D_j = \frac{n - j + 1}{\frac{n}{2}(n+1)}(C - S_n)$$

$$D_2 = \frac{3 - 2 + 1}{\frac{3}{2}(3+1)}(15{,}000 - 1000) = \$4667$$

The answer is (d).

A-63. The second-year sinking-fund depreciation, based on 8% interest, is nearest to

(a) \$3000 (d) \$4500
(b) \$3500 (e) \$5000
(c) \$4000

Solution

$$\begin{aligned} D_2 &= (15{,}000 - 1000)(A/F,8\%,3)(F/P,8\%,1) \\ &= 14{,}000(0.3080)(1.08) = \$4657 \end{aligned}$$

The answer is (d).

A-64. An individual, who has a 28% incremental income tax rate, is considering purchasing a \$1000 taxable corporation bond. As the bondholder he will receive \$100 a year in interest and his \$1000 back when the bond becomes due in six years. This individual's after-tax rate of return from the bond is nearest to

(a) 6% (d) 9%
(b) 7% (e) 10%
(c) 8%

Solution

Twenty-eight percent of the \$100 interest income must be paid in taxes. The balance of \$72 is the after-tax income. Thus the after-tax rate of return = 72/1000 = 0.072 = 7.2%. The answer is (b).

A-65. A \$20,000 investment in equipment will produce \$6000 of net annual benefits for the next eight years. The equipment will be depreciated by straight-line depreciation over its eight year useful life. The equipment has no salvage value. Assuming a 34% income tax rate, the after-tax rate of return for this investment is closest to

(a) 6%
(b) 8%
(c) 10%
(d) 12%
(e) 18%

Solution

Year	Before-tax cash flow	SL deprec	Taxable income	34% Income taxes	After-tax cash flow
0	–\$20,000				–\$20,000
1-8	+6000	2500	3500	1190	+4810

$$D_j = (P - S)/n = \frac{20{,}000 - 0}{8} = 2500$$

$$\text{PW of cost} = \text{PW of benefits}$$

$$20{,}000 = 4810\,(P/A,i\%,8)$$

$$(P/A,i\%,8) = 20{,}000/4810 = 4.158$$

From interest tables, i = 18%. The answer is (e).

A-66. An individual bought a one-year savings certificate for \$10,000, and it pays 6%. He has a taxable income that puts him at the 28% incremental income tax rate. His after-tax rate of return on this investment is closest to

(a) 2%
(b) 3%
(c) 4%
(d) 5%
(e) 6%

Solution

Additional taxable income = 0.06(10,000) = 600

Additional income tax = 0.28(600) = 168

After-tax income = 600 – 168 = 432

After-tax rate of return = 432/10,000 = 0.043 = 4.3%

Alternate Solution

After-tax rate of return = (1 – Incremental tax rate)(Before-tax rate of return)

$$= (1 - 0.28)(0.06) = 0.043 = 4.3\%$$

The answer is (c).

A-67. A tool costing $300 has no salvage value. Its resulting before-tax cash flow is shown in the following partially completed cash flow table.

Year	Before-tax cash flow	Effect on SOYD Deprec	Effect on taxable income	Income taxes	After-tax cash flow
0	–$300				
1	+100				
2	+150				
3	+200				

The tool is to be depreciated over three years using Sum-Of-Years-Digits depreciation. The income tax rate is 50%. The after-tax rate of return is nearest to

(a) 8%
(b) 10%
(c) 12%
(d) 15%
(e) 20%

Solution

Year	Before-tax cash flow	Effect on SOYD Deprec	Effect on taxable income	50% Income taxes	After-tax cash flow
0	–$300				–$300
1	+100	–150	–50	+25	+125
2	+150	–100	+50	–25	+125
3	+200	–50	+150	–75	+125

For the after-tax cash flow:

$$\text{PW of cost} = \text{PW of benefits}$$

$$300 = 125(P/A,i\%,3)$$

$$(P/A,i\%,3) = 300/125 = 2.40$$

From the interest tables, we find that i (the after-tax rate of return) is close to 12%. The answer is (c).

A-68. An engineer is considering the purchase of an annuity that will pay $1000 per year for ten years. The engineer feels he should obtain a 5% rate of return on the annuity after considering the effect of an estimated 6% inflation per year. The amount he would be willing to pay to purchase the annuity is closest to

(a) $1500 (d) $6000

(b) $3000 (e) $7500

(c) $4500

Solution

$d = i + f + if = 0.05 + 0.06 + 0.05(0.06) = 0.113 = 11.3\%$

$$P = A(P/A, 11.3\%, 10) = 1000\left[\frac{(1+0.113)^{10}-1}{0.113(1+0.113)^{10}}\right] = 1000\left[\frac{1.9171}{0.3296}\right] = \$5816$$

The answer is (d).

A-69. An automobile costs $20,000 today. You can earn 12% tax free on an "auto purchase account." If you expect the cost of the auto to increase by 10% per year, the amount you would need to deposit in the account to provide for the purchase of the auto five years from now is closest to

(a) $12,000 (d) $18,000

(b) $14,000 (e) $20,000

(c) $16,000

Solution

$$\text{Cost of auto 5 years hence } (F) = P(1 + \text{inflation rate})^n$$
$$= 20{,}000(1 + 0.10)^5 = 32{,}210$$

Amount to deposit now to have $32,210 available 5 years hence:

$P = F(P/F, i\%, n) = 32{,}210(P/F, 12\%, 5) = 32{,}210(0.5674) = \$18{,}276$

The answer is (d).

A-70. An engineer purchases a building lot for $40,000 cash and plans to sell it after five years. If he wants an 18% before-tax rate of return, after taking the 6% annual inflation rate into account, the selling price must be nearest to

(a) $55,000 (d) $100,000

(b) $65,000 (e) $125,000

(c) $75,000

Solution

$$\text{Selling price } (F) = 40{,}000(F/P, 18\%, 5)(F/P, 6\%, 5)$$
$$= 40{,}000(2.288)(1.338) = \$122{,}500$$

The answer is (e).

A-71. A piece of equipment with a list price of $450 can actually be purchased for either $400 cash or $50 immediately plus four additional annual payments of $115.25. All values are in dollars of current purchasing power. If the typical customer considered a 5% interest rate appropriate, the inflation rate at which the two purchase alternatives are equivalent is nearest to

(a) 5%
(b) 6%
(c) 8%
(d) 10%
(e) 12%

Solution

$$\text{PW of cash purchase} = \text{PW of installment purchase}$$

$$400 = 50 + 115.25(P/A,\text{d}\%,4)$$

$$(P/A,\text{d}\%,4) = 350/115.25 = 3.037$$

From the interest tables, $d = 12\%$.

$$d = i + f + i(f)$$

$$0.12 = 0.05 + f + 0.05f$$

$f = 0.07/1.05 = 0.0667 = 6.67\%$

The answer is (b).

A-72. A man wants to determine whether to invest $1000 in a friend's speculative venture. He will do so if he thinks he can get his money back. The probabilities of the various outcomes at the end of one year are:

Result	Probability
$2000 (double his money)	0.3
1500	0.1
1000	0.2
500	0.3
0 (lose everything)	0.1

His expected outcome if he invests the $1000 is closest to

(a) $800
(b) $900
(c) $1000
(d) $1100
(e) $1200

Solution

The expected income = 0.3(2000) + 0.1(1500) + 0.2(1000) + 0.3(500) + 0.1(0) = $1100. The answer is (d).

A-73. The amount you would be willing to pay for an insurance policy protecting you against a one in twenty chance of losing $10,000 three years from now, if interest is 10%, is closest to

(a) $175 (d) $1500

(b) $350 (e) $2000

(c) $1000

Solution PW of benefit = (10,000/20)(P/F,10%,3)

= 500(0.7513) = $376

The answer is (b).

½% Compound Interest Factors ½%

	Single Payment		Uniform Payment Series				Uniform Gradient		
n	Compound Amount Factor Find *F* Given *P* *F/P*	Present Worth Factor Find *P* Given *F* *P/F*	Sinking Fund Factor Find *A* Given *F* *A/F*	Capital Recovery Factor Find *A* Given *P* *A/P*	Compound Amount Factor Find *F* Given *A* *F/A*	Present Worth Factor Find *P* Given *A* *P/A*	Gradient Uniform Series Find *A* Given *G* *A/G*	Gradient Present Worth Find *P* Given *G* *P/G*	n
1	1.005	.9950	1.0000	1.0050	1.000	0.995	0	0	1
2	1.010	.9901	.4988	.5038	2.005	1.985	0.499	0.991	2
3	1.015	.9851	.3317	.3367	3.015	2.970	0.996	2.959	3
4	1.020	.9802	.2481	.2531	4.030	3.951	1.494	5.903	4
5	1.025	.9754	.1980	.2030	5.050	4.926	1.990	9.803	5
6	1.030	.9705	.1646	.1696	6.076	5.896	2.486	14.660	6
7	1.036	.9657	.1407	.1457	7.106	6.862	2.980	20.448	7
8	1.041	.9609	.1228	.1278	8.141	7.823	3.474	27.178	8
9	1.046	.9561	.1089	.1139	9.182	8.779	3.967	34.825	9
10	1.051	.9513	.0978	.1028	10.228	9.730	4.459	43.389	10
11	1.056	.9466	.0887	.0937	11.279	10.677	4.950	52.855	11
12	1.062	.9419	.0811	.0861	12.336	11.619	5.441	63.218	12
13	1.067	.9372	.0746	.0796	13.397	12.556	5.931	74.465	13
14	1.072	.9326	.0691	.0741	14.464	13.489	6.419	86.590	14
15	1.078	.9279	.0644	.0694	15.537	14.417	6.907	99.574	15
16	1.083	.9233	.0602	.0652	16.614	15.340	7.394	113.427	16
17	1.088	.9187	.0565	.0615	17.697	16.259	7.880	128.125	17
18	1.094	.9141	.0532	.0582	18.786	17.173	8.366	143.668	18
19	1.099	.9096	.0503	.0553	19.880	18.082	8.850	160.037	19
20	1.105	.9051	.0477	.0527	20.979	18.987	9.334	177.237	20
21	1.110	.9006	.0453	.0503	22.084	19.888	9.817	195.245	21
22	1.116	.8961	.0431	.0481	23.194	20.784	10.300	214.070	22
23	1.122	.8916	.0411	.0461	24.310	21.676	10.781	233.680	23
24	1.127	.8872	.0393	.0443	25.432	22.563	11.261	254.088	24
25	1.133	.8828	.0377	.0427	26.559	23.446	11.741	275.273	25
26	1.138	.8784	.0361	.0411	27.692	24.324	12.220	297.233	26
27	1.144	.8740	.0347	.0397	28.830	25.198	12.698	319.955	27
28	1.150	.8697	.0334	.0384	29.975	26.068	13.175	343.439	28
29	1.156	.8653	.0321	.0371	31.124	26.933	13.651	367.672	29
30	1.161	.8610	.0310	.0360	32.280	27.794	14.127	392.640	30
36	1.197	.8356	.0254	.0304	39.336	32.871	16.962	557.564	36
40	1.221	.8191	.0226	.0276	44.159	36.172	18.836	681.341	40
48	1.270	.7871	.0185	.0235	54.098	42.580	22.544	959.928	48
50	1.283	.7793	.0177	.0227	56.645	44.143	23.463	1 035.70	50
52	1.296	.7716	.0169	.0219	59.218	45.690	24.378	1 113.82	52
60	1.349	.7414	.0143	.0193	69.770	51.726	28.007	1 448.65	60
70	1.418	.7053	.0120	.0170	83.566	58.939	32.468	1 913.65	70
72	1.432	.6983	.0116	.0166	86.409	60.340	33.351	2 012.35	72
80	1.490	.6710	.0102	.0152	98.068	65.802	36.848	2 424.65	80
84	1.520	.6577	.00961	.0146	104.074	68.453	38.576	2 640.67	84
90	1.567	.6383	.00883	.0138	113.311	72.331	41.145	2 976.08	90
96	1.614	.6195	.00814	.0131	122.829	76.095	43.685	3 324.19	96
100	1.647	.6073	.00773	.0127	129.334	78.543	45.361	3 562.80	100
104	1.680	.5953	.00735	.0124	135.970	80.942	47.025	3 806.29	104
120	1.819	.5496	.00610	.0111	163.880	90.074	53.551	4 823.52	120
240	3.310	.3021	.00216	.00716	462.041	139.581	96.113	13 415.56	240
360	6.023	.1660	.00100	.00600	1 004.5	166.792	128.324	21 403.32	360
480	10.957	.0913	.00050	.00550	1 991.5	181.748	151.795	27 588.37	480

1% Compound Interest Factors 1%

	Single Payment		Uniform Payment Series				Uniform Gradient		
n	Compound Amount Factor Find *F* Given *P* *F/P*	Present Worth Factor Find *P* Given *F* *P/F*	Sinking Fund Factor Find *A* Given *F* *A/F*	Capital Recovery Factor Find *A* Given *P* *A/P*	Compound Amount Factor Find *F* Given *A* *F/A*	Present Worth Factor Find *P* Given *A* *P/A*	Gradient Uniform Series Find *A* Given *G* *A/G*	Gradient Present Worth Find *P* Given *G* *P/G*	n
1	1.010	.9901	1.0000	1.0100	1.000	0.990	0	0	1
2	1.020	.9803	.4975	.5075	2.010	1.970	0.498	0.980	2
3	1.030	.9706	.3300	.3400	3.030	2.941	0.993	2.921	3
4	1.041	.9610	.2463	.2563	4.060	3.902	1.488	5.804	4
5	1.051	.9515	.1960	.2060	5.101	4.853	1.980	9.610	5
6	1.062	.9420	.1625	.1725	6.152	5.795	2.471	14.320	6
7	1.072	.9327	.1386	.1486	7.214	6.728	2.960	19.917	7
8	1.083	.9235	.1207	.1307	8.286	7.652	3.448	26.381	8
9	1.094	.9143	.1067	.1167	9.369	8.566	3.934	33.695	9
10	1.105	.9053	.0956	.1056	10.462	9.471	4.418	41.843	10
11	1.116	.8963	.0865	.0965	11.567	10.368	4.900	50.806	11
12	1.127	.8874	.0788	.0888	12.682	11.255	5.381	60.568	12
13	1.138	.8787	.0724	.0824	13.809	12.134	5.861	71.112	13
14	1.149	.8700	.0669	.0769	14.947	13.004	6.338	82.422	14
15	1.161	.8613	.0621	.0721	16.097	13.865	6.814	94.481	15
16	1.173	.8528	.0579	.0679	17.258	14.718	7.289	107.273	16
17	1.184	.8444	.0543	.0643	18.430	15.562	7.761	120.783	17
18	1.196	.8360	.0510	.0610	19.615	16.398	8.232	134.995	18
19	1.208	.8277	.0481	.0581	20.811	17.226	8.702	149.895	19
20	1.220	.8195	.0454	.0554	22.019	18.046	9.169	165.465	20
21	1.232	.8114	.0430	.0530	23.239	18.857	9.635	181.694	21
22	1.245	.8034	.0409	.0509	24.472	19.660	10.100	198.565	22
23	1.257	.7954	.0389	.0489	25.716	20.456	10.563	216.065	23
24	1.270	.7876	.0371	.0471	26.973	21.243	11.024	234.179	24
25	1.282	.7798	.0354	.0454	28.243	22.023	11.483	252.892	25
26	1.295	.7720	.0339	.0439	29.526	22.795	11.941	272.195	26
27	1.308	.7644	.0324	.0424	30.821	23.560	12.397	292.069	27
28	1.321	.7568	.0311	.0411	32.129	24.316	12.852	312.504	28
29	1.335	.7493	.0299	.0399	33.450	25.066	13.304	333.486	29
30	1.348	.7419	.0287	.0387	34.785	25.808	13.756	355.001	30
36	1.431	.6989	.0232	.0332	43.077	30.107	16.428	494.620	36
40	1.489	.6717	.0205	.0305	48.886	32.835	18.178	596.854	40
48	1.612	.6203	.0163	.0263	61.223	37.974	21.598	820.144	48
50	1.645	.6080	.0155	.0255	64.463	39.196	22.436	879.417	50
52	1.678	.5961	.0148	.0248	67.769	40.394	23.269	939.916	52
60	1.817	.5504	.0122	.0222	81.670	44.955	26.533	1 192.80	60
70	2.007	.4983	.00993	.0199	100.676	50.168	30.470	1 528.64	70
72	2.047	.4885	.00955	.0196	104.710	51.150	31.239	1 597.86	72
80	2.217	.4511	.00822	.0182	121.671	54.888	34.249	1 879.87	80
84	2.307	.4335	.00765	.0177	130.672	56.648	35.717	2 023.31	84
90	2.449	.4084	.00690	.0169	144.863	59.161	37.872	2 240.56	90
96	2.599	.3847	.00625	.0163	159.927	61.528	39.973	2 459.42	96
100	2.705	.3697	.00587	.0159	170.481	63.029	41.343	2 605.77	100
104	2.815	.3553	.00551	.0155	181.464	64.471	42.688	2 752.17	104
120	3.300	.3030	.00435	.0143	230.039	69.701	47.835	3 334.11	120
240	10.893	.0918	.00101	.0110	989.254	90.819	75.739	6 878.59	240
360	35.950	.0278	.00029	.0103	3 495.0	97.218	89.699	8 720.43	360
480	118.648	.00843	.00008	.0101	11 764.8	99.157	95.920	9 511.15	480

1½% Compound Interest Factors 1½%

n	Single Payment: Compound Amount Factor, Find F Given P, F/P	Single Payment: Present Worth Factor, Find P Given F, P/F	Uniform Payment Series: Sinking Fund Factor, Find A Given F, A/F	Uniform Payment Series: Capital Recovery Factor, Find A Given P, A/P	Uniform Payment Series: Compound Amount Factor, Find F Given A, F/A	Uniform Payment Series: Present Worth Factor, Find P Given A, P/A	Uniform Gradient: Gradient Uniform Series, Find A Given G, A/G	Uniform Gradient: Gradient Present Worth, Find P Given G, P/G	n
1	1.015	.9852	1.0000	1.0150	1.000	0.985	0	0	1
2	1.030	.9707	.4963	.5113	2.015	1.956	0.496	0.970	2
3	1.046	.9563	.3284	.3434	3.045	2.912	0.990	2.883	3
4	1.061	.9422	.2444	.2594	4.091	3.854	1.481	5.709	4
5	1.077	.9283	.1941	.2091	5.152	4.783	1.970	9.422	5
6	1.093	.9145	.1605	.1755	6.230	5.697	2.456	13.994	6
7	1.110	.9010	.1366	.1516	7.323	6.598	2.940	19.400	7
8	1.126	.8877	.1186	.1336	8.433	7.486	3.422	25.614	8
9	1.143	.8746	.1046	.1196	9.559	8.360	3.901	32.610	9
10	1.161	.8617	.0934	.1084	10.703	9.222	4.377	40.365	10
11	1.178	.8489	.0843	.0993	11.863	10.071	4.851	48.855	11
12	1.196	.8364	.0767	.0917	13.041	10.907	5.322	58.054	12
13	1.214	.8240	.0702	.0852	14.237	11.731	5.791	67.943	13
14	1.232	.8118	.0647	.0797	15.450	12.543	6.258	78.496	14
15	1.250	.7999	.0599	.0749	16.682	13.343	6.722	89.694	15
16	1.269	.7880	.0558	.0708	17.932	14.131	7.184	101.514	16
17	1.288	.7764	.0521	.0671	19.201	14.908	7.643	113.937	17
18	1.307	.7649	.0488	.0638	20.489	15.673	8.100	126.940	18
19	1.327	.7536	.0459	.0609	21.797	16.426	8.554	140.505	19
20	1.347	.7425	.0432	.0582	23.124	17.169	9.005	154.611	20
21	1.367	.7315	.0409	.0559	24.470	17.900	9.455	169.241	21
22	1.388	.7207	.0387	.0537	25.837	18.621	9.902	184.375	22
23	1.408	.7100	.0367	.0517	27.225	19.331	10.346	199.996	23
24	1.430	.6995	.0349	.0499	28.633	20.030	10.788	216.085	24
25	1.451	.6892	.0333	.0483	30.063	20.720	11.227	232.626	25
26	1.473	.6790	.0317	.0467	31.514	21.399	11.664	249.601	26
27	1.495	.6690	.0303	.0453	32.987	22.068	12.099	266.995	27
28	1.517	.6591	.0290	.0440	34.481	22.727	12.531	284.790	28
29	1.540	.6494	.0278	.0428	35.999	23.376	12.961	302.972	29
30	1.563	.6398	.0266	.0416	37.539	24.016	13.388	321.525	30
36	1.709	.5851	.0212	.0362	47.276	27.661	15.901	439.823	36
40	1.814	.5513	.0184	.0334	54.268	29.916	17.528	524.349	40
48	2.043	.4894	.0144	.0294	69.565	34.042	20.666	703.537	48
50	2.105	.4750	.0136	.0286	73.682	35.000	21.428	749.955	50
52	2.169	.4611	.0128	.0278	77.925	35.929	22.179	796.868	52
60	2.443	.4093	.0104	.0254	96.214	39.380	25.093	988.157	60
70	2.835	.3527	.00817	.0232	122.363	43.155	28.529	1 231.15	70
72	2.921	.3423	.00781	.0228	128.076	43.845	29.189	1 279.78	72
80	3.291	.3039	.00655	.0215	152.710	46.407	31.742	1 473.06	80
84	3.493	.2863	.00602	.0210	166.172	47.579	32.967	1 568.50	84
90	3.819	.2619	.00532	.0203	187.929	49.210	34.740	1 709.53	90
96	4.176	.2395	.00472	.0197	211.719	50.702	36.438	1 847.46	96
100	4.432	.2256	.00437	.0194	228.802	51.625	37.529	1 937.43	100
104	4.704	.2126	.00405	.0190	246.932	52.494	38.589	2 025.69	104
120	5.969	.1675	.00302	.0180	331.286	55.498	42.518	2 359.69	120
240	35.632	.0281	.00043	.0154	2 308.8	64.796	59.737	3 870.68	240
360	212.700	.00470	.00007	.0151	14 113.3	66.353	64.966	4 310.71	360
480	1 269.7	.00079	.00001	.0150	84 577.8	66.614	66.288	4 415.74	480

2% Compound Interest Factors 2%

n	Single Payment: Compound Amount Factor, Find F Given P, F/P	Single Payment: Present Worth Factor, Find P Given F, P/F	Uniform Payment Series: Sinking Fund Factor, Find A Given F, A/F	Uniform Payment Series: Capital Recovery Factor, Find A Given P, A/P	Uniform Payment Series: Compound Amount Factor, Find F Given A, F/A	Uniform Payment Series: Present Worth Factor, Find P Given A, P/A	Uniform Gradient: Gradient Uniform Series, Find A Given G, A/G	Uniform Gradient: Gradient Present Worth, Find P Given G, P/G	n
1	1.020	.9804	1.0000	1.0200	1.000	0.980	0	0	1
2	1.040	.9612	.4951	.5151	2.020	1.942	0.495	0.961	2
3	1.061	.9423	.3268	.3468	3.060	2.884	0.987	2.846	3
4	1.082	.9238	.2426	.2626	4.122	3.808	1.475	5.617	4
5	1.104	.9057	.1922	.2122	5.204	4.713	1.960	9.240	5
6	1.126	.8880	.1585	.1785	6.308	5.601	2.442	13.679	6
7	1.149	.8706	.1345	.1545	7.434	6.472	2.921	18.903	7
8	1.172	.8535	.1165	.1365	8.583	7.325	3.396	24.877	8
9	1.195	.8368	.1025	.1225	9.755	8.162	3.868	31.571	9
10	1.219	.8203	.0913	.1113	10.950	8.983	4.337	38.954	10
11	1.243	.8043	.0822	.1022	12.169	9.787	4.802	46.996	11
12	1.268	.7885	.0746	.0946	13.412	10.575	5.264	55.669	12
13	1.294	.7730	.0681	.0881	14.680	11.348	5.723	64.946	13
14	1.319	.7579	.0626	.0826	15.974	12.106	6.178	74.798	14
15	1.346	.7430	.0578	.0778	17.293	12.849	6.631	85.200	15
16	1.373	.7284	.0537	.0737	18.639	13.578	7.080	96.127	16
17	1.400	.7142	.0500	.0700	20.012	14.292	7.526	107.553	17
18	1.428	.7002	.0467	.0667	21.412	14.992	7.968	119.456	18
19	1.457	.6864	.0438	.0638	22.840	15.678	8.407	131.812	19
20	1.486	.6730	.0412	.0612	24.297	16.351	8.843	144.598	20
21	1.516	.6598	.0388	.0588	25.783	17.011	9.276	157.793	21
22	1.546	.6468	.0366	.0566	27.299	17.658	9.705	171.377	22
23	1.577	.6342	.0347	.0547	28.845	18.292	10.132	185.328	23
24	1.608	.6217	.0329	.0529	30.422	18.914	10.555	199.628	24
25	1.641	.6095	.0312	.0512	32.030	19.523	10.974	214.256	25
26	1.673	.5976	.0297	.0497	33.671	20.121	11.391	229.196	26
27	1.707	.5859	.0283	.0483	35.344	20.707	11.804	244.428	27
28	1.741	.5744	.0270	.0470	37.051	21.281	12.214	259.936	28
29	1.776	.5631	.0258	.0458	38.792	21.844	12.621	275.703	29
30	1.811	.5521	.0247	.0447	40.568	22.396	13.025	291.713	30
36	2.040	.4902	.0192	.0392	51.994	25.489	15.381	392.036	36
40	2.208	.4529	.0166	.0366	60.402	27.355	16.888	461.989	40
48	2.587	.3865	.0126	.0326	79.353	30.673	19.755	605.961	48
50	2.692	.3715	.0118	.0318	84.579	31.424	20.442	642.355	50
52	2.800	.3571	.0111	.0311	90.016	32.145	21.116	678.779	52
60	3.281	.3048	.00877	.0288	114.051	34.761	23.696	823.692	60
70	4.000	.2500	.00667	.0267	149.977	37.499	26.663	999.829	70
72	4.161	.2403	.00633	.0263	158.056	37.984	27.223	1 034.050	72
80	4.875	.2051	.00516	.0252	193.771	39.744	29.357	1 166.781	80
84	5.277	.1895	.00468	.0247	213.865	40.525	30.361	1 230.413	84
90	5.943	.1683	.00405	.0240	247.155	41.587	31.793	1 322.164	90
96	6.693	.1494	.00351	.0235	284.645	42.529	33.137	1 409.291	96
100	7.245	.1380	.00320	.0232	312.230	43.098	33.986	1 464.747	100
104	7.842	.1275	.00292	.0229	342.090	43.624	34.799	1 518.082	104
120	10.765	.0929	.00205	.0220	488.255	45.355	37.711	1 710.411	120
240	115.887	.00863	.00017	.0202	5 744.4	49.569	47.911	2 374.878	240
360	1 247.5	.00080	.00002	.0200	62 326.8	49.960	49.711	2 483.567	360
480	13 429.8	.00007		.0200	671 442.0	49.996	49.964	2 498.027	480

4% Compound Interest Factors 4%

n	Single Payment: Compound Amount Factor, Find F Given P, F/P	Single Payment: Present Worth Factor, Find P Given F, P/F	Uniform Payment Series: Sinking Fund Factor, Find A Given F, A/F	Uniform Payment Series: Capital Recovery Factor, Find A Given P, A/P	Uniform Payment Series: Compound Amount Factor, Find F Given A, F/A	Uniform Payment Series: Present Worth Factor, Find P Given A, P/A	Uniform Gradient: Gradient Uniform Series, Find A Given G, A/G	Uniform Gradient: Gradient Present Worth, Find P Given G, P/G	n
1	1.040	.9615	1.0000	1.0400	1.000	0.962	0	0	1
2	1.082	.9246	.4902	.5302	2.040	1.886	0.490	0.925	2
3	1.125	.8890	.3203	.3603	3.122	2.775	0.974	2.702	3
4	1.170	.8548	.2355	.2755	4.246	3.630	1.451	5.267	4
5	1.217	.8219	.1846	.2246	5.416	4.452	1.922	8.555	5
6	1.265	.7903	.1508	.1908	6.633	5.242	2.386	12.506	6
7	1.316	.7599	.1266	.1666	7.898	6.002	2.843	17.066	7
8	1.369	.7307	.1085	.1485	9.214	6.733	3.294	22.180	8
9	1.423	.7026	.0945	.1345	10.583	7.435	3.739	27.801	9
10	1.480	.6756	.0833	.1233	12.006	8.111	4.177	33.881	10
11	1.539	.6496	.0741	.1141	13.486	8.760	4.609	40.377	11
12	1.601	.6246	.0666	.1066	15.026	9.385	5.034	47.248	12
13	1.665	.6006	.0601	.1001	16.627	9.986	5.453	54.454	13
14	1.732	.5775	.0547	.0947	18.292	10.563	5.866	61.962	14
15	1.801	.5553	.0499	.0899	20.024	11.118	6.272	69.735	15
16	1.873	.5339	.0458	.0858	21.825	11.652	6.672	77.744	16
17	1.948	.5134	.0422	.0822	23.697	12.166	7.066	85.958	17
18	2.026	.4936	.0390	.0790	25.645	12.659	7.453	94.350	18
19	2.107	.4746	.0361	.0761	27.671	13.134	7.834	102.893	19
20	2.191	.4564	.0336	.0736	29.778	13.590	8.209	111.564	20
21	2.279	.4388	.0313	.0713	31.969	14.029	8.578	120.341	21
22	2.370	.4220	.0292	.0692	34.248	14.451	8.941	129.202	22
23	2.465	.4057	.0273	.0673	36.618	14.857	9.297	138.128	23
24	2.563	.3901	.0256	.0656	39.083	15.247	9.648	147.101	24
25	2.666	.3751	.0240	.0640	41.646	15.622	9.993	156.104	25
26	2.772	.3607	.0226	.0626	44.312	15.983	10.331	165.121	26
27	2.883	.3468	.0212	.0612	47.084	16.330	10.664	174.138	27
28	2.999	.3335	.0200	.0600	49.968	16.663	10.991	183.142	28
29	3.119	.3207	.0189	.0589	52.966	16.984	11.312	192.120	29
30	3.243	.3083	.0178	.0578	56.085	17.292	11.627	201.062	30
31	3.373	.2965	.0169	.0569	59.328	17.588	11.937	209.955	31
32	3.508	.2851	.0159	.0559	62.701	17.874	12.241	218.792	32
33	3.648	.2741	.0151	.0551	66.209	18.148	12.540	227.563	33
34	3.794	.2636	.0143	.0543	69.858	18.411	12.832	236.260	34
35	3.946	.2534	.0136	.0536	73.652	18.665	13.120	244.876	35
40	4.801	.2083	.0105	.0505	95.025	19.793	14.476	286.530	40
45	5.841	.1712	.00826	.0483	121.029	20.720	15.705	325.402	45
50	7.107	.1407	.00655	.0466	152.667	21.482	16.812	361.163	50
55	8.646	.1157	.00523	.0452	191.159	22.109	17.807	393.689	55
60	10.520	.0951	.00420	.0442	237.990	22.623	18.697	422.996	60
65	12.799	.0781	.00339	.0434	294.968	23.047	19.491	449.201	65
70	15.572	.0642	.00275	.0427	364.290	23.395	20.196	472.479	70
75	18.945	.0528	.00223	.0422	448.630	23.680	20.821	493.041	75
80	23.050	.0434	.00181	.0418	551.244	23.915	21.372	511.116	80
85	28.044	.0357	.00148	.0415	676.089	24.109	21.857	526.938	85
90	34.119	.0293	.00121	.0412	827.981	24.267	22.283	540.737	90
95	41.511	.0241	.00099	.0410	1 012.8	24.398	22.655	552.730	95
100	50.505	.0198	.00081	.0408	1 237.6	24.505	22.980	563.125	100

6% Compound Interest Factors 6%

n	Single Payment: Compound Amount Factor, Find F Given P, F/P	Single Payment: Present Worth Factor, Find P Given F, P/F	Uniform Payment Series: Sinking Fund Factor, Find A Given F, A/F	Uniform Payment Series: Capital Recovery Factor, Find A Given P, A/P	Uniform Payment Series: Compound Amount Factor, Find F Given A, F/A	Uniform Payment Series: Present Worth Factor, Find P Given A, P/A	Uniform Gradient: Gradient Uniform Series, Find A Given G, A/G	Uniform Gradient: Gradient Present Worth, Find P Given G, P/G	n
1	1.060	.9434	1.0000	1.0600	1.000	0.943	0	0	1
2	1.124	.8900	.4854	.5454	2.060	1.833	0.485	0.890	2
3	1.191	.8396	.3141	.3741	3.184	2.673	0.961	2.569	3
4	1.262	.7921	.2286	.2886	4.375	3.465	1.427	4.945	4
5	1.338	.7473	.1774	.2374	5.637	4.212	1.884	7.934	5
6	1.419	.7050	.1434	.2034	6.975	4.917	2.330	11.459	6
7	1.504	.6651	.1191	.1791	8.394	5.582	2.768	15.450	7
8	1.594	.6274	.1010	.1610	9.897	6.210	3.195	19.841	8
9	1.689	.5919	.0870	.1470	11.491	6.802	3.613	24.577	9
10	1.791	.5584	.0759	.1359	13.181	7.360	4.022	29.602	10
11	1.898	.5268	.0668	.1268	14.972	7.887	4.421	34.870	11
12	2.012	.4970	.0593	.1193	16.870	8.384	4.811	40.337	12
13	2.133	.4688	.0530	.1130	18.882	8.853	5.192	45.963	13
14	2.261	.4423	.0476	.1076	21.015	9.295	5.564	51.713	14
15	2.397	.4173	.0430	.1030	23.276	9.712	5.926	57.554	15
16	2.540	.3936	.0390	.0990	25.672	10.106	6.279	63.459	16
17	2.693	.3714	.0354	.0954	28.213	10.477	6.624	69.401	17
18	2.854	.3503	.0324	.0924	30.906	10.828	6.960	75.357	18
19	3.026	.3305	.0296	.0896	33.760	11.158	7.287	81.306	19
20	3.207	.3118	.0272	.0872	36.786	11.470	7.605	87.230	20
21	3.400	.2942	.0250	.0850	39.993	11.764	7.915	93.113	21
22	3.604	.2775	.0230	.0830	43.392	12.042	8.217	98.941	22
23	3.820	.2618	.0213	.0813	46.996	12.303	8.510	104.700	23
24	4.049	.2470	.0197	.0797	50.815	12.550	8.795	110.381	24
25	4.292	.2330	.0182	.0782	54.864	12.783	9.072	115.973	25
26	4.549	.2198	.0169	.0769	59.156	13.003	9.341	121.468	26
27	4.822	.2074	.0157	.0757	63.706	13.211	9.603	126.860	27
28	5.112	.1956	.0146	.0746	68.528	13.406	9.857	132.142	28
29	5.418	.1846	.0136	.0736	73.640	13.591	10.103	137.309	29
30	5.743	.1741	.0126	.0726	79.058	13.765	10.342	142.359	30
31	6.088	.1643	.0118	.0718	84.801	13.929	10.574	147.286	31
32	6.453	.1550	.0110	.0710	90.890	14.084	10.799	152.090	32
33	6.841	.1462	.0103	.0703	97.343	14.230	11.017	156.768	33
34	7.251	.1379	.00960	.0696	104.184	14.368	11.228	161.319	34
35	7.686	.1301	.00897	.0690	111.435	14.498	11.432	165.743	35
40	10.286	.0972	.00646	.0665	154.762	15.046	12.359	185.957	40
45	13.765	.0727	.00470	.0647	212.743	15.456	13.141	203.109	45
50	18.420	.0543	.00344	.0634	290.335	15.762	13.796	217.457	50
55	24.650	.0406	.00254	.0625	394.171	15.991	14.341	229.322	55
60	32.988	.0303	.00188	.0619	533.126	16.161	14.791	239.043	60
65	44.145	.0227	.00139	.0614	719.080	16.289	15.160	246.945	65
70	59.076	.0169	.00103	.0610	967.928	16.385	15.461	253.327	70
75	79.057	.0126	.00077	.0608	1 300.9	16.456	15.706	258.453	75
80	105.796	.00945	.00057	.0606	1 746.6	16.509	15.903	262.549	80
85	141.578	.00706	.00043	.0604	2 343.0	16.549	16.062	265.810	85
90	189.464	.00528	.00032	.0603	3 141.1	16.579	16.189	268.395	90
95	253.545	.00394	.00024	.0602	4 209.1	16.601	16.290	270.437	95
100	339.300	.00295	.00018	.0602	5 638.3	16.618	16.371	272.047	100

8% Compound Interest Factors 8%

	Single Payment		Uniform Payment Series				Uniform Gradient		
	Compound Amount Factor	Present Worth Factor	Sinking Fund Factor	Capital Recovery Factor	Compound Amount Factor	Present Worth Factor	Gradient Uniform Series	Gradient Present Worth	
n	Find F Given P F/P	Find P Given F P/F	Find A Given F A/F	Find A Given P A/P	Find F Given A F/A	Find P Given A P/A	Find A Given G A/G	Find P Given G P/G	n
1	1.080	.9259	1.0000	1.0800	1.000	0.926	0	0	1
2	1.166	.8573	.4808	.5608	2.080	1.783	0.481	0.857	2
3	1.260	.7938	.3080	.3880	3.246	2.577	0.949	2.445	3
4	1.360	.7350	.2219	.3019	4.506	3.312	1.404	4.650	4
5	1.469	.6806	.1705	.2505	5.867	3.993	1.846	7.372	5
6	1.587	.6302	.1363	.2163	7.336	4.623	2.276	10.523	6
7	1.714	.5835	.1121	.1921	8.923	5.206	2.694	14.024	7
8	1.851	.5403	.0940	.1740	10.637	5.747	3.099	17.806	8
9	1.999	.5002	.0801	.1601	12.488	6.247	3.491	21.808	9
10	2.159	.4632	.0690	.1490	14.487	6.710	3.871	25.977	10
11	2.332	.4289	.0601	.1401	16.645	7.139	4.240	30.266	11
12	2.518	.3971	.0527	.1327	18.977	7.536	4.596	34.634	12
13	2.720	.3677	.0465	.1265	21.495	7.904	4.940	39.046	13
14	2.937	.3405	.0413	.1213	24.215	8.244	5.273	43.472	14
15	3.172	.3152	.0368	.1168	27.152	8.559	5.594	47.886	15
16	3.426	.2919	.0330	.1130	30.324	8.851	5.905	52.264	16
17	3.700	.2703	.0296	.1096	33.750	9.122	6.204	56.588	17
18	3.996	.2502	.0267	.1067	37.450	9.372	6.492	60.843	18
19	4.316	.2317	.0241	.1041	41.446	9.604	6.770	65.013	19
20	4.661	.2145	.0219	.1019	45.762	9.818	7.037	69.090	20
21	5.034	.1987	.0198	.0998	50.423	10.017	7.294	73.063	21
22	5.437	.1839	.0180	.0980	55.457	10.201	7.541	76.926	22
23	5.871	.1703	.0164	.0964	60.893	10.371	7.779	80.673	23
24	6.341	.1577	.0150	.0950	66.765	10.529	8.007	84.300	24
25	6.848	.1460	.0137	.0937	73.106	10.675	8.225	87.804	25
26	7.396	.1352	.0125	.0925	79.954	10.810	8.435	91.184	26
27	7.988	.1252	.0114	.0914	87.351	10.935	8.636	94.439	27
28	8.627	.1159	.0105	.0905	95.339	11.051	8.829	97.569	28
29	9.317	.1073	.00962	.0896	103.966	11.158	9.013	100.574	29
30	10.063	.0994	.00883	.0888	113.283	11.258	9.190	103.456	30
31	10.868	.0920	.00811	.0881	123.346	11.350	9.358	106.216	31
32	11.737	.0852	.00745	.0875	134.214	11.435	9.520	108.858	32
33	12.676	.0789	.00685	.0869	145.951	11.514	9.674	111.382	33
34	13.690	.0730	.00630	.0863	158.627	11.587	9.821	113.792	34
35	14.785	.0676	.00580	.0858	172.317	11.655	9.961	116.092	35
40	21.725	.0460	.00386	.0839	259.057	11.925	10.570	126.042	40
45	31.920	.0313	.00259	.0826	386.506	12.108	11.045	133.733	45
50	46.902	.0213	.00174	.0817	573.771	12.233	11.411	139.593	50
55	68.914	.0145	.00118	.0812	848.925	12.319	11.690	144.006	55
60	101.257	.00988	.00080	.0808	1 253.2	12.377	11.902	147.300	60
65	148.780	.00672	.00054	.0805	1 847.3	12.416	12.060	149.739	65
70	218.607	.00457	.00037	.0804	2 720.1	12.443	12.178	151.533	70
75	321.205	.00311	.00025	.0802	4 002.6	12.461	12.266	152.845	75
80	471.956	.00212	.00017	.0802	5 887.0	12.474	12.330	153.800	80
85	693.458	.00144	.00012	.0801	8 655.7	12.482	12.377	154.492	85
90	1 018.9	.00098	.00008	.0801	12 724.0	12.488	12.412	154.993	90
95	1 497.1	.00067	.00005	.0801	18 701.6	12.492	12.437	155.352	95
100	2 199.8	.00045	.00004	.0800	27 484.6	12.494	12.455	155.611	100

10% Compound Interest Factors 10%

	Single Payment		Uniform Payment Series				Uniform Gradient		
	Compound Amount Factor	Present Worth Factor	Sinking Fund Factor	Capital Recovery Factor	Compound Amount Factor	Present Worth Factor	Gradient Uniform Series	Gradient Present Worth	
n	Find F Given P F/P	Find P Given F P/F	Find A Given F A/F	Find A Given P A/P	Find F Given A F/A	Find P Given A P/A	Find A Given G A/G	Find P Given G P/G	n
1	1.100	.9091	1.0000	1.1000	1.000	0.909	0	0	1
2	1.210	.8264	.4762	.5762	2.100	1.736	0.476	0.826	2
3	1.331	.7513	.3021	.4021	3.310	2.487	0.937	2.329	3
4	1.464	.6830	.2155	.3155	4.641	3.170	1.381	4.378	4
5	1.611	.6209	.1638	.2638	6.105	3.791	1.810	6.862	5
6	1.772	.5645	.1296	.2296	7.716	4.355	2.224	9.684	6
7	1.949	.5132	.1054	.2054	9.487	4.868	2.622	12.763	7
8	2.144	.4665	.0874	.1874	11.436	5.335	3.004	16.029	8
9	2.358	.4241	.0736	.1736	13.579	5.759	3.372	19.421	9
10	2.594	.3855	.0627	.1627	15.937	6.145	3.725	22.891	10
11	2.853	.3505	.0540	.1540	18.531	6.495	4.064	26.396	11
12	3.138	.3186	.0468	.1468	21.384	6.814	4.388	29.901	12
13	3.452	.2897	.0408	.1408	24.523	7.103	4.699	33.377	13
14	3.797	.2633	.0357	.1357	27.975	7.367	4.996	36.801	14
15	4.177	.2394	.0315	.1315	31.772	7.606	5.279	40.152	15
16	4.595	.2176	.0278	.1278	35.950	7.824	5.549	43.416	16
17	5.054	.1978	.0247	.1247	40.545	8.022	5.807	46.582	17
18	5.560	.1799	.0219	.1219	45.599	8.201	6.053	49.640	18
19	6.116	.1635	.0195	.1195	51.159	8.365	6.286	52.583	19
20	6.728	.1486	.0175	.1175	57.275	8.514	6.508	55.407	20
21	7.400	.1351	.0156	.1156	64.003	8.649	6.719	58.110	21
22	8.140	.1228	.0140	.1140	71.403	8.772	6.919	60.689	22
23	8.954	.1117	.0126	.1126	79.543	8.883	7.108	63.146	23
24	9.850	.1015	.0113	.1113	88.497	8.985	7.288	65.481	24
25	10.835	.0923	.0102	.1102	98.347	9.077	7.458	67.696	25
26	11.918	.0839	.00916	.1092	109.182	9.161	7.619	69.794	26
27	13.110	.0763	.00826	.1083	121.100	9.237	7.770	71.777	27
28	14.421	.0693	.00745	.1075	134.210	9.307	7.914	73.650	28
29	15.863	.0630	.00673	.1067	148.631	9.370	8.049	75.415	29
30	17.449	.0573	.00608	.1061	164.494	9.427	8.176	77.077	30
31	19.194	.0521	.00550	.1055	181.944	9.479	8.296	78.640	31
32	21.114	.0474	.00497	.1050	201.138	9.526	8.409	80.108	32
33	23.225	.0431	.00450	.1045	222.252	9.569	8.515	81.486	33
34	25.548	.0391	.00407	.1041	245.477	9.609	8.615	82.777	34
35	28.102	.0356	.00369	.1037	271.025	9.644	8.709	83.987	35
40	45.259	.0221	.00226	.1023	442.593	9.779	9.096	88.953	40
45	72.891	.0137	.00139	.1014	718.905	9.863	9.374	92.454	45
50	117.391	.00852	.00086	.1009	1 163.9	9.915	9.570	94.889	50
55	189.059	.00529	.00053	.1005	1 880.6	9.947	9.708	96.562	55
60	304.482	.00328	.00033	.1003	3 034.8	9.967	9.802	97.701	60
65	490.371	.00204	.00020	.1002	4 893.7	9.980	9.867	98.471	65
70	789.748	.00127	.00013	.1001	7 887.5	9.987	9.911	98.987	70
75	1 271.9	.00079	.00008	.1001	12 709.0	9.992	9.941	99.332	75
80	2 048.4	.00049	.00005	.1000	20 474.0	9.995	9.961	99.561	80
85	3 299.0	.00030	.00003	.1000	32 979.7	9.997	9.974	99.712	85
90	5 313.0	.00019	.00002	.1000	53 120.3	9.998	9.983	99.812	90
95	8 556.7	.00012	.00001	.1000	85 556.9	9.999	9.989	99.877	95
100	13 780.6	.00007	.00001	.1000	137 796.3	9.999	9.993	99.920	100

12% Compound Interest Factors 12%

n	F/P	P/F	A/F	A/P	F/A	P/A	A/G	P/G	n
1	1.120	.8929	1.0000	1.1200	1.000	0.893	0	0	1
2	1.254	.7972	.4717	.5917	2.120	1.690	0.472	0.797	2
3	1.405	.7118	.2963	.4163	3.374	2.402	0.925	2.221	3
4	1.574	.6355	.2092	.3292	4.779	3.037	1.359	4.127	4
5	1.762	.5674	.1574	.2774	6.353	3.605	1.775	6.397	5
6	1.974	.5066	.1232	.2432	8.115	4.111	2.172	8.930	6
7	2.211	.4523	.0991	.2191	10.089	4.564	2.551	11.644	7
8	2.476	.4039	.0813	.2013	12.300	4.968	2.913	14.471	8
9	2.773	.3606	.0677	.1877	14.776	5.328	3.257	17.356	9
10	3.106	.3220	.0570	.1770	17.549	5.650	3.585	20.254	10
11	3.479	.2875	.0484	.1684	20.655	5.938	3.895	23.129	11
12	3.896	.2567	.0414	.1614	24.133	6.194	4.190	25.952	12
13	4.363	.2292	.0357	.1557	28.029	6.424	4.468	28.702	13
14	4.887	.2046	.0309	.1509	32.393	6.628	4.732	31.362	14
15	5.474	.1827	.0268	.1468	37.280	6.811	4.980	33.920	15
16	6.130	.1631	.0234	.1434	42.753	6.974	5.215	36.367	16
17	6.866	.1456	.0205	.1405	48.884	7.120	5.435	38.697	17
18	7.690	.1300	.0179	.1379	55.750	7.250	5.643	40.908	18
19	8.613	.1161	.0158	.1358	63.440	7.366	5.838	42.998	19
20	9.646	.1037	.0139	.1339	72.052	7.469	6.020	44.968	20
21	10.804	.0926	.0122	.1322	81.699	7.562	6.191	46.819	21
22	12.100	.0826	.0108	.1308	92.503	7.645	6.351	48.554	22
23	13.552	.0738	.00956	.1296	104.603	7.718	6.501	50.178	23
24	15.179	.0659	.00846	.1285	118.155	7.784	6.641	51.693	24
25	17.000	.0588	.00750	.1275	133.334	7.843	6.771	53.105	25
26	19.040	.0525	.00665	.1267	150.334	7.896	6.892	54.418	26
27	21.325	.0469	.00590	.1259	169.374	7.943	7.005	55.637	27
28	23.884	.0419	.00524	.1252	190.699	7.984	7.110	56.767	28
29	26.750	.0374	.00466	.1247	214.583	8.022	7.207	57.814	29
30	29.960	.0334	.00414	.1241	241.333	8.055	7.297	58.782	30
31	33.555	.0298	.00369	.1237	271.293	8.085	7.381	59.676	31
32	37.582	.0266	.00328	.1233	304.848	8.112	7.459	60.501	32
33	42.092	.0238	.00292	.1229	342.429	8.135	7.530	61.261	33
34	47.143	.0212	.00260	.1226	384.521	8.157	7.596	61.961	34
35	52.800	.0189	.00232	.1223	431.663	8.176	7.658	62.605	35
40	93.051	.0107	.00130	.1213	767.091	8.244	7.899	65.116	40
45	163.988	.00610	.00074	.1207	1 358.2	8.283	8.057	66.734	45
50	289.002	.00346	.00042	.1204	2 400.0	8.304	8.160	67.762	50
55	509.321	.00196	.00024	.1202	4 236.0	8.317	8.225	68.408	55
60	897.597	.00111	.00013	.1201	7 471.6	8.324	8.266	68.810	60
65	1 581.9	.00063	.00008	.1201	13 173.9	8.328	8.292	69.058	65
70	2 787.8	.00036	.00004	.1200	23 223.3	8.330	8.308	69.210	70
75	4 913.1	.00020	.00002	.1200	40 933.8	8.332	8.318	69.303	75
80	8 658.5	.00012	.00001	.1200	72 145.7	8.332	8.324	69.359	80
85	15 259.2	.00007	.00001	.1200	127 151.7	8.333	8.328	69.393	85
90	26 891.9	.00004		.1200	224 091.1	8.333	8.330	69.414	90
95	47 392.8	.00002		.1200	394 931.4	8.333	8.331	69.426	95
100	83 522.3	.00001		.1200	696 010.5	8.333	8.332	69.434	100

18% Compound Interest Factors 18%

n	F/P	P/F	A/F	A/P	F/A	P/A	A/G	P/G	n
1	1.180	.8475	1.0000	1.1800	1.000	0.847	0	0	1
2	1.392	.7182	.4587	.6387	2.180	1.566	0.459	0.718	2
3	1.643	.6086	.2799	.4599	3.572	2.174	0.890	1.935	3
4	1.939	.5158	.1917	.3717	5.215	2.690	1.295	3.483	4
5	2.288	.4371	.1398	.3198	7.154	3.127	1.673	5.231	5
6	2.700	.3704	.1059	.2859	9.442	3.498	2.025	7.083	6
7	3.185	.3139	.0824	.2624	12.142	3.812	2.353	8.967	7
8	3.759	.2660	.0652	.2452	15.327	4.078	2.656	10.829	8
9	4.435	.2255	.0524	.2324	19.086	4.303	2.936	12.633	9
10	5.234	.1911	.0425	.2225	23.521	4.494	3.194	14.352	10
11	6.176	.1619	.0348	.2148	28.755	4.656	3.430	15.972	11
12	7.288	.1372	.0286	.2086	34.931	4.793	3.647	17.481	12
13	8.599	.1163	.0237	.2037	42.219	4.910	3.845	18.877	13
14	10.147	.0985	.0197	.1997	50.818	5.008	4.025	20.158	14
15	11.974	.0835	.0164	.1964	60.965	5.092	4.189	21.327	15
16	14.129	.0708	.0137	.1937	72.939	5.162	4.337	22.389	16
17	16.672	.0600	.0115	.1915	87.068	5.222	4.471	23.348	17
18	19.673	.0508	.00964	.1896	103.740	5.273	4.592	24.212	18
19	23.214	.0431	.00810	.1881	123.413	5.316	4.700	24.988	19
20	27.393	.0365	.00682	.1868	146.628	5.353	4.798	25.681	20
21	32.324	.0309	.00575	.1857	174.021	5.384	4.885	26.300	21
22	38.142	.0262	.00485	.1848	206.345	5.410	4.963	26.851	22
23	45.008	.0222	.00409	.1841	244.487	5.432	5.033	27.339	23
24	53.109	.0188	.00345	.1835	289.494	5.451	5.095	27.772	24
25	62.669	.0160	.00292	.1829	342.603	5.467	5.150	28.155	25
26	73.949	.0135	.00247	.1825	405.272	5.480	5.199	28.494	26
27	87.260	.0115	.00209	.1821	479.221	5.492	5.243	28.791	27
28	102.966	.00971	.00177	.1818	566.480	5.502	5.281	29.054	28
29	121.500	.00823	.00149	.1815	669.447	5.510	5.315	29.284	29
30	143.370	.00697	.00126	.1813	790.947	5.517	5.345	29.486	30
31	169.177	.00591	.00107	.1811	934.317	5.523	5.371	29.664	31
32	199.629	.00501	.00091	.1809	1 103.5	5.528	5.394	29.819	32
33	235.562	.00425	.00077	.1808	1 303.1	5.532	5.415	29.955	33
34	277.963	.00360	.00065	.1806	1 538.7	5.536	5.433	30.074	34
35	327.997	.00305	.00055	.1806	1 816.6	5.539	5.449	30.177	35
40	750.377	.00133	.00024	.1802	4 163.2	5.548	5.502	30.527	40
45	1 716.7	.00058	.00010	.1801	9 531.6	5.552	5.529	30.701	45
50	3 927.3	.00025	.00005	.1800	21 813.0	5.554	5.543	30.786	50
55	8 984.8	.00011	.00002	.1800	49 910.1	5.555	5.549	30.827	55
60	20 555.1	.00005	.00001	.1800	114 189.4	5.555	5.553	30.846	60
65	47 025.1	.00002		.1800	261 244.7	5.555	5.554	30.856	65
70	107 581.9	.00001		.1800	597 671.7	5.556	5.555	30.860	70
75	246 122.1				1 367 339.2	5.556	5.555	30.862	75
100	15 424 131.9				85 689 616.2	5.556	5.555	30.864	100

Exam Files

Professors around the country have opened their exam files and revealed their examination problems and solutions. These are actural exam problems with the complete solutions prepared by the same professors who wrote the problems. Exam Files are currently available for these topics:

Calculus I
Calculus II
Calculus III
Circuit Analysis
College Algebra
Differential Equations
Dynamics
Engineering Economic Analysis
Fluid Mechanics
Linear Algebra
Materials Science
Mechanics of Materials
Organic Chemistry
Physics I Mechanics
Physics III Electricity and Magnetism
Probability and Statistics
Statics
Thermodynamics

For a description of all available **Exam Files**, or to order them, ask at your college or technical bookstore, or call **1-800-800-1651** or write to:

Engineering Press
P.O. Box 1
San Jose, CA 95103-0001